73rd World Foundry Congress (WFC 2018)

AA006353

Creative Foundry

Young Researchers' Seminar

Krakow, Poland
23 - 27 September 2018

ISBN: 978-1-5108-8042-9

Printed from e-media with permission by:

Curran Associates, Inc.
57 Morehouse Lane
Red Hook, NY 12571

Some format issues inherent in the e-media version may also appear in this print version.

Copyright© (2018) by World Foundry Organization Ltd
All rights reserved.

Printed by Curran Associates, Inc. (2019)

For permission requests, please contact World Foundry Organization Ltd
at the address below.

World Foundry Organization Ltd
Winton House, Lyonshall
Kington, Herefordshire
HR5 3JP, United Kingdom

Phone: +44 (0) 1544 340332
Fax: +44 (0) 1544 340332

www.thewfo.com

Additional copies of this publication are available from:

Curran Associates, Inc.
57 Morehouse Lane
Red Hook, NY 12571 USA
Phone: 845-758-0400
Fax: 845-758-2633
Email: curran@proceedings.com
Web: www.proceedings.com

73rd World Foundry Congress

"Creative Foundry"

23rd–27th September 2018, Kraków, Poland

Congress Proceedings

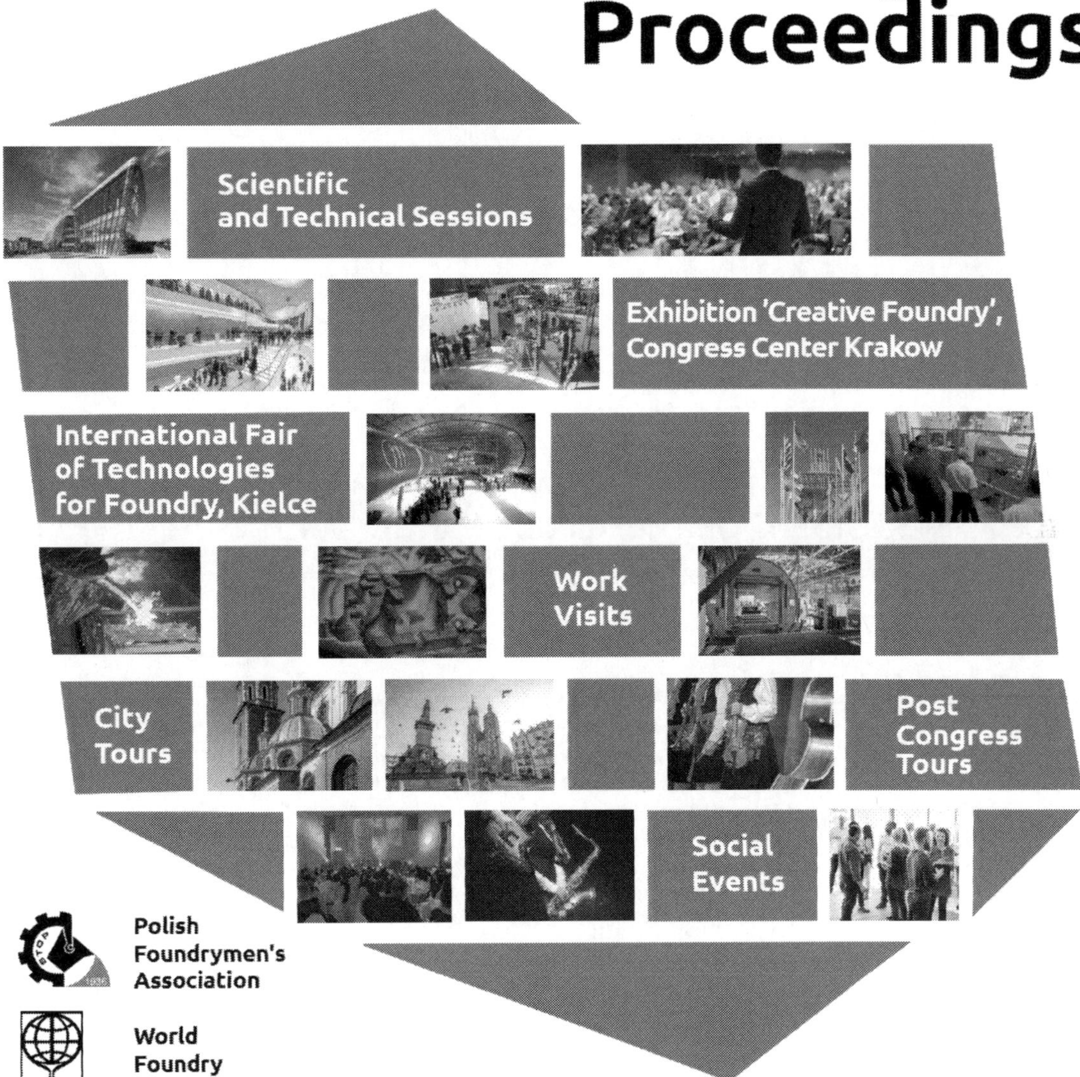

Polish Foundrymen's Association

World Foundry Organizatuon

Young Researchers' Seminar

Organization

Executive of WFO

President	M. Fenyes (United Kingdom)
Vice President	U. Denizci (Turkey)
Immediate Past President	M.H. Kim (Korea)
Past President	X. Azpiri Gonzalez (Spain)
Traesurer	J.S. Suchy (Poland)

Members:
Y. Lou (China)
D. Dotson (United States of America)
U. Denizci (Turkey)
C. Kuhlgatz (Germany)
S. Panchal (India)
P.M. Cabbanne (France)
T. Torigoe (Japan)

General Secretary	A. Turner (United Kingdom)
Assistant Secretary	L. Postle (United Kingdom)
Media Manager	J. Call (USA)
Administration Secretary	Mr. J. J. Gozalez (Spain)

THE HONORARY PATRONAGE
OF THE PRESIDENT OF THE REPUBLIC OF POLAND
ANDRZEJ DUDA

Honorary Committee of the 73rd World Foundry Congress

Jarosław GOWIN	– Minister of Science and Higher Education
Jacek KRUPA	– The Marshal of the Małopolska Region
Jacek MAJCHROWSKI	– Major of the City of Krakow
Tadeusz SŁOMKA	– Rector of University of Science and Technology (AGH)
Ewa MAŃKIEWICZ-CUDNY	– President of NOT, The Polish Federation of Engineering Associatons
Mark FENYES	– President of World Foundry Organization

PUBLISHER/WYDAWCA:
STOWARZYSZENIE TECHNICZNE ODLEWNIKÓW POLSKICH/POLISH FOUNDRYMEN'S ASSOCIATION

Organizational Committee of the 73rd World Foundry Congress

Józef Szczepan SUCHY – Chairman
Tadeusz FRANASZEK – President of Polish Foundrymen's Association
Katarzyna LISZKA – General Secretary of Polish Foundrymen's Association
Andrew TURNER – General Secretary of World Foundry Organization

Michał BALICKI
Aldona GARBACZ-KLEMPKA
Aleksandra GRABARCZYK
Beata GRABOWSKA
Marcin GÓRNY
Joanna MADEJ
Jerzy Józef SOBCZAK
Michał SZUCKI
Emilia WILDHIRT
Dorota WILK-KOŁODZIEJCZYK
Jerzy ZYCH
Janusz LELITO
Magdalena JASIŃSKA
Magdalena PIETROŃ

Program Committee of the 73rd World Foundry Congress

Tadeusz FRANASZEK – President of Polish Foundrymen's Association
Józef Szczepan SUCHY – AGH University of Science and Technology
Rafał DAŃKO – Dean of Faculty of Foundry Engineering/AGH University of Science and Technology
Jerzy SOBCZAK – Foundry Research Institute, Krakow
Andrzej MOCHOŃ – Chairman of the Board/Kielce Trade Fairs
Przemysław TRAWA – Chairman of the Board/Poznań International Fair

Scientific Committee of the 73rd World Foundry Congress

Prof. Natalia Sobczak – Poland (Chairperson, Guest Editor)
Prof. Rajiv Asthana – USA (Vice-Chairman, Guest Editor)
Prof. Jan Szajnar – Poland (Vice-Chairman, Guest Editor)

Dr. Zdenka Zovko Brodarac – Croatia
Prof. Ludmil Drenchev – Bulgaria
Dr. Patricia Caballero – Spain
Dr. Rafał Cygan – Poland
Dr Donatella Giuranno – Poland/Italy
Prof. Edward Guzik – Poland
Prof. Milan Horacek – Czech Republic
Prof. Jolanta Janczak-Rusch – Switzerland
Prof. Jan Jezierski – Poland
Prof. Zenon Ignaszak – Poland
Prof. Savko Malinov – UK
Dr. Antton Meléndez – Spain
Prof. Sergey Mileiko – Russia
Prof. Primoz Mrvar – Slovenia
Prof. Anita Olszówka-Myalska – Poland
Prof. Hideo Nakae – Japan
Prof. Adel Nofal – Egipt
Prof. Tadeusz Pacyniak – Poland
Prof. Alberto Passerone – Italy
Prof. Marcin Perzyk – Poland
Prof. Yuriy Plevachuk – Ukraine
Dr. Wojciech Polkowski – Poland (Guest Editor)
T. Prucha – US (Guest Editor)
R.M. Purgert – US
Prof. Peter Schumacher - Austria
Dr. Mrityunjay Singh – USA
Prof. Augustin Sladek – Slovakia
Prof. Jerzy Józef Sobczak – Poland (Guest Editor)
Prof. Duru Stefanescu – US
Prof. Boris Straumal – Russia
Prof. Józef Szczepan Suchy – Poland (Guest Editor)
Dr. Lou Yanchun – China
Dr Vincent Vignal – France
Prof Gotthard Wolf – Germany
Prof. Pawel Zięba – Poland (Guest Editor)

Preface

Marcin Górny
Young Researcher's Seminar Organizer

The 73rd World Foundry Congress will be held again in Poland. After 27 years, on September 23–27 in Krakow, we will hosting representatives of universities, research units and industry from over 40 countries. Once again, within WFC, the Young Researchers' Seminar will be held. The participants of this Young Researchers' Seminar are students, doctorate candidates, young scientists and engineers up to 35 years old. I would like to express my appreciation to all authors, who came forward to participate in this event. We will have 100 participants, representing Austria, Germany, Czech Republic, Sweden, Slovenia, China, Japan, South Africa, Pakistan, Iraq, Switzerland, Croatia, India, France, Hungary, Italy, Slovakia, Finland, Brazil, Turkey, Republic of Korea and Poland. During the Young Researchers' Seminar special Guest Lectures will be delivered. Kazutoshi Kimura President of Kimura Foundry CO., Ltd, from JAPAN, will present two lectures: 'Kimura Foundry Co., Ltd. Overview' and 'The state of art and foresight of world casting production'. The lecture entitled: 'The role of the Faculty of Foundry Engineering AGH-UST in educating engineers for casting industry' will be delivered by professor Rafał Dańko - Dean of the Faculty of Foundry Engineering, AGH University of Science and Technology, POLAND. The Young Researchers' Seminar will consist of two parts: oral presentations from 12 countries and the YOUNG RESEARCHERS' POSTER SESSION. The special Committee consisting of outstanding representatives of science and industry will select the Best Young Researcher Paper, which will be awarded by PGO SA.

The Young Researchers' Seminar is an excellent occasion for students, doctorate candidates and young scientists to present their achievements to a wide group of researchers and industry representatives from all over the world. They have the possibility to participate in plenary and technical sessions during the 73rd WFC, sharing of the most recent achievements from widely understood foundry industry, gathered in five thematic blocks i.e. Materials, Technologies, Digitalisation, Management and Ecology.

Scientific sessions and social events within the World Foundry Congress, provide an opportunity for young researchers to establish cooperation with the world's most important scientific centres and foundry plants.

WFC is also a celebration of foundry men, accompanied with inspiring exchanges of thoughts, ideas and research projects.

In this place I would like to express my gratitude to the Organisational Committee of the Young Researchers' Seminar of 73rd WFC for organising this meeting. I would like to thank all participants of this Young Researchers' Seminar for taking part in the world celebration of foundry men. I wish all of you fruitful deliberations, creative discussions and unforgettable impressions.

Contents

Materials:

24 – Steeldesign and activated infiltration of metal-matrix composites for wearing parts
P. Rähmer, TU Bergakademie Freiberg, Foundry Institute, Germany 13

27 – Oxidation resistant aluminum alloyed simo cast iron for high temperature application
N. Scheidhauer, Technical University Bergakademie Freiberg, Foundry Department, Germany 15

35 – Modeling of microstructure in hypoeutectic high chromium cast iron
D. Siekaniec, AGH University of Science and Technology, Faculty of Foundry Engineering, Poland 17

37 – Possibility on estimation of fatigue limit using X-ray CT apparatus in spheroidal graphite cast iron
K. Tanaka, Tokyo City University, Graduate school of Engineering, Japan 19

48 – Influence of solution hardening on microstructure and mechanical properties of Al-2.5Mg-0.7Li alloy
F. Kozina, University of Zagreb Faculty of Metallurgy, Department for Process Metallurgy, Croatia 21

50 – Modification of non-metallic inclusions in steel by application of deoxodizing agents with alkaline metals (calcium and barium)
J. Cerny, COMTES FHT, Czech Republic 23

77 – New possibilities in the thermal analysis of moulding materials
J. Svidró, Jönköping University School of Engineering, Department of Materials and Manufacturing, Sweden 25

136 – Development of mathematical relationships for calculating of material-dependent-flowability of green moulding sand
D. Abdulamer, IMKF. TU- Bergakademie Freiberg, Germany 27

147 – Spinel forming systems (Mg-/Fe-/Mn-Al-O) as functional coating materials on carbon-bonded filters for steel melt filtration
B. Bock, TU Bergakademie Freiberg, Institute of Ceramic, Glass and Construction Materials, Germany 29

160 – Development of new heat – resistant steel to avoid the detrimental Z-phase precipitation at high temperatures
W. Pasini, Federal University of Rio Grande do Sul, Metallurgy Department - School of Engineering, Brazil 31

170 – The deviation from eutectic composition in boundary layer for eutectic growth: a phase-field study
Z. Tu, Huazhong University of Science and Technology, State Key Laboratory of Materials Processing and Die & Mould Technology, China 33

173 – Effect of cooling rate on constituent particle formation in high-speed twin-roll cast Al-Mn based alloy strips
R. Song, Tokyo Institute of Technology, Department of Metallurgy and Ceramics Science, Japan 35

190 – Suppressing of ultra-high temperature wetting between molten Si and SiC by using h-BN spray coatings
W. Polkowski, Foundry Research Institute, Poland 37

196 – Indicator of the efficient protection against the humidity W_{so}, as a new parameter characterising protective coatings
N. Kaźnica, AGH University of Science and Technology, Faculty of Foundry Engineering, Poland 39

207 – The influence of remelting on the properties of AlSi9Cu3 alloy with higher iron content
M. Matejka, University of Žilina, Department of Technological Engineering, Slovakia 41

215 – Study of damping properties of composite materials on a metallic matrix
M. Poręba, AGH University of Science and Technology, Faculty of Foundry Engineering, Poland 43

216 – Tribological properties of composite materials on the metallic matrix
B. Gospodarczyk, AGH University of Science and Technology, Faculty of Foundry Engineering, Poland 45

219 – Dynamic measurement and FEM analysis of restraint force from mold on gray cast iron castings and contraction of castings during cooling in green sand mold
M. Kaneko, Waseda University, Modern mechanical engineering, Japan 47

250 – Deformations in chemically bonded moulding sands
A. Grabarczyk, AGH University of Science and Technology, Faculty of Foundry Engineering, Poland 49

254 – Austempering in Zamak bath: influence of austempering time on the properties of ductile cast iron
W. M. Pasini, Federal University of Rio Grande do Sul, Metallurgy Department - School of Engineering, Brazil 51

267 – Influence of AlP in solidification sequence of Al-10%Si-0.3%Mg
Y. Zhao, University of Toyama, Graduate School of Science and Engineering for Education, Japan 53

268 – Effect of AlP in solidification sequence of Al-6%Mg-3%Si
A. Osugi, University of Toyama, Graduate School of Science and Engineering for Research, Japan 55

269 – Influence of P content in eutectic crystallization of Al-10%Si-0.8%Mg
H. Kazuta, University of Toyama, Graduate School of Science and Engineering for Education, Japan 57

278 – Infruence of tellurium addition to spheroidal graphite cast iron on the number of graphite particles
R. Saito, Graduate School of Kansai University, Department of Chemistry and Materials Engineering, Japan 59

286 – Influence of Ti and RE on primary crystallization and wear resistance of chromium cast iron
M. Dojka, Silesian University of Technology, Department of Foundry Engineering, Poland 61

298 – *Corrosion resistance of Ti30Nb alloy in Hank's solution*
J. Ryba, AGH University of Science and Technology, Faculty of Foundry Engineering, , Poland 63
299 – *Magnesium alloys for biomedical applications of implants*
A. Fijołek, AGH University of Science and Technology, Faculty of Foundry Engineering, Poland 65
300 – *Modeling of the kinetics of carbonitrides precipitation process in HSLA steels using cellular automata method*
P. Marynowski, AGH University of Science and Technology, Poland 67
301 – *Grain refinement of Al-2%Cu alloy using by vibrating mold*
Y. Yoshitake, National Institute of Technology, Support Center of Education and Research, Kurume College, Japan 69
310 – *Settling effect of Ti on melt quality and mechanical properties in 356 cast alloy*
Ö. Gürsoy, Istanbul University, Faculty of Engineering, Turkey 71
317 – Hot tearing tendency and tensile property change of Ni and Cu added A356 casting alloys
K. Yildirim, Giesserei Institut RWTH, Germany 73
319 – *Efficiency of Ti addition in A356 alloy under different holding time*
M. Gurtaran, Bursa Technical University, Material Science And Engineering, Turkey 75
325 – *Effect of thermal exposure on high temperature mechanical properties of secondary AlSi7Cu3Mg alloys*
A. De Mori, University of Padova, Department of Management and Engineering, Italy 77
330 – *Increasing the lifetime of die casting molds made of the hot work tool steel H13 by means of a local surface heat treatment with the electron-beam*
T. Schuchardt, TU Braunschweig, Institute of Joining and Welding, Germany 79
336 – *Effect of quenching mediums on mechanical properties of Al-Cu alloy*
M. Gurtaran, Bursa Technical University, Metallurgical and Materials Engineering, Turkey 81
360 – *The examination of the modification effect of strontium in AlSi8Cu3 foundry alloy*
M. Tokár, University of Miskolc, Foundry Institute, Hungary 83
362 – *Manganese in Compacted Graphite Iron*
B. Kurowska, Łódź University of Technology, Department of Materials Engineering and Production Systems, Poland 85
381 – *Influence of evaluated temperature of intercritical treatment on mechanical properties of austempered ductile iron*
P. Nawrocki, Warsaw Univeristy of Technology, Faculty of Production Engineering, Poland 87
383 – *Microstrutural and mechanical characterization of high-aluminium zinc alloy modified with 2.0% TiB_2*
M. Roberto Bellé, Federal University of Rio Grande do Sul, Metallurgy Department, School of Engineering, Brazil 89
390 – *Effect of Al, Cr and Mo addition on mechanical properties of Nb-Si based alloys*
T. Uemura, Kindai University, Faculty of Mechanical Engineering, Japan 91
406 – *Quality index - a way of illustrating diversified properties of a heat-treated Al-Zn-Mg-Cu aluminum alloys depending on homogenization parameters*
Z. Kwak, AGH University od Science and Technology, Faculty of Foundry Engineering, Poland 93
409 – *Non-destructive testing methods in the analysis of metal artefacts*
A. Sosnowska, AGH University od Science and Technology, Faculty of Management, Poland 95
428 – *Effect of predeformation and semisolid isothermal treatment time on the globular grains in MRI230 alloy during SIMA process*
M. Roberto Bellé, Federal University of Rio Grande do Sul, Metallurgy Department - School of Engineering, Brazil 97
443 – *Wettability, reactivity and interfaces in Gd/Y_2O_3 system*
P. Turalska, Foundry Research Institute, Poland 99
446 – *High-temperature interaction of molten high chromium cast iron with ZrO_2- Al_2O_3 ceramic*
M. Kolev, Institute of Metal Science, Equipment and Technologies with Hydroaerodynamics Centre, Bulgaria 101
458 – The effect of boron content on wetting kinetics in Si-B alloy/h-BN system
W. Polkowski, Foundry Research Institute, Poland 103

Technology:
15 – *Technical capabilities for low-temperature waste heat recovery in foundries*
T. Ludwig, Technical University Freiberg, Foundry Department, Germany 107
32 – *Design options to increase the service life of cast accessories for thermal and chemical treatment furnaces*
A. Bajwoluk, West Pomeranian University of Technology, Faculty of Mechanical Engineering and Mechatronics, Poland 109
33 – *Suppressing the basket deformation process during heat treatment*
A. Bajwoluk, West Pomeranian University of Technology, Faculty of Mechanical Engineering and Mechatronics, Poland 111

53 – A novel measurement method to study the thermal aspects of moulding mixture decomposition
J. T. Svidró, Jönköping University School of Engineering, Department of Materials and Manufacturing, Sweden 113
56 – Development of casting genome for automotive piston casting: a study
R. Ozah, North Eastern Regional Institute of Science and Technology, Mechanical Engineering, India 115
63 – Investigation of the impact of roughness on adhesion forces and filtration efficiency in a water-based model system with ceramic foam filters
D. Hoppach, TU Freiberg, Insitute of Mechanical Process Engineering and Mineral Processing, Germany 117
93 – Comparative study of waste foundry sand reclamation techniques
M. Khan, Indian Institute of Technology Bombay, Chemical Engineering, India 119
157 – Exchangeable carbon-bonded alumina foam filter systems for the continuous casting of steel – development and testing
T. Wetzig, TU Bergakademie Freiberg, Institute of Ceramics, Glass and Construction Materials, Germany 123
169 – Simulation of sand filling process and analysis of flow behavior with different binder ratio
L. Tong, Huazhong University of Science and Technology, State Key Laboratory of Materials Processing and Die & Mould Technology, China 125
176 – Optimal gating system design of steel casting by fruit fly optimization algorithm based on casting simulation technology
T. Wang, Huazhong University of Science and Technology, State Key Laboratory of Materials Processing and Die &Mould Technology, China 127
200 – Bayesian inference based optimization of process parameters for chemically bonded molding system
H. Khandelwal, MPM Infosoft Pvt. Ltd., India 129
212 – Coupled methods of thermal analysis in the study of materials used in the foundry industry
A. Roczniak, AGH University of Science and Technology, Faculty of Foundry Engineering, Poland 131
217 – Gating system design optimization for investment casted part
M. Bruna, University of Žilina, Department of Technological Engineering, Slovakia 133
239 – Development of a method for measuring the ejector forces in the die casting process
S. Krischke, TU Braunschweig, Institute of Joining and Welding, High-Pressure Die Casting, Germany 135
260 – Teaching-and-playback approach based on pouring flow rate in tilting-ladle-type pouring machine
T. Yajima, University of Yamanashi, Department of Mechanical Engineering, Japan 137
266 – Optimization of the runner shape considering the molten metal flow of product in die casting
T. Okuno, Mie University, Faculty of Engineering, Department of Mechanical Engineering, Japan 139
282 – Multi-scale topologically optimized components made by casting and additive manufacturing
K. Jalava, Aalto University, Department of Mechanical Engineering, Finland 141
294 – Product design and manufacturing in foundry industry using augmented virtual reality technology
Y. Pandya, AGH University of Science and Technology, Faculty of Foundry Engineering, , Poland 143
295 – Study of viscosity of water- and alcohol-based protective coatings
E. Wildhirt, AGH University of Science and Technology, Faculty of Foundry Engineering, Poland 145
320 – Influence of manufacturing process parameters on the stability of layers printed from ceramic powders using additive technology
A. Kania, AGH University of Science and Technology, Faculty of Foundry Engineering, Poland 147
337 – Foundry engineering and art. Casting of the bronze statuette inspired by Salvador Dali
B. Chachurska, AGH University of Science and Technology, Faculty of Foundry Engineering, Poland 149
343 – The use of experimental and computer aided methods in reconstruction of casting process of metal artifacts from the bronze age
D. Gruszka, AGH University of Science and Technology, Faculty of Foundry Engineering, Poland 151
419 – Influence of alloy additives and casting parameters on microstructure and mechanical properties of silicon bronzes
D. Witasiak, AGH University od Science and Technology, Faculty of Foundry Engineering, Poland 153
447 – Analysis and evaluation of the effect of manganese content on the properties of aluminum alloy EN AC 46000
M. Pasternak, AGH University od Science and Technology, Faculty of Foundry Engineering, Poland 155
449 – Analysis of the influence of the degree of wear of the pressing piston in cold chamber pressure machines on the quality of aluminium castings
G. Piwowarczyk, AGH University od Science and Technology, Faculty of Foundry Engineering, Poland 157

Digitalization:

165 – Adaptative product manufacturing technique for foundry industry using IoT technology an industry 4.0 initiative
Y. Pandya, AGH University of Science and Technology, Faculty of Foundry Engineering, Poland 161
185 – Casting EMPS (ERP\MES\PDM\SCADA) integrated system architecture and its application in aeronautic titanium alloy foundry enterprise
J. Zhou, Huazhong University of Science & Technology, State Key Laboratory of Materials Processing and Die & Mould Technology, China 163

198 – Application of 3D printing to produce an investment casting model of hip-joint endoprosthesis
M. Skorupska, AGH University of Science and Technology, Faculty of Foundry Engineering, Poland 165
201 – Numerical modelling of SHSB metal matrix composite solidification
A. Wojtyla, AGH University of Science and Technology, Faculty of Foundry Engineering, Poland 167
332 – Implementation of 3D printing in dentistry to improve the production of dental prosthetics restoration
D. Dulian, AGH University of Science and Technology, Faculty of Foundry Engineering, Poland 169
335 – Numerical simulation of the impact of casting wall thickness and initial mould temperature on the ductile cast iron microstructure
G. Witek, AGH University of Science and Technology , Faculty of Foundry Engineering , Poland 171
424 – Vision of the digitalized foundry -with cyper-physical systems to fully automatic cutting process for casting
M. Füglister, University of Applied Science and Arts Northwestern, Institute of Automation, School of Engineering, Switzerland 173

Ecology:

145 – Thermal reclamation of foundry sands using repurposed sand dryer equipment
T. Sappinen, Aalto University, School of Engineering, Finland 177

Management:

45 – A six sigma framework for eliminating defects in sand casting foundries
K. Nyembwe, University of Johanneburg, South Africa 181

MATERIALS

The 73rd World Foundry Congress
„Creative Foundry"
23 - 27 September 2018, Krakow, Poland

STEELDESIGN AND ACTIVATED INFILTRATION OF METAL-MATRIX COMPOSITES FOR WEARING PARTS

Paul Rähmer [1]*, Gotthard Wolf [1], Claudia Dommaschk [1]

[1]TU Bergakademie Freiberg, Faculty of Materials Science and Materials Technology (Foundry Institute), Bernhard-von-Cotta-Straße 4, 09599 Freiberg, Germany
**Corresponding address: e-mail: paul.raehmer@gi.tu-freiberg.de*

Key words: metal-matrix composite, infiltration, ceramics, TRIP-steel, quenching and partitioning

1. Introduction

Metal-matrix composites (MMC) unite the outstanding properties of metals, for example austenitic-martensitic cast steels, with those of oxides. To increase the wear resistance, varying contents of hard particles like zirconium or aluminum oxide can be added to the metal-matrix. If the hardness and the wear resistance are increased, the toughness of the MMC often decreases. Therefore TRIP-steels (transformation-induced plasticity) are investigated, which offer a matrix material with a high strength and outstanding ductility. These properties resulting from a strain-induced martensitic phase transformation from austenite to martensite.

2. Steeldesign

The Collaborative Research Center 799 investigates intensively TRIP-steels. During abrasion these steels show a continuous hardening process at the surface by achieving the required force for the transformation from austenite to martensite.

Due to increase the hardness and strength of the steel nickel gets substituted with results in decreased material cost. The composition of the used steel is shown in table 1.

***Table 1.* Chemical composition of the investigated steels in wt.-% (rest Fe)**

C	Si	Mn	Cr	Ni	N
0.3	1.0	2.0-4.0	13.0	1.0	0.1

Manganese is an austenite promoting element, so that it is responsible for the content of residual austenite after the quenching and partitioning (Q&P) process.

The aim of the Q&P process is to generate properties needed for the 3rd generation Advanced High Strength Steel with an ultimate tensile strength above 1200/1300 MPa. After the austenitization heat treatment follows a rapid cooling in liquid nitrogen to produce martensite between martensite start and martensite finish temperature. After the cooling starts the partitioning process at 450 °C to raise the stability of austenite and to perform M_3C precipitates in the martensite phase [1]. To quantify the ferromagnetic martensite, a Metis MSAT magnetization device is used.

3. Interface steel-ceramic

The interface between the metal-matrix and ceramic is substantial for the practical application of the MMC in wearing parts. For example the thermal properties and the chemical character of ceramics like aluminium or zirconium oxide and steel vary widely, so that a promoter is required, which triggers reactions in the metal/ceramic interface [2]. Initial investigations show, that titanate compounds improve the interface. Before casting, the aluminium oxide particles are coated with a titanium oxide slurry. After coating, the ceramics will be heated up and filled into the mould before casting. With support of the casting heat some spinells are shaped between the steel and the ceramics. The spinells are stable, because of elements like manganese or chromium, whereas aluminium titanate will be not stable at room temperature [3].

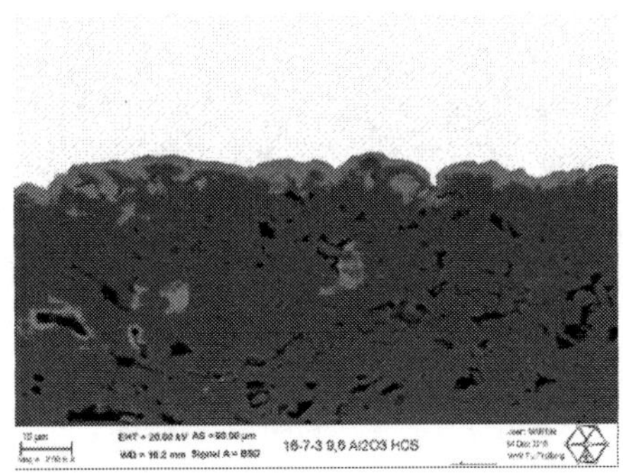

Fig. 1.* *REM-picture of an steel/ceramic interface with spinell [3]

4. Three-body-abrasive test

To quantify the abrasion resistance of the MMC, three-body-abrasive tests are performed. Abrasiv particles, for example silica sand, are scattered in front of the infiltrated sample, whereas the sample is pressed on a rotating NiHard4 surface with a load of 0.5 MPa. Every 1000 rotations the weight loss is calculated and evaluated [4].

5. Conclusion

The Enhancement of the production of MMC is in the focus of this research. The TRIP-steel is adjusted for the use in wear parts. Furthermore the interface steel-ceramics is investigated and improved. To make practical statements to the abrasive behaviour three-body-abrasive tests are performed.

Acknowledgments

The authors gratefully acknowledge the financial support of the German Research Foundation for funding the research project under the Collaborative Research Center 799.

References

[1] Wendler M., Ullrich C., Hauser M. et al. (2017). Quenching and partitioning (Q&P) processing of fully austenitic stainless steels. Acta Materialia. 133, 346-345.

[2] Weigelt C., Aneziris C.G., Ehinger D. et al. (2015). Effect of zirconia and aluminium titanate on the mechanical properties of transformation-induced plasticity-matrix composite materials. Journal of Composite Materials. 49, 3567-3579.

[3] Wagler H. (2017). Surface coating of ceramic particles to enhance infiltration. Diploma thesis. TU Bergakademie Freiberg. Foundry Institute.

[4] Acker A., Martin S., Meltke K. et al. Casting of Fe-CrMnNi and ZrO2-Based Metal-Matrix Composites and Their Wear Properties. Steel research international. 87. 1111-1117.

The 73rd World Foundry Congress
„Creative Foundry"
23 - 27 September 2018, Krakow, Poland

OXIDATION RESISTANT ALUMINUM ALLOYED SIMO CAST IRON FOR HIGH TEMPERATURE APPLICATION

Nico Scheidhauer [1*], **Claudia Dommaschk** [2], **Gotthard Wolf** [3]

[1 2 3] *Technical University Freiberg, Foundry Department, Bernhard-von-Cotta-Str. 4, 09599 Freiberg, Germany,*
**Corresponding address: e-mail: Nico.Scheidhauer@gi.tu-freiberg.de*

Key words: oxidation resistance, oxide layer thickness, ferritic SiMo cast iron, aluminum addition, austenitic D5S

1. Introduction

The investigation of materials for high temperature application in combustion engines, based on the material costs and the mechanical, physical and chemical properties, is promoted.

The present work will focus on an aluminum alloyed ferritic SiMo4.1 cast iron to find an economical alternative for D5S austenitic cast iron. The new alloy has to resist temperatures up to 950°C without lattice transformation. Furthermore, it has to have sufficient mechanical properties.

To reach an Ac1 temperature of 950°C, a high silicon and aluminum content is necessary. At the same time, those elements improve the durability of the protective oxide layer but they also increase the embrittlement of the material. Furthermore, aluminum disrupts the spherical graphite formation. To prevent those negative effects, alloy optimization was done. The present work will focus mainly on the remarkable oxidation resistance of aluminum and silicon alloyed cast iron.

2. Basic Principles

Oxidation resistance of exhaust gas components is essential for the lifetime of the casting because:

- it avoids the reduction of the bearing cross section below a critical level
- damages at the casting and moving parts can be prevented by less spallation of oxide particles
- the optimized air flow and heat transfer conditions in exhaust gas components are preserved.

In this case, the efficiency and the maximum loading pressure will stay at a high level.

In the literature [1,2] two mechanisms of scale growth are reported by Wagner's model. The first one is promoted by diffusion of metal ions to the interface between oxide scale and the surrounding atmosphere. The second one instead, is characterized by the diffusion of oxygen through the oxide layer to the metal. Generally, oxidation processes are mixed forms of both mechanisms.

If an alloy is oxidized the scale composition is dependent on the single Gibbs free energies of the referring elements but also on their parabolic oxidation rate constant (k_p) and content. As S. Gosh et. al [3] found out at first a FeO/Fe2O3 layer with low density is formed and ruptured by vaporizing CO gas. The reason for that is the low aluminum and silicon content of the cast iron and the higher k_p value of the iron oxides FeO and Fe2O3. Further oxidation leads to the formation of a more and more dense oxide layer with higher Si and/or Al contents. This partially sintered layer acts as a diffusion barrier because of its passivating character and the lower k_p values. [1,2,4]

3. Experimental Procedure

The material development was supported by phase simulation with JMatPro. The best composition of aluminum alloyed cast iron was tested in detail. Concurrently to this, a basic research was done to determine the influence of silicon and aluminum. The structure of the whole research is shown in **Table 1**.

Table 1. Experimental research

Basic research	Material development
• Lamellar graphite • 3,5%C | <4,5%Si | <9%Al	• Vermicular graphite • ~3%C | >2,5%Si | >4%Al | Mo, Nb, ...
➢ Analysis of oxide layers, Ac1 temperature, hardness	➢ Analysis of relevant mechanical and physical properties

The investigation of the oxidation resistance was done by:

- weight measurement with an accuracy of ±0,1mg at cylindric specimens with a surface of 9.425cm²
- analyzing the thickness of oxide layer crosscuts embedded in gap-free hardening resin, the analysed surface was ground to 1200P and cleaned with ethanol before oxidation.

Both tests were done with specimens that were oxidized for 96h at 820°C in static air atmosphere.

4. Results

It was tried to rate the oxidation resistance by weight gain measurements. Two crucial factors, see **Fig. 1**, leaded to incomparable results:

- very low oxide scale thickness and weight gain
- weight loss due to random graphite vaporization depending on the graphite morphology.

These effects were already reported by M. J. Kadhim [5]. The layer thickness measurement instead shows the typical parabolic growth mechanism of oxide scales [2]. For highly hot corrosion resistant cast irons, this measurement methodology is necessary.

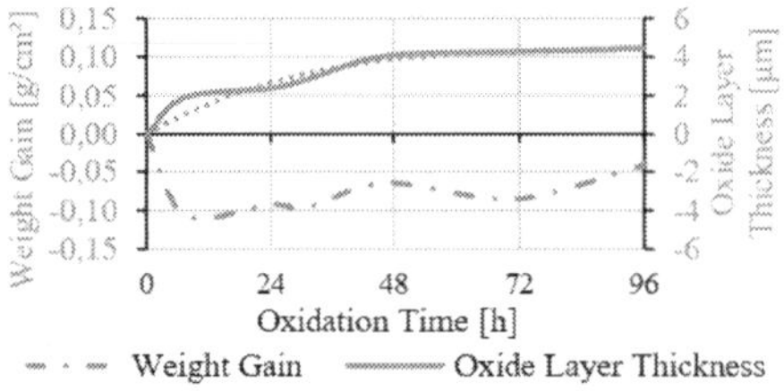

Fig. 1. Al alloyed SiMo cast iron weight gain measurement compared to oxide layer thickness

Generally, the oxidation resistance of cast iron increases with the silicon and aluminum content. The layer thickness decreases from 2376µm to less than 3µm for high silicon and aluminum alloyed cast irons, which is shown in **Fig. 2**. In this picture also an almost linear dependency can be seen between 1 and 3.5wt.-% Si with a decrease in scale thickness of 790µm per percent silicon and between 1 and 6wt.-% aluminum with a decrease of 350µm per percent aluminum.

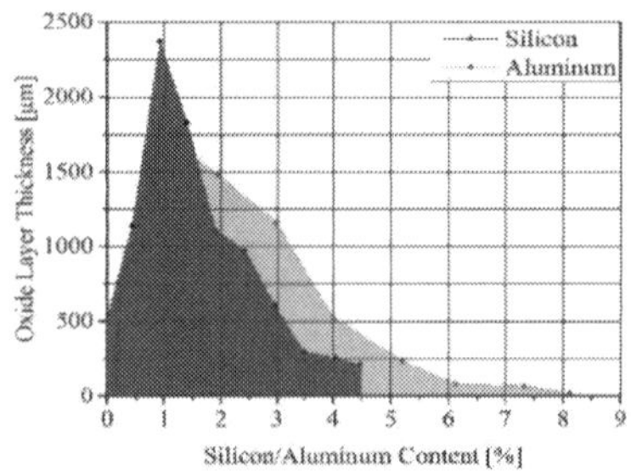

Fig. 2. Oxide layer thickness dependency of cast iron on silicon and aluminum content

The reduction in oxidation speed can be explained by more dense inner oxide layers that are formed by Si/Al-Fe-oxides, as it can be seen in **Fig. 3**. The higher impact of silicon on the oxidation resistance can be explained by the higher oxide to metal volume, the so-called Pilling-Bedworth ratio (PBR). If $PBR > 1$, protective oxide layers can be formed. This results in a higher density of silicon oxide $(PBR = 2{,}15)$ containing scales, then oxide layers which contain aluminum oxide $(PBR = 1{,}28)$. [2]

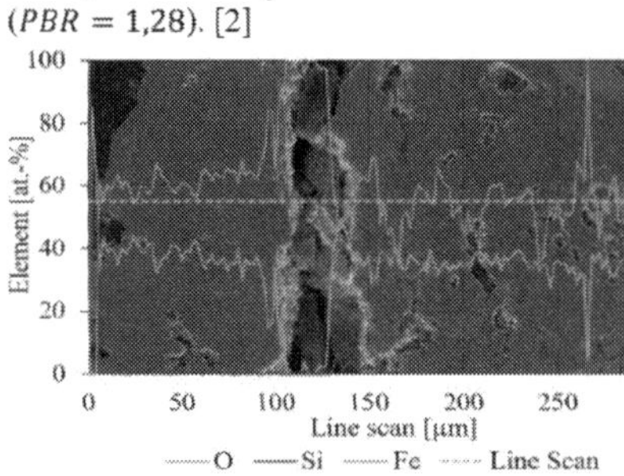

Fig. 3. Line scan of the oxide layer of a grey cast iron with 3.15wt.-%C and 4.05wt.-%Si (average layer thickness 256µm)

Fig. 3 shows the EDX analysis of an oxide layer of grey cast iron with 4.05wt.-% Si. The outer layer is almost 100% iron oxide. When the analysis is going closer to the metallic matrix, the silicon content increases in numerous layers. Those layers slow down the whole oxidation process. The crack through the oxide layers appears at the surface of biggest difference in oxidic composition for silicon alloyed as well as aluminum alloyed cast irons. The reason for that is a differnce in the expansion coefficients of the different layers [6].

In **Table. 2** the comparison of scale thickness and spalling behavior between the austenitic cast iron D3S, D5S and a CGI-SiMoAl is shown. The ferritic cast iron has much better oxidation resistance and spallation behavior. In that way, the surface roughness does not increase and physical damage to the casting itself, can be prevented by less detached oxide particles.

Table 2. Oxidation behavior of D3S, D5S and CGI-SiMoAl

Material	Scale Width [µm]	Spallation [$\frac{g}{cm^2}$]
Austenitic D3S	192.3	0.043
Austenitic D5S	171.5	0.030
Ferritic SiMoAl	8.7	0.000

5. Conclusion and Outlook

In conclusion high silicon and aluminum alloyed cast iron is attractive for usage in high temperature application. Although there are some deficits to mention. At first, it is hard to control and to keep the graphite morphology in close borders and secondly, the brittleness of this material at room temperature is a grave subject to handle and to control.

In summary, the benefits of the material exceed the deficits:

- extremely high oxidation and spallation resistance
- material costs are one fourth of the costs of an austenitic cast iron because of the nickel savings
- good mechanical high temperature properties due to kappa-phase precipitation.

Subjects to deal with during future investigation:

- focus on mechanical properties of the alloys
- EBSD and growth behavior analysis of the oxide layer

References

[1] I. Anžel. (2007). High Temperature Oxidation of Metals and Alloys. Metalurgija - Journal of Metallurgy. 13, 325-336.

[2] R. Bürgel, H. J. Maier & T. Niendorf. (2011). Oxidation. Handbuch Hochtemperatur-Werkstofftechnik (267-293). Vieweg+Teubner Verlag.

[3] S. Ghosh, A. Prodhan, N. Mohanty & A. K. Chakrabarti. (1996). A Study on the Oxidation Characteristics of Cast Irons Containing Aluminum. Oxidation of Metals. 45, 109-131.

[4] C. Bartels, R. Gerhards, H. Hanselka, & K. Herfurth. (2007) Gusseisen mit Kugelgraphit: Herstellung - Eigenschaften - Anwendung. Konstruieren + Giessen. 32, 1-98.

[5] M. J. Kadhim. (2010). Microstructure and High Temperature Oxidation Behavior of Aluminum Alloyed Cast Irons. Modern Applied Science. 4, 130-135.

[6] M. Takeda, T. Onishi, S. Nakakubo & S. Fujimoto. (2009). Physical Properties of Iron-Oxide Scales on Si-Containing Steels at High Temperature. Materials Transactions. 50 (9), 2242-2246

The 73rd World Foundry Congress
„Creative Foundry"
23 - 27 September 2018, Krakow, Poland

MODELING OF MICROSTRUCTURE IN HYPOEUTECTIC HIGH CHROMIUM CAST IRON

Dorota Siekaniec*[1], Dariusz Kopyciński[1], Andrzej Szczęsny[1], Edward Guzik[1], Andrzej Burbelko[1], Adam Nowak[2]

[1]AGH University of Science and Technology, Department of Foundry Engineering, Al. Mickiewicza 30, 30-059 Kraków, Poland
[2] Odlewnie Polskie S.A. based in Starachowice, al. Wyzwolenia 70, 27-200 Starachowice, Poland
**Corresponding address: e-mail: dorotkasiekaniec@gmail.com*

Key words: High Chromium Cast Iron, Titanium, Modeling, Law of Growth, Microstructure,

The present investigation shows the results of modeling the microstructure of hypoeutectic High Chromium Cast Iron. To achieve this, a series of tests and experiments were carried out. It consisted of conducting the melts and analyze the obtained microstructure using the EBSD technique. Based on this results performed a statistical analysis of the influence the degree of undercooling (ΔT) on the amount of primary austenite grains (N_V). All of these studies led to design the model of microstructure of primary austenite grains in the hypoeutectic High Chromium Cast Iron in ProCast.

1. Introduction

The High Chromium Cast Irons is an effective material for castings with the required high wear resistance. The applications of that material include parts for steel making plants, power plants, mineral industry plants, and others. Grain orientation and shape, their properties as well as the type and size of the phases have a major impact on the properties of modern engineering materials. Therefore, it is important to study, analyze and control the way of alloys crystallization. [1-5]

2. Methodology

The melts were carried out in industrial conditions. Liquid metal overheated to 1600°C in induction furnace. The liquid metal poured from furnace to ladle. A cooling curve was recorded for each melt. The inoculant was in powdered form with grain size less than 1 mm. The inoculation was carried out on the stream of liquid metal. The following amount of inoculants have been used; 0.17%Fe-Ti, 0.33%Fe-Ti and 0.66%Fe-Ti. The sample for testing was cut from the casts.

The sample for EBSD analysis was prepared and tested in the Institute of Metallurgy and Materials Science Polish Academy of Sciences, using high resolution scanning electron microscope with EBSD system.

Statistical analysis was performed based on received data. Modeling was made in Computer Modeling Laboratory of Crystallization Processes, using ProCast software.

3. Results and discussion

A degree of undercooling was determined for each melt from the cooling curve. Results are presented in the Table 1.

Table 1. Degree of undercooling vs Fe-Ti additions

Sample, wt%	Degree of undercooling, K
0. 0 Fe-Ti	18.5
0.17 Fe-Ti	17.5
0.33 Fe-Ti	9.0
0.66 Fe-Ti	6.0

Based on the IPF maps (Fig. 1) obtained by analysis using the EBSD technique, the amount of primary austenite grains was calculated for each melt.

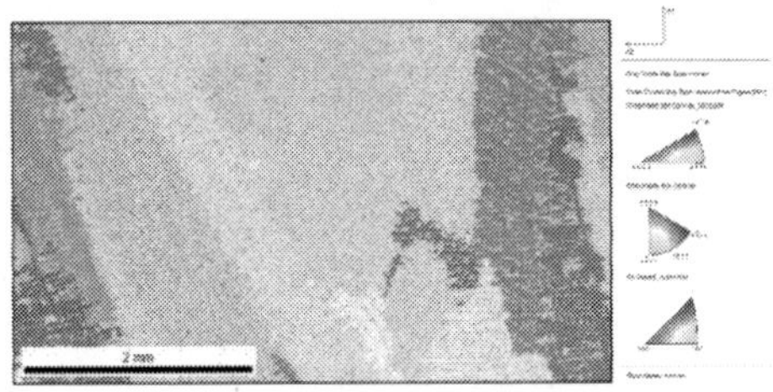

Fig. 1. IPF map with marked primary grains (a) and reference triangle (b) for initial sample

The next step was to perform statistical analysis and present the law of the growth of primary austenite grains as a function $N_V = f(\Delta T)$.

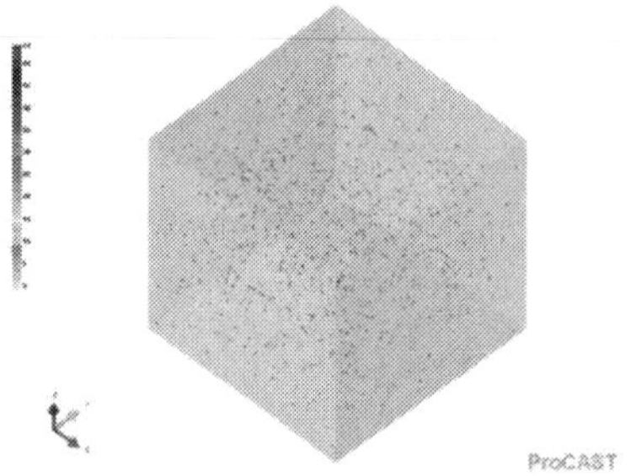

Fig. 2. The distribution of nucleus in the casting for initial sample

The diagram of the growth was modeled based on the experimental data. Fig. 2 shows the distribution of nucleus in the casting for initial sample. Based on these studies, a model for High Chromium Cast Iron was created.

4. Conclusions

The presented results show a model of shaping the cast iron microstructure. It was created and verified with experimental data. It has been established that it is possible to predict and, consequently, influence the number of grains in the casting. This is importance because microstructure influences the properties of the cast. This type of study is innovative and uses the latest research techniques

Acknowledgments
The study was co-financed by NCBiR as a targeted project No. PBS/B5/44/2015

References

[1] D. Siekaniec, D. Kopyciński, E. Guzik, E. Tyrała, A. Nowak. (2017). Effect of different inoculants on impact toughness in high chromium cast iron. Archives of Foundry Engineering. 17(2), 85–88.

[2] D. Siekaniec, D. Kopyciński (2017) Phase analysis and crystallographic orientation of high chromium cast iron grain using EBSD technique. Journal of Casting & Materials Engineering. 1(1), 15–19.

[3] M. Faryna (2012). Diffraction of backscattered electrons in a scanning electron microscope. Cracow: Wydawnictwo AGH.

[4] A. Studnicki, (2005).Crystallization temperatures of chromium cast iron in function cooling rate. Archives of Foundry Engineering. 48/15, 371-378.

[5] M. Rappaz, Ch.-A. Gandin. (1993) Probabilistic modelling of microstructure formation in solidification processes. Acta Metallurgica et Materialia. 41(2) 345-360.

The 73rd World Foundry Congress
„Creative Foundry"
23 - 27 September 2018, Krakow, Poland

POSSIBILITY ON ESTIMATION OF FATIGUE LIMIT USING X-RAY CT APPARATUS IN SPHEROIDAL GRAPHITE CAST IRON

Kaho Tanaka[1], **Naoto Shiraki**[2*], **Akira Sugawara**[1], **Ryosuke Fujimoto**[3],

Masami Tomizawa[4] **and Takumi Hara**[4]

[1] *Tokyo City University, Graduate school of Engineering, Setagaya, Tokyo, Japan*
[2] *Tokyo City University, Faculty of Engineering, Setagaya, Tokyo, Japan*
[3]*Toshiba Machine Co.,LTD., Numazu, Shizuoka, Japan*
[4]*Toshiba IT & Control Systems Corporation, Shinjuku, Tokyo, Japan*
**Corresponding address: nshiraki@tcu.ac.jp*

Key words: Spheroidal graphite cast iron, Fatigue limit, High-resolution X-ray CT, Fracture origin, Defect size

In this study, the axial fatigue test specimen was machined from large size ingot of spheroidal graphite cast iron. Axial fatigue test was also carried out using an electric-hydraulic fatigue tester (made by Shimadzu Corp. Ltd.) with a capacity of ±19.6 kN. The test was conducted using a sinusoidal load control system with a load repetition frequency of 17Hz, and stress ratio, R=-1. The maximum number of repetitions for the fatigue test was 1×107 cycles. The test was carried out at room temperature. The approximation of S-N diagrams obtained by the axial fatigue test and the fatigue limits were calculated in conformity with JSMS-SD-6-02. The fracture origins and their peripheral areas of fractured all specimens after the fatigue test were observed using a scanning electron microscope.The defects in specimen were detected using X-rays CT and estimated fatigue limit by equation of Murakami considering defect position and stress distribution. And we compare fatigue limit which was estimated from fracture analysis using the equation with the fatigue limit of experimental results. The possibility on estimation of the fatigue limit was investigated by the non-destructive technique.

1. Introduction

It is desired to establish a safe and simple method using for strength design because experimental determination of fatigue limit spends enormous time and effort [1]. Murakami et al. proposed the 4-parameters method to estimate the fatigue limit in consideration of the hardness of the matrix and defect size in a material, and showed that the fatigue limit can be evaluated with a good accuracy for a material with defects whose sizes are less than 1mm [2].

Hardness of the matrix can be obtained conveniently, but for defect size, many fatigue specimens are required and also the statistical analyzing of defect size from fracture surface observation is necessary.

In this study, the fatigue limit was estimated by equation of Murakami considering the defect position and stress distribution. The fatigue limit obtained by carrying out axial fatigue test and the fatigue limit which was estimated from defect size, which was detected by X-ray CT apparatus with high resolution was compared. The possibility on estimation of the fatigue limit was investigated by the non-destructive technique.

2. Sample material

Sample material targeted chemical components shown in Table.1 and rare-earth elements in spheroidizing agent was performed casting by adding so as to be 200 ppm in the melt. Figure 1 shows the shape of the axial load fatigue test specimen followed by JIS Z2273 [3]. The specimens were taken from h500×w500×t500 mm size of ingot.

Table 1. Chemical composition of material used
(mass%)

Material	C	Si	Mn	P	Mg	S
FCD400	3.4~ 3.5	2.5	0.9	≤0.03	0.05	≤0.01

3. Experimental methods

15 axial fatigue test specimens were observed before fatigue test, using X-ray CT apparatus with high resolution, TOSCANER- 33000μFD-Z manufactured by Toshiba IT & control systems Co.,LTD. The observation range is 36 mm including the parallel part of the specimens indicated by hatching in Fig.1.

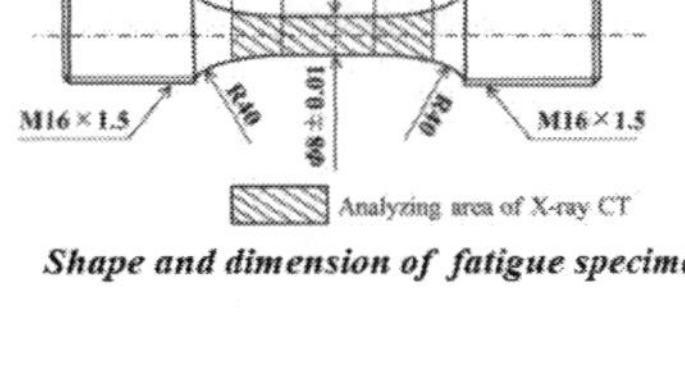

Fig. 1. ***Shape and dimension of fatigue specimen***

In the axial fatigue test, electro-hydraulic servo pulsar was used with a loading frequency of 17 Hz and stress ratio R=-1 at room temperature in the atmosphere. The number of cycle to discontinue test was set to 1×107 cycles. The fracture origins were observed by scanning electron microscope (SEM).

4. Experimental Results and Discussion

Defects can be detected larger than 0.02mm^3 with a volume in the parallel position in axial fatigue specimen observed by X-ray CT apparatus.. Defects which were detected using the software with the apparatus was classified by the volume. And largest defect size (cross sectional direction) was determined from largest volume defect in parallel position. In consideration of the position (surface or inside), fatigue limit of each specimen was estimated.

It is known that the fatigue limit of a material with a defect (crack) can be estimated using the 4-parameter model [4].

Equation proposed by Murakami et al. can be applied at region mainly ruled by the resistance against the propagation of a small defect (crack). The fatigue limit could be accurately estimated using Eq.(1) considering stress distribution for the material with defects of the size smaller than 1,000μm[2].

$$\sigma_w = \frac{\alpha(HV + 120)(r - \sqrt{area})}{(r \times \sqrt{area}^{1/6})} \quad (1)$$

HV is the Vickers hardness of matrix, r is a radius of parallel part of specimen, α is a coefficient, which depends on the position of a defect.

$\sqrt{area}_{max}$ is square root of the rectangle approximation of the defect when the cross sectional area of the defect having the largest volume is the maximum. Estimated by equation (1), the fatigue limit was 122 MPa. As a result of the experimental fatigue limit is about 130 MPa. It can be said that highly precise estimation is possible by this method.

Fig.2 shows the maximum defects in test specimens observed by high-resolution X-ray CT and secondary electron image of fractured surfaces observed by SEM. It was confirmed that the fracture origins was micro-shrinkage and was the same as the maximum defect observed by X-ray CT.

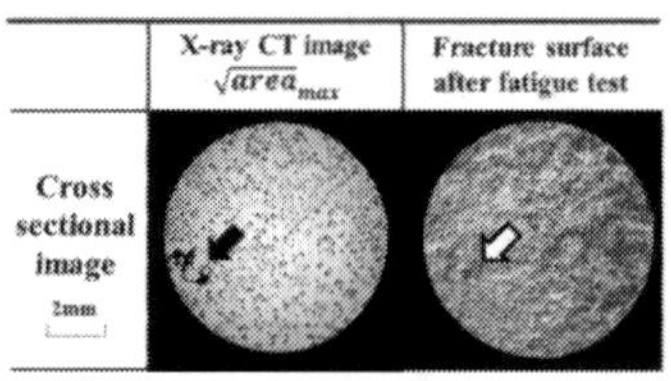

Fig. 2. Cross section images of specimen at maximum defect (fracture origin)

However, in the case of a few specimens, the fracture origin didn't correspond to a detected largest defect by X-ray CT. In such of case, it seems to be impossible to estimate fatigue limit correctly.

So, the volume of the defect included in each specimen, large, 10 of dominance was picked out. Fig.3 is the plotting the results of defect size of the largest section, √area of 15 specimens by X-ray CT on gumbel paper. From the results, the defect size (call as $\sqrt{area}_{F=50\%(n=1-10)}$) at the cumulative distribution function F = 50% was obtained.

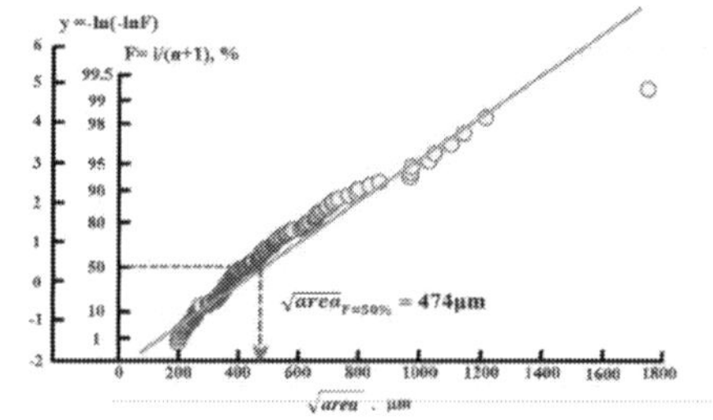

Fig. 3. The plotting the results of defect size of the largest section, √area of 15 specimens on gumbel paper

$\sqrt{area}_{F=50\%(n=1-10)}$ was calculated to be 474μm. And using this $\sqrt{area}_{F=50\%(n=1-10)}$, the fatigue limit was estimated from equation (1). When fatigue limit is estimated, since coefficient α converted the position of the defect as a surface, α is estimated as 1.43. As a result, Estimated from this method, the fatigue limit was 139 MPa.

In the two kinds of estimation method, it's possible to estimate the high precision. It seems to be possible that estimation method of the fatigue limit by X-ray CT apparatus with high resolution.

5. Conclusions

An axial load fatigue testing was carried out and specimen observed by high-resolution X-ray CT.

Obtained conclusions as follows.

(1) X-ray CT is possible to observe internal defects of spheroidal graphite cast iron.

(2) Estimation method of fatigue limit by Eq.(1) using size of the defects observed by X-ray CT and hardness of the matrix was an useful method.

References

[1] K.Ohmichi : Strength of fracture Materials, (Ohmsha,1985)p.194.

[2] Y.Murakami : Metal Fatigue, Effects of Small Defects and Nonmetallic Inclusions, (Yokendo, 1993)p.43.

[3] JIS Z 2273 : General Rules for Fatigue Testing of Metals, (1978)

[4] Y.Sugiyama, K.Asami and S.Matsuoka : Transactions of JSME(A)(in Japanese)588(1992)2287-2292.

The 73rd World Foundry Congress
„Creative Foundry"
23 - 27 September 2018, Krakow, Poland

INFLUENCE OF SOLUTION HARDENING ON MICROSTRUCTURE AND MECHANICAL PROPERTIES OF Al-2.5 Mg-0.7 Li ALLOY

Franjo Kozina[1], Zdenka Zovko Brodarac[1], Mitja Petrič[2]

[1] *University of Zagreb Faculty of Metallurgy, Aleja narodnih heroja 3, Sisak, Croatia*
[2] *University of Ljubljana, Faculty of Natural Sciences and Engineering, Aškerčeva cesta 12, Ljubljana, Slovenia*
**Corresponding address: fkozin@simet.hr*

Key words: Al-2.5Mg-0.7Li alloy, solution hardening, microstructure, thermo-mechanical testing, mechanical properties

ABSTRACT

Although utilization of lightweight aluminum alloys in transportation industry reduced harmful emissions and improves fuel economy, their further application is limited by low stiffness, low Young modulus and sensitivity to elevated temperatures. It was found that specific strength properties of aluminum alloys can be improved by Li additions. The beneficial effect is achieved by precipitation of hardening Al_3Li precipitate.

The microstructural and mechanical properties of Al-2.5 Mg-0.7 Li alloy were investigated in as cast and solution hardened condition. The microstructure constituents and thermo-mechanical properties for both conditions were determined and compared.

1. Introduction

Dependence of environmental conservation on transportation industry placed automotive manufacturers under great pressure to reduce fuel consumption and emissions. Weight reduction is considered to be the most cost-effective option to reduce harmful emissions and improve fuel economy [1]. The Al alloys were introduced as a response to the weight reduction demand. The downsizing effect of foundry Al-Si, Al-Si-Cu, Al-Si-Mg alloys is achieved by precipitation of complex intermetallic phases [2].

However, the sensitivity to elevated temperatures [3], low Young modulus of elasticity and additional weight savings indicated the need for chemical composition redesign. Modifying chemical composition with Li additions enables direct weight savings, improved elastic modulus and specific strength properties. The additions of Li enable the formation of potent hardening precipitates, in particular metastable Al_3Li precipitate [4].

An Al-2.5 Mg-0.7 Li alloy was investigated in as cast and solution hardened condition in order to compare microstructure and mechanical properties. The microstructure in as cast condition consists of following constituents: primary aluminum (α_{Al}), intermetallic phases based on Li (metastable Al_3Li and stabile AlLi phases), ternary phase (Al_2LiMg) and secondary eutectic (Al_3Mg_2). Since the hardening is mainly achieved by precipitation of metastable Al_3Li phase, solution hardening was used in order to dissolve other intermetallic precipitates and bulk α_{Al} matrix. Identification of solution hardening parameters was performed based on thermodynamic behaviour of an alloy.

2. Materials and methods

The Al alloy was synthesized by melting in the induction melting furnace under protective atmosphere of argon. The synthesized alloy was cast into a permanent steel mold without protective atmosphere.

In order to identify temperatures significant for phase transformation and precipitations, Simultaneous Thermal Analysis (STA) was applied. The temperature of solution hardening was estimated on the base of Differential Scanning Calorimetry (DSC) results. The micro constituents in as cast condition were identified using X-Ray Diffraction (XRD).

Number of grains per unit area was determined for samples in in as cast condition, after solution hardening and after compression testing. Olympus SZ 11 with digital camera Promicra was used.

Thermo-mechanical properties were measured using Gleeble 1500D machine. Samples 1 and 2 were tested in as cast condition and at room temperature. The samples 3 and 4 were solution hardened and tested at elevated temperatures. The micro-hardness was measured on samples in as cast condition and after solution hardening using LEICA VMHT 302801 machine.

3. Results and discussion

The heat treatment regime consisted of solution hardening at 520°C followed by quenching in water and artificial aging at 190°C during thermo-mechanical testing.

The number of grains pre unit area (N_A) is given in Table 1.

Table 1. Number of grains per unit area [grains/mm]

Condition	Before deformation	After Deformation
As cast	0,51	0,37
Solution Hardened	0,64	0,57

The solution hardening led to the recrystallization of primary microstructure. The number of grains was reduced respectively. During thermo-mechanical testing of primary as cast microstructure, the number of grains significantly increased. In contrary, heat treated microstructure was not deformed at all during the same testing parameters.

The microstructure of samples before and after thermo-mechanical testing is given in Fig.1

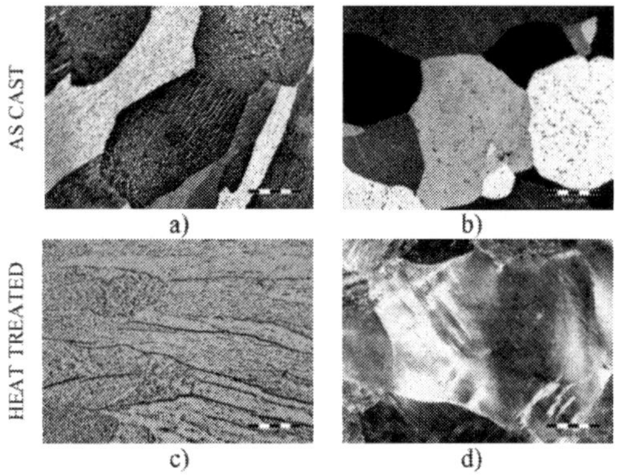

Fig. 1. The microstructure of: a) sample 1 - as cast; b) sample 2 – heat treated, c) sample 3 – as cast, thermo-mechanically tested, d) sample 4 – heat treated, thermo-mechanically tested.

The thermo-mechanical testing of as cast microstructure (Fig.1a) led to the significant texture development (Fig.1c). Thermo-mechanical testing of solution hardened and aged microstructure (Fig.1b) did not cause any directed microstructural change (Fig.1c).

The results of thermo-mechanical testing are given in Table 2.

Table 2. The results of thermo-mechanical testing

Sample No.	Strain rate, mm/min	Re, MPA	Rm, MPa	Deformation, %
1	100	375.02	543.70	46.4
2	230	155.17	289.89	41.5
3	100	180.32	273.47	58.2
4	100	178.20	260.17	55.3

The sample 1 had the highest tensile and yield strength values. Increasing strain rate up to 230 mm/min caused the drop in both values (sample 2). Samples 3 and 4 had a significantly lower strength properties compared to the first two samples. It is assumed that higher strength properties in as cast condition occurred as a result of Al_3Li precipitates. These Li based intermetallic precipitates interact with dislocations progressing from the grain boundaries. The assumption is supported by texture development during thermo-mechanical testing (Fig.1b). The solution hardening caused dissolution of Al_3Li precipitates enabling free movement of dislocations. Furthermore, the mobility of dislocations was increased due to the increase in testing temperature.

The micro-hardness measurement results are shown in Table 3.

Table 3. Results of Micro-hardness measurements

Condition	HV		
As cast	92.0	92.4	93.4
Solution Hardened	102.5	103.2	103.3

The solution hardening led to the increase of micro-hardness values.

4. Conclusions

The influence of solution hardening on microstructure and mechanical properties of Al-2.5 Mg-0.7 Li alloy were investigated. The solution hardening was performed in order to induce the metastable Al_3Li hardening precipitation and dissolving of ternary Al_2LiMg intermetallic phase and secondary eutectic Al_3Mg_2 phase and their incorporation in α_{Al} solid solution.

However, samples tested in as cast condition showed the higher strength properties (Table 2). The texture developing during deformation was a result of interaction between Al_3Li precipitates and dislocations progressing from the grain boundaries (Fig.1c). Although solution hardening has had a beneficial influence on micro-hardness values (Table 3), while at the same time having a weak influence on strength development. During deformation dislocations were moving freely, without limitation due to precipitations, resulting in low strength properties (Table 2). This is supported by lack of texture after deformation (Fig.1d).

Acknowledgements

This research and paper are supported by the funds of the University of Zagreb within the framework of Financial Support for Research "Design and Characterization of Innovative Engineering Alloys", Code: TP167.

References

[1] E. Aghion, B. Bronfin, D. Eliezer. (2001). The role of the magnesium industry in protecting the environment. Journal of Materials Processing Technology. 117(3),381-385.

[2] S. Miller, L. Zhuang, J. Bottema, A. Wittebrood, P. Smet, A. Haszler, A. Vieregge. (2014). Recent development in aluminium alloys for the automotive industry. Materials Science and Engineering. 280(37), 7-49.

[3] M. Uludağ, R. Çetin, D. Dispinar, M. Tiryakioğlu. (2017) Characterization of the Effect of Melt Treatments on Melt Quality in Al-7wt %Si-Mg Alloys. Materials. 7(5), 157.

[4] R. J. Rioja, J. Liu. The Evolution of Al-Li Based Products for Aerospace and Space industry. (2012). Metallurgical and Materials Transactions A., 43 A, 3325-3337.

The 73rd World Foundry Congress
„Creative Foundry"
23 - 27 September 2018, Krakow, Poland

MODIFICATION OF NON-METALLIC INCLUSIONS IN STEEL BY APPLICATION OF DEOXIDIZING AGENTS WITH ALKALINE METALS (CALCIUM AND BARIUM)

Jan Cerny [1]*, Pavel Ludvik [1], Martina Sipova [1] and Josef Odehnal [2]

[1] *COMTES FHT a.s, Prumyslova 995, 334 41 Dobrany, Czech Republic*
[2] *University of West Bohemia in Pilsen, Faculty of Mechanical Engineering, Univerzitni 22, 301 00 Pilsen, Czech Republic*
**Corresponding address: e-mail: jan.cerny@comtesfht.cz*

Key words: Steel refining, Non-metallic inclusions, Barium, Calcium, Vacuum induction furnace

1. Introduction

One way how to enhance steel mechanical and other properties is to decrease or modify non-metallic inclusions (NMIs). NMIs come like a result from several physical-chemical reactions during casting and subsequent solidification of liquid metal. They disturb the homogenous structure and can act as a crack imitation or cause pitting in rolling bearing. The modification of NMIs mainly takes place in final stage of steel making process - refining process. The content of detrimental inclusions and further harmful elements is reduced. In addition, the shape and chemical composition of NMIs is changed. Nowadays is widely applied aluminium, silicon and calcium (or their mixture) as deoxidizing/modification agents for metallurgy processes. According to articles [1-3] the modification with less investigated alloys containing barium in addition could be more favourable than refining only by above-mentioned elements. Generally can be concluded from [1-3] that the barium influenced the microstructure. It was observed reduction of NMIs content and change their shape. Simultaneously higher mechanical properties (impact strength and tensile strength) were achieved. Reached results were but very imperceptible that was main reason for focus on this area.

2. Experimental task

The objective of the experiment was studying the impact of deoxidizing/modifying agents containing alkaline earth elements (calcium, barium and their mixture) on steel quality. There were evaluated three kinds of modifiers CaSiBa, FeSiBa and FeSiCa with chemical composition given in **Table 1**.

Table 1. Chemical composition of modifiers (all. in wt. %, rest element Fe)

	Ca	Ba	Si
CaSiBa	14	14	61
FeSiBa	-	13	60
FeSiCa	17	-	55

These modifiers were provided in the shape of cored wire. The powder made up the core that was wrapped by steel plate. At the first stage, the main purpose was melting and casting of differently deoxidized ingots for further testing. The experimental melt of low alloy steel was casted in the ceramic crucible with lining made up mainly from Al_2O_3 in vacuum induction furnace in company COMTES FHT [4,5]. The batch consisted of several ferroalloys with very high purity, together amount of 250 kg. The power input of furnace was approximately 200 kW at the start of melting process and afterwards 100 kW for holding on the temperature. The level of vacuum was 10 kPa. After final alloying the melt was deoxidized in crucible by 200 g pure Al (0.8 kg/ton). At temperature 1600 °C, the liquid metal was casted into four ingot moulds with circle cross section. The first ingot was kept in base condition like reference, deoxidized only by Al in crucible. The rest three ingots were further modified by cored wires that represented 2-step modification. The cored wires were inserted directly into the ingot moulds before casting. The weight of pure modifiers were chosen same 100 g (1.6 kg/ton). The exact dimensions of moulds – diameter 107/117 mm and length approximately 500 mm. The exothermic lining and powder were applied on the upper part of ingot to solidification carry out here up to the end. Schematically is the whole experiment depicted in **Fig. 1.**

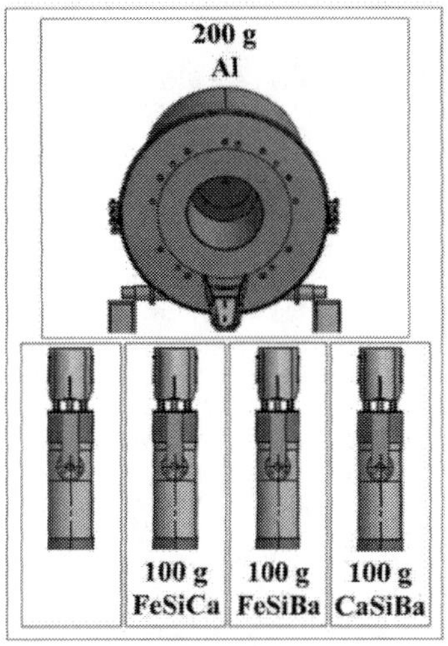

Fig. 1. Scheme of the whole experiment

Ingots (weight 60 kg) were subsequently cut for further testing. The chemical composition including content of gases was measured by optical emission spectrometry and by the method of inert gas fusion analysis. For measurement barium content was applied the method of atomic absorption spectroscopy (AAS). The determination of content of non-metallic inclusions was evaluated according to the standard ISO 4967 by means of optical microscopy (OM). The samples for documentation of morphology and chemical composition of inclusions were then observed by scanning electron microscope (SEM) together with energy dispersive analyser (EDX). The specimens were prepared by standard metallographic way.

3. Results

Firstly, it is important to note that in this abstract are presented only two materials – material deoxidized by Al and the second one by both alkaline metals - CaSiBa. Ingots in casted state did not show cracks or any surface inhomogenity. The casted structure was dendritic segregated – **Fig. 2**. This effect was caused by fast solidification rate. According the SW simulation the whole ingot should be solid in 7 minutes. The microstructure was made up from bainite (darker areas) and martensite (lighter areas). The determination the lighter area was a martensite performed by means of measurement of microhardness. In lighter aras was measured higher hardness than in darker areas.

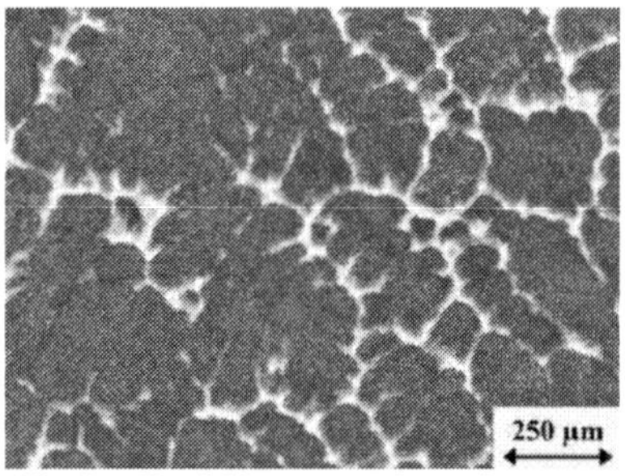

Fig. 2. Dendritic segregated casted structure

The type of deoxidation had almost no impact on chemical composition. The only effect was higher calcium content in material modified by CaSiBa. The content of oxygen remained in margin for error. (Al – 27 ± 4 ppm and CaSiBa – 30 ± 4 ppm). The barium was not detect in both ingots; more exactly the detection limit of AAS was 20 ppm.

The size or more precisely content of non-metallic inclusions stayed on the same level (mostly were presented globular oxides and aluminate type inclusions). This effect was slightly in contrary to experiments mentioned in [2,3] where was observed NMIs decrease after modification by agents with barium. It was probably influenced by casting process that was varied. In articles was refining process in ladle furnace where was cored wire applied.

Despite the same NMIs content, their chemical composition and morphology as a shape differed – **Fig. 3**.

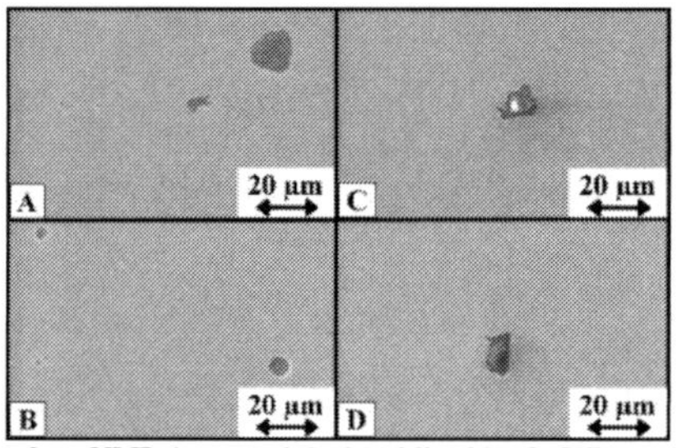

Fig. 3. NMIs in materials deoxidized by CaSiBa and Al

The material deoxidized only by Al contained especially randomly located particles comprising aluminium and oxygen (**Fig. 3 C-D**). These aluminates had irregular shape with sharp edges. The size was approximately 10-20 μm. In contrast, in material deoxidized by CaSiBa were most present particles with globular shape (**Fig. 3 A-B**) consisting of calcium, aluminium and oxygen. This effect – spheroidization of non-metallic particles was in conformity with article [2]. However, the barium was not founded by EDX in these particles.

4. Conclusion

In COMTES FHT were casted four ingots that were differently deoxidized at the end of melting process. Subsequently the impact of modifiers with alkaline metals (barium and calcium) was aim of the study. From up to now observations of microstructure can be concluded that modifier CaSiBa had favourable effect on inclusion morphology – spheroidization of NMIs. In complete version will be presented all materials (deoxidized by Al, CaSiBa, FeSiCa and FeSiBa) in details, in addition influence of modifiers on mechanical properties such a tensile/impact strength will be discussed.

References

[1] MUKAI, K. & HAN, Q.. (1999). Application of Barium-bearing Alloys in Steelmaking. ISIJ International. 39 (7), 625-636.

[2] DERDA, W., LAWENDA, P. & PARDELA, P.. (2015). The Use of Barium and Calcium Alloys in the Ladle Furnace Treatment of Selected Grades of Continuously Cast Steels. Archives of Metallurgy and Materials. 60 (1), 281-288.

[3] ISSAGULOV, A. Z., KVON, SV. S., KULIKOV, V. Y. & AITBAYEV, N. B.. (2016). Studying Microstructure of Heat Resistant Steel Deoxidized by Barium Ferrosilicon. Metalurgija. 55 (3), 388-390.

[4] MARTINEK, P., DUCHEK, M., KONOPIK, P. & PODANY, P.. (2015). Laboratory Melting, Casting and Forging of Manganese TWIP Steel. Journal of Achievements in Materials and Manufacturing Engineering. 71 (1), 39-44.

[5] CERNY, J., LUDVIK, P., SIPOVA, M. & ODEHNAL, J.. (2017). Influence of Deoxidizing and Modifying Agent CaSiBa on Microstructure of Steel with 0.3% C. Slevarenstvi. 65 (9-10), 281-285.

The 73rd World Foundry Congress
„Creative Foundry"
23 - 27 September 2018, Krakow, Poland

NEW POSSIBILITIES IN THE THERMAL ANALYSIS OF MOULDING MATERIALS

Judit Svidró[1]*, Attila Diószegi [1]

[1]*Jönköping University, School of Engineering, Department of Materials and Manufacturing, P.O Box 1026 SE-551 11 Jönköping, Sweden*
**Corresponding address: e-mail: judit.svidro@ju.se*

Key words: casting, foundry sand, Fourier thermal analysis, heat absorption, moulding material

Moulding material related studies within research activities concerning foundry technology have always been limited, despite the significant effect of moulding mixtures on the quality of cast parts. One reason behind this trend is the difficulty in interpreting results of such complex systems like moulds and cores. This paper provides a new possibility to study the heat absorption performance of materials used as moulding media in metal casting processes. By further developing the Fourier thermal analysis method of cores and moulds introduced by earlier authors, the investigation of unbonded sand has become available. Heat absorption properties of the components can be hereby separated and studied respectively. Thermal analyses were performed on sphere shaped, resin bonded cores with various binder levels and on unbonded sand samples. The temperature data collected in two points of the samples were then used for the calculation of novel thermophysical properties. The results revealed not only quantitative but qualitative differences in the characteristics of binder decomposition processes, providing deeper understanding on the thermal behavior of moulding materials. The outcome of the research provides more accurate data, which is the key for improved simulation of casting processes.

1. Introduction

Besides traditional experiments to determine mechanical properties of cores and moulds, research interest has been focusing on the thermophysical behaviour of the materials as well [1-3]. They are crucial for the prediction of the temperature field, therefore determine every parameter that affects the simulation result. Researchers make intensive efforts to find solutions for reproducible, in-situ methods with improved accuracy, and thus there are many experimental layouts can be found in the literature [4,5].

However, interpretation of the results related to moulding mixtures is complicated since the number of influencing factors is too high and it is difficult to determine the relation between cause and effect. Moulds and cores consist of base aggregates, binders and additives, constructing a granulous, porous system in which the refractory grains are connected to each other with binder bridges. During the casting process, the properties of each component is changing as the temperature increases, resulting in unpredictable behaviour of the mould. The measured parameter is thus an average of the whole system and can not be referred to the components directly. To obtain comprehensive understanding of moulding materials, the problem of separating complex results need to be solved.

Thermal analysis is a well-known and applied technique to monitor the solidification behavior of cast alloys. Svidro et. al [6] have developed a novel application of the method to gain knowledge about the thermophysical properties of moulding mixtures [7]. This paper introduces the recent updates obtained on the approach.

2. Materials and measurement methods

The temperature of sphere shaped samples with 40 mm diameter was measured during immersing them into molten aluminum. The two measuring points were in the horizontal center plane, one in the middle of the specimen and the other was 10 mm from the edge. Since unbonded sand samples would collapse without binder, thin walled containers made of aluminum-oxide were used for the experiments. The heat absorbing properties of pure silica sand was studied and used as reference point for furan hot-box mixtures with various resin levels.

The composition of studied materials is shown in Table 1. The base sand was the same type of foundry purpose silica for all mixtures as the unbonded sample. The mixing time was 1+1 minutes with the hardener and then the resin respectively, the curing temperature was 200 °C. To avoid differences in the experimental conditions, resin bonded cores were also coated with the aluminum-oxide layer. The bulk density was set to equal for all four specimens, 1,37±0,02 g/cm^3. The temperature of the aluminum melt was 680±10 °C.

Table 1. Composition of the studied samples

Binder content – based on the weight of the sand, %	Hardener content – based on the weight of the resin, %
1	33
2	33
3	33
0	0

The sphere shape has essential importance in the calculation of thermophysical properties, since the temperature gradient is homogenous in this geometry.

Temperature data were evaluated by an iteration algorithm based on heat conduction equation [6].

3. Results and discussion

The main results of the calculations are the rates of heat absorption for unbonded and bonded samples. The comparison of the pure sand and the mixture with 2% resin is shown in Fig. 1. Peaks indicate heat absorbing processes like moisture evaporation, resin decomposition and the allotropic transition of quartz. In the unbonded sand, there are only two significant processes can be observed. The first one is related to the evaporation of the water content of the sand above 100 °C. The second peak is apparently due to the α-quartz to β-quartz transition at 573 °C. The heat absorption of the mixture is more complex, with several peak values which show multistage decomposition of the resin.

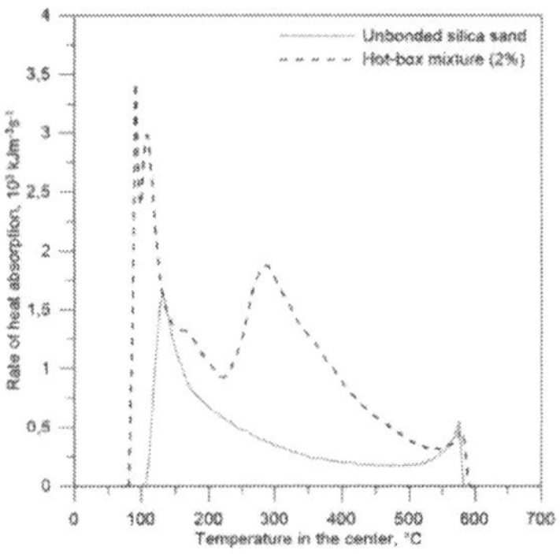

Fig. 1. Rate of heat absorption of unbonded and resin bonded samples versus temperature

By means of the unbonded sand results, it is possible to determine the binder decomposition process of the hot-box system directly. Subtracting the data regarding pure sand from the results of the mixture, the heat absorption rate of the polymerized binder can be obtained (Fig. 2).

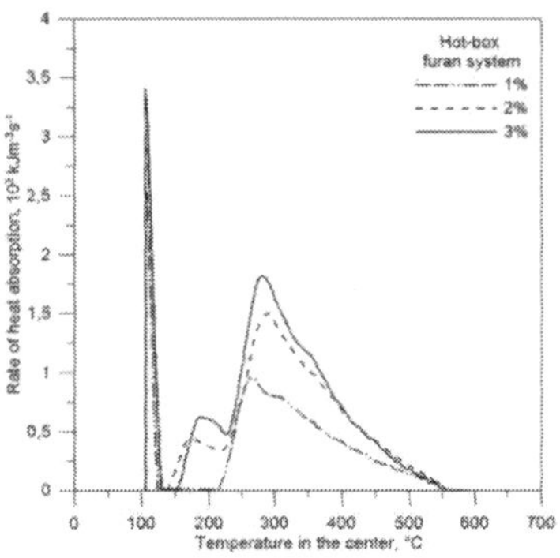

Fig. 2. Rate of heat absorption of hot-box furan systems versus temperature

The curves also show, that the energy needed for the decomposition is increasing by the addition of resin. However, the values of the particular maximum peaks are not in linear relationship. Furthermore, below a certain level of resin content, the second step of the decomposition around 200 °C has no significant influence on the heat absorbing properties because it is overlapped by the effect of the base sand in that temperature range. Regardless of the resin level, degradation of hot-box furan system is completed by 550 °C.

4. Conclusion

The unique thermal analysis technique used for the examination of moulds and cores has been refined. The opportunity to investigate unbonded aggregates provides new approach to the evaluation of thermal analysis results.

Studying different levels of binder highlighted, that although higher resin content causes higher heat absorption, affects the heat absorbing characteristic as well. By collecting the thermophysical properties of the main ingredients, it is possible to build up any kind of moulding mixture combinations, component by component. With the help of such database, not only more accurate simulation is achievable, but customized needs can be fulfilled.

Acknowledgments

The present work was financed by the Swedish Knowledge Foundation. Cooperating parties in the project were Jönköping University, Scania CV AB and Volvo Powertrain Production Gjuteriet AB. Participating persons from these institutions and companies are acknowledged.

References

[1] G. Malherbe et al. (2012). Measurement of thermal conductivity of granular materials over a wide range of temperatures. Comparison with theoretical models. 6th European Thermal Sciences Conference (Eurotherm 2012). Journal of Physics: Conference Series 395. IOP Publishing

[2] B. Grabowska et al. (2016) Thermal analysis in foundry technology. Journal of Thermal Analysis and Calorimetry. Volume 126 (Issue 1), 245-50.

[3] B. Grabowska et al. (2017) Thermal analysis in foundry technology. Journal of Thermal Analysis and Calorimetry. Volume 130 (Issue 1), 301-9.

[4] G. Solenicki, I. Budic, D. Ciglar. (2010) Determination of thermal conductivity in foundry mould mixtures. Metalurgija. Volume 49 (Issue 1), 3-7.

[5] J. Zych, J. Mocek. (2015) Destruction of moulding sands with chemical binders caused by the thermal radiation of liquid metal. Archives of Foundry Engineering. Volume 15 (Issue 4), 95-100.

[6] JT. Svidró, A. Diószegi & J. Tóth. (2014). The novel application of Fourier thermal analysis in foundry technologies. Journal of Thermal Analysis and Calorimetry. Volume 115 (Issue 1), 331–8.

[7] JT. Svidró et al. (2017) The effect of different binder levels on the heat absorption capacity of moulding mixtures made by the phenolic urethane cold-box process. Volume 130 (Issue 3), 1769-77.

DEVELOPMENT OF MATHEMATICAL RELATIONSHIPS FOR CALCULATING OF MATERIAL-DEPENDENT-FLOWABILITY OF GREEN MOULDING SAND

Dheya Abdulamer[1,2]

[1]*IMKF. TU - Bergakademie Freiberg, Agricolastr.1, 09599 Freiberg, Germany,*
[2]*ERETC. University Of Technology, Baghdad, Iraq*
[*]Corresponding address: Dheya.Abdulamer@imkf.tu-freiberg.de

Key words: Green sand mould, Stationary old methods, Continuous pressure, contact area, material-dependent-flowability

Flowability has a great effect on the performance of moulding process of green sand mixtures. Flowability is an ability of green sand to flow during mould making process. It is an important property for obtaining mould without bridge formation and without loss in energy consumed for the moulding process.

The old stationary methods are predetermined tests give an indication of flowability before an actual moulding process. Our method promotes the technician in the foundry for measuring of material-dependent-flowability of the green sand mixture during compaction process, then determine the effective compaction time.

The new method used two sensors that were developed by Prof. Bast [1]. These sensors were integrated into the template similar to the template of Orlov test [2]. In Fig. 1 is shown green sand flowability sensor.

Fig. 1 Flowability sensor

As is shown in figure 2, during the moulding process, the signals outgoing from the sensors appear on the computer as:

The (S_u) yellow curve is displacement value of the bottom sensor and the (S_o) blue curve is displacement value of the upper sensor.

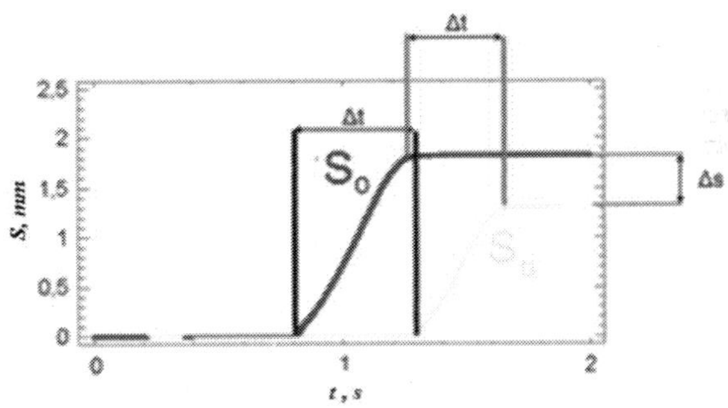

Fig. 2 Sensor signals

Several mathematical relationships have been developed for calculating of material-dependent-flowability (FB (t)) of green moulding sand. These relationships are:

$$\boldsymbol{FB(t)} = \frac{S_u(t)}{S_o(t)} * \left[\frac{\sin\left(\cos^{-1}\left(\frac{S_o(t)}{r}\right)\right)}{\sin\left(\cos^{-1}\left(\frac{S_u(t)}{r}\right)\right)}\right]^2 * \mathbf{100}, \% \qquad (1)$$

$$FB(t) = (58.79\, P_u(t) + \frac{P_u(t)}{P_o(t)}) * 100, \% \qquad (2)$$

$$FB(t) = \frac{\int_{t_i}^{t_1} S_u(t)dt}{\int_{t_i}^{t_1} S_o(t)dt} * 100, \% \qquad (3)$$

$$FB(t) = \frac{\int_0^{t_{end}} \tanh(m_u * t) * S_{u\,end}}{\int_0^{t_{end}} \tanh(m_o * t) * S_{o\,end}} * 100, \% \qquad (4)$$

where:

r	The radius of the sensor.
P_u	Contact pressure within the bottom sensor.
P_o	Contact pressure within the upper sensor.
t_i	Time of the initial signal.
t_{end}	Time of the end signal.
m_u	Th slope of the bottom curve.
m_o	The slope of the upper curve.
$S_{u\,end}$	The end value of the bottom curve.
$S_{o\,end}$	The end value of the upper curve.

In the first formula, flowability measurement based on the spring stiffness of the sensor [3], to determine the continuous pressure that generated in the contact areas between the formed sand, the bottom sensor and the upper sensor, as is shown in Fig. 3.

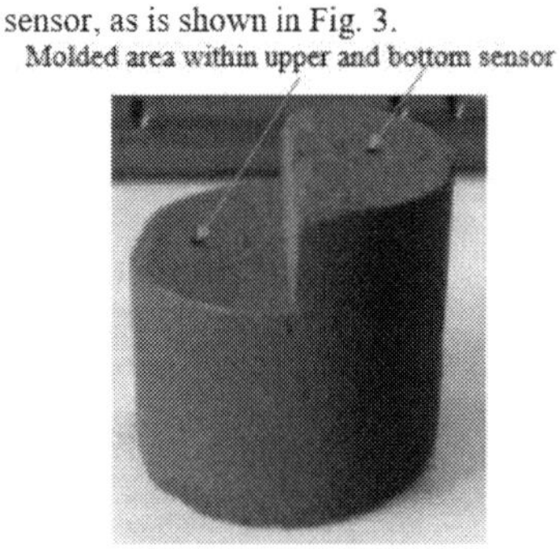

Fig. 3 Flowabiliy sample

Fig. 4 shows material-dependent-flowability that has been calculated based on the first formula. By this figure, in addition to determine the effective compactability, it can be also determine the effective compaction time without loss in energy and the reduction of the cost of the moulding process.

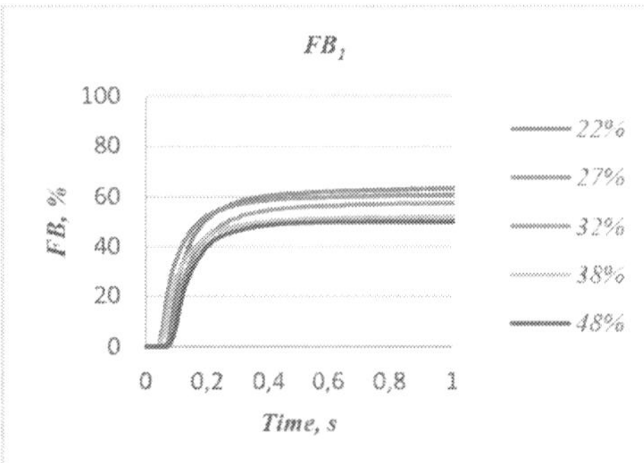

Fig. 4 FB_1, Flowability percentage vs compaction time

In the second formula [1], flowability expression is based on the relationship between the bottom and upper sensor must be 1.

The third formula relies on the limited integration of polynomial regression for measuring flowability, as is shown in Fig. 5.

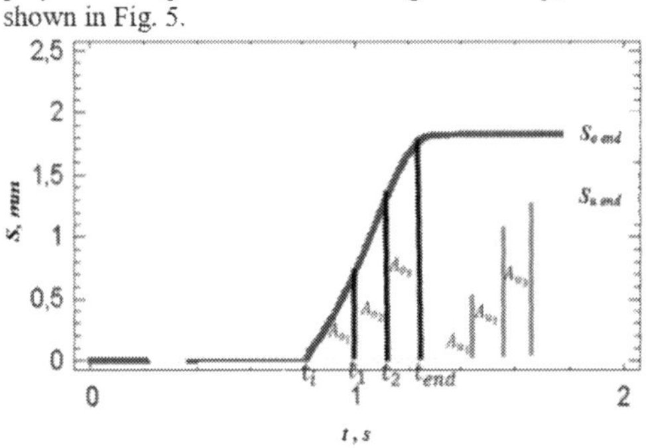

Fig. 5 Division areas under the curves

The mathematical function tanh can be used as an expression of curves of the sensor signals [1]. The slope of the bottom curve and slope of the upper curve can be determined by using genfit function through Mathcad software, and then used limited integration method for calculating the flowability.

Material-dependent-flowability that has been calculated based on the sensor signals shows good conformity with the flowability behaviour that obtained from Orlov method. There is no big difference between flowabiliy that obtained from four mathematical functions, as is shown in Fig. 6. Thereby; all these mathematical methods can be used for determination of the flowability.

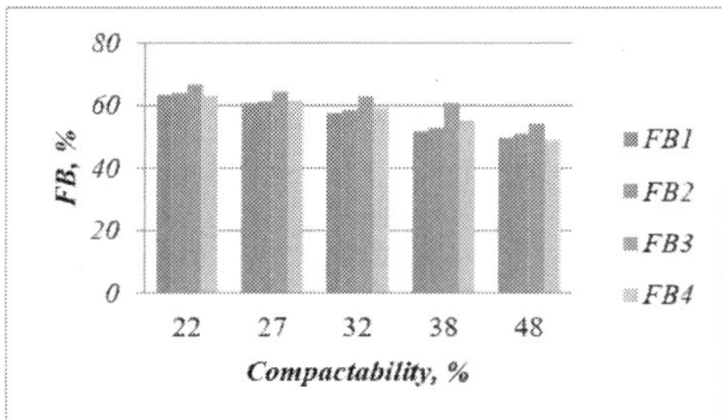

Fig. 6 Comparison final values of the flowability % FB_1, FB_2, FB_3 and FB_4

Acknowledgements

I have to say many thanks to DAAD, my supervisors Prof. Hentschel and Dr. Kadauw for giving me this opportunity for studying in TU-Bergakademie Freiberg. I have to introduce my deep thank Prof. Bast and to the best technician Mr. Engmann for giving me wise advice.

References

[1] J. Bast. (2013). A new method for measurement of flowability of green moulding sand, Archives of metallurgy and materials. Vol. 58, Issue 3.

[2] Masahiro MORI and Sennosuke INUI. (1972). Green Mold Characteristics of Bentonites and Their Deterioration on Repeated Usage,(in Japanese), The Journal of the Japan Foundrymen's Society, Vol. 44, No. 8 p. 659-666.

[3] Andrej Malaschkin. (2002). Entwicklung einer Methode zur kontinuierlichen Qualitätskontrolle beim Nassgussformverfahren mit automatischen Formanlagen. Dissertation, TU Bergakademie Freiberg.

The 73rd World Foundry Congress
„Creative Foundry"
23 - 27 September 2018, Krakow, Poland

SPINEL FORMING SYSTEMS (Mg-/Fe-/Mn-Al-O) AS FUNCTIONAL COATING MATERIAL ON CARBON-BONDED FILTERS FOR STEEL MELT FILTRATION

Benjamin Bock [1*], Anne Schmidt [1], Edyta Sniezek [2], Steffen Dudczig [1], Gert Schmidt [1], Jacek Szczerba [2], Christos G. Aneziris [1]

[1]TU Bergakademie Freiberg, Institute of Ceramic, Glass and Construction Materials, Agricolastraße 17, 09599 Freiberg, Germany,
[2]AGH University of Science and Technology, Faculty of Materials Science and Ceramics, al. A. Mickiewicza 30, 30-059 Krakow, Poland
**Corresponding address: e-mail: benjamin.bock@ikgb.tu-freiberg.de*

Key words: steel melt filtration, carbon-bonded alumina, spinel coating, steel cleanliness

1. Introduction

Non-metallic inclusions considerably impair the properties of casted steel products and hence reduce their quality. While large, exogenous inclusions can be removed by slag or casting systems [1], the removal of endogenous inclusions, exhibiting a size of smaller than 10 µm, is more difficult. For this purpose, new filter systems have been developed to bind these inclusions on the filter surface and thus achieve a higher purification degree of molten steel. According to Aneziris et al. [2], carbon-bonded alumina proves as promising filter material. Since steel melts are generally deoxidized by aluminium, a high amount of alumina inclusions is formed. Due to its mode of action as "reactive" [3] and "active" [4] filter material, carbon-bonded alumina interacts with a steel melt, that leads to the development of an in-situ formed, polycrystalline layer on the filter surface, as shown in Figure 1.

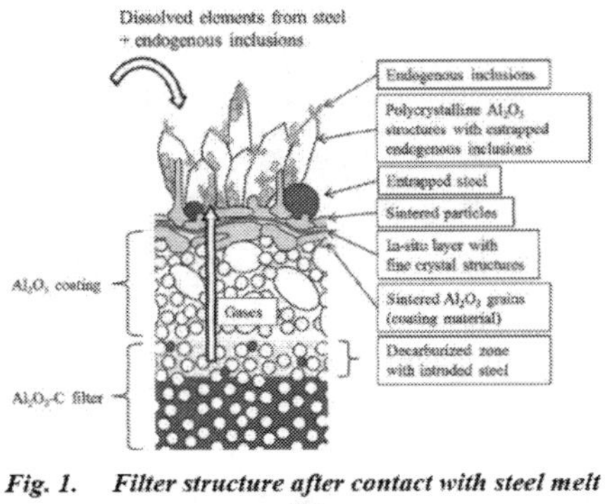

Fig. 1. Filter structure after contact with steel melt [5]

On the one hand, alumina inclusions are entrapped in this structure during the development and on the other hand are deposited on the new layer due to the attraction forces between these media. [6] Recent studies [5,7,8] revealed that the application of a functional coating, exhibiting a "reactive" and "active" behavior, increases considerably the filtration efficiency. In order to investigate the impact of spinels as coating material for the steel melt filtration, five different spinel compounds ($MgAl_2O_4$, $FeAl_2O_4$, $MnAl_2O_4$, $Fe_{0.5}Mg_{0.5}Al_2O_4$, and $Fe_{0.5}Mn_{0.5}Al_2O_4$) were applied on carbon-bonded alumina filters and analyzed after contact with molten steel.

2. Materials and Methods

Based on the patent of Schwartzfelder and Sommers [9], polyurethane filters (20x20x150 mm³ and 10 ppi) were covered with carbon-bonded alumina slurries, optimized by Emmel et al. [8], in two steps. Firstly, the filters were impregnated and secondly spray coated with 24 hours drying at room temperature after every procedure. Subsequently, the filters were sintered at 800 °C.

The sintered filters were then spray coated with a carbonaceous spinel slurry, containing one of the mentioned spinel compounds (93.75 wt%), Carbores®P (6.25 wt%), deionized water (42.00 wt% regarding the solid content) as well as appropriate additives, and were afterwards sintered. Both sintering at 800 °C and 1400 °C corresponded to those of Emmel et al. [8].

Thereupon, the coated filters were immersed in a 42CrMo4-steel melt for 10 s with the aid of a steel cast simulator. All other test parameters were the same as described by Dudczig at el. [6]. The characterization of the filter surfaces before and after the immersion attempts were carried out by light microscopy and SEM. Also, the phase development of the coating materials was investigated before and after sintering by XRD. In order to estimate the filtration efficiency, solidified steel samples were taken after each immersion test and analyzed by means of ASPEX and spark emission spectrometer to evaluate remaining inclusions and the chemical composition.

3. Results and Discussion

After the filters were sintered at 1400 °C, a color change of the filter surfaces was observed. The XRD investigation evinced that a partial decomposition of the spinel compounds, except of $MgAl_2O_4$-C, proceeded. As result of this process, the functional coating did not only consist of spinels, but also include the oxidic components of their components (Al_2O_3, Fe_2O_3, and MnO_2). According to Woodland and Woods [10], this decomposition occurs at high temperatures and low oxygen partial pressures, which were prevailed during sintering.

Due to the interactions between filter and steel melt, a vitreous, in-situ formed layer with a multi-crystalline structure arose on the filter surfaces during the immersion tests, resulting from a carbothermal reaction [11]. The EDX analysis prefigured that these media mainly consisted of alumina, which is in agreement with the data of Schmidt et al. [5] and Dudczig et al. [6]. However, the number and size these crystals distinguished between the individual carbonaceous spinel coatings. Figure 2 presents that this multi-crystalline structure was most distinctive in case of $Fe_{0.5}Mn_{0.5}Al_2O_4$-C, followed by $FeAl_2O_4$-C and $MnAl_2O_4$-C. In comparison, $MgAl_2O_4$-C and $Fe_{0.5}Mg_{0.5}Al_2O_4$-C revealed smaller crystal sizes and distributions. Due to the decomposition of the spinel compounds, $Fe_{0.5}Mn_{0.5}Al_2O_4$-C, $FeAl_2O_4$-C, and $MnAl_2O_4$-C encompassed higher amounts of alumina that could lead to elevated dissolution and precipitation processes of alumina, originating from the filter coating, and intensified the formation of the multi-crystalline structures. Moreover, it was expected that Fe and MnO, emerged by the reduction of Fe_2O_3 and MnO_2, acted as catalysts during the carbothermal reaction. Hence, the development of the multi-crystalline structures in case of Fe- and Mn-containing spinel coatings could be accelerated.

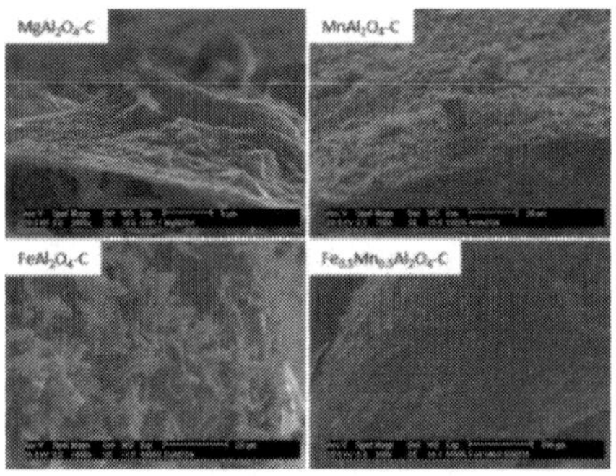

Fig. 2. Filter surfaces after immersion test

Although, the steel melt was deoxidized with aluminium, the spark emission spectrometry determined a lower aluminium content after filtration. This observation was traced back to an increment of the aluminium activity due to an increasing amount of dissolved carbon [12] and thus alumina was formed by the reaction with dissolved oxygen. This is consistent with the results of the ASPEX investigations, which identified alumina as the main inclusion type. However, the alumina inclusions were reduced to 40 – 65 % by filtration, whereby the $Fe_{0.5}Mn_{0.5}Al_2O_4$-C coated filter attained the best results (40 % remaining alumina inclusions).

4. Conclusion

In order to improve the steel cleanliness, carbon-bonded alumina filters with different, carbonaceous spinel coatings were immersed for 10 s in an Al-deoxidized 42CrMo4 steel melt. All tested filters exhibited an interaction with the steel melt, whereby filters with a Fe- and Mn-containing coating, especially $Fe_{0.5}Mn_{0.5}Al_2O_4$, obtained a higher filtration efficiency.

Acknowledgment

The authors would like to thank the German Research Foundation (DFG) for the financial support of this work in frame of the Collaborative Research Center CRC 920.

References

[1] Dominik Krewerth, Tim Lippmann & Anja Weidner. (2016). Influence of non-metallic inclusions on fatigue life in the very high cycle fatigue regime. International Journal of Fatigue. 84. 40-52.

[2] Christos G. Aneziris, Undine Fischer & Marcus Emmel. (2012). Multifunctional Filter Systems for the steel-melt-filtration. Keramische Zeitschrift. 64 (2). 124-128.

[3] Christos G. Aneziris. (2014). DE 102011109684. Stahl- und Aluminiumschmelzefilter aus reaktiven Keramiken.

[4] Christos G. Aneziris. (2014). DE 102011109681. Stahl-schmelzfilter und Verfahren zu ihrer Herstellung.

[5] Anne Schmidt, Anton Salomon & Steffen Dudczig. (2017). Functionalized Carbon-Bonded Filters with an Open Porous Alumina Coating: Impact of Time on Interactions and Steel Cleanliness. Advanced Engineering Materials. 19 (9). 1700170.

[6] Steffen Dudczig, Christos G. Aneziris & Marcus Emmel. (2014). Characterization of carbon-bonded alumina filters with active or reactive coatings in a steel casting simulator. Ceramics International. 40 (PB). 16727-16742.

[7] Enrico Storti, Steffen Dudczig & Marcus Emmel. (2015). Functional Coatings on Carbon-Bonded Ceramic Foam Filters for Steel Melt Filtration. Steel Research. 87 (8). 1030-1037.

[8] Marcus Emmel & Christos G. Aneziris. (2013). Functionalization of carbon-bonded alumina filters through the application of active oxide coatings for steel melt filtration. Journal of Materials Research. 28 (17). 2234-2242.

[9] Schwartzenfelder. (1963). US 3090094. Method of making porous ceramic articles.

[10] Alan B. Woodland & Bernard J. Wood. (1990). The breakdown of hercynite at low f_{O_2}. American Mineralogist. 75. 1342-1348.

[11] Tilo Zienert, Steffen Dudczig & Olga Fabrichnaya, (2014). Interface reactions between liquid iron and alumina-carbon refractory filter materials. Ceramics International. 41. 2089-2098.

[12] Jacques Poirier. (2015). A review: Influence of refractories on steel quality. Metallurgical Research Technology. 112 (4). 410.

DEVELOPMENT OF NEW HEAT-RESISTANT STEEL TO AVOID THE DETRIMENTAL Z-PHASE PRECIPITATION AT HIGH TEMPERATURES

Willian Martins Pasini[1,2]*, Matheus Roberto Bellé[1], Túlio Sérgio Nascimento[2], Leonardo Pereira[1], Régis Fabiano do Amaral[1], Vinicius Karlinski de Barcellos[1]

[1]*Federal University of Rio Grande do Sul, Engineering School, Foundry Laboratory, Bento Gonçalves Ave. 9500, Postal code: 15.021, Porto Alegre, Brazil.*
[2] *Federal Institute of Science, Technology and Education of Rio Grande do Sul, Avelino Antônio de Souza Street 1730 Caxias do Sul, Brazil,*
**Corresponding address: e-mail: willian.pasini@ufrgs.br*

***Key words**: Heat-resistant steel, Z-phase precipitation, Long-term creep-resistance, Thermodynamic modelling*

1. Introduction

Traditional coal-fired power plants have been emitting environmentally damaging gases, at high levels compared to other electric power generation options. Adoption of Ultra Supercritical (USC) power plants with increased steam parameters significantly improves efficiency, which reduces gas emissions [1]. For adoption of USC power plants, it is necessary an improvement in creep performance of martensitic steels to allow the use of components operating at 650°C in the next generation of USC power plants. High chromium ferritic-martensitic steels have been employed due to their good thermal properties and low cost, as compared to austenitic steels and Ni-based superalloys [1,2].

The long term creep strength of the 9–12%Cr martensitic steels relies primarily on precipitation strengthening by VN and/or Nb(C,N). These precipitates are small and finely distributed in the microstructure and have a very low coarsening rate at the intended service temperature. However, they are not thermodynamically stable and will, in time, be replaced by a similar nitride known as the Z-phase, Cr(V, Nb)N, which is the cause of the microstructure instabilities. The Z-phase precipitation will not contribute to the strengthening and the consumption of the (V, Nb)N nitrides during their growth, causing a significant drop in creep strength [2].

In the present research, which is based on the chemical composition of C12A steel (ASTM A217), an alternative alloy is suggested to avoid the detrimental Z-phase precipitation. For that, the vanadium content in this steel alloy was intentionally increased and thermodynamically designed to evaluate phase constituents and the chemical composition. Afterwards, a small-scale ingot was casted for microstructural characterization and mechanical properties evaluation.

2. Experimental

MatCalc® software with a validated thermodynamic database for calculation of phase equilibria has been used to design the heat treatment routs and to evaluate the phase stability in the complex multi-component systems. The phase fractions were simulated for each alloy, presented in Table 1, as function of the temperature. After, the value of phase fractions was compared with conventional C12A steel.

Table 1. Chemical composition of simulated alloys (%wt)

Steel	C	Cr	Mo	W	V
C12A	0.1	9.0	1	-	0.2
10Cr1W1V	0.12	10.0	-	1.0	1.0
	Si	Mn	N	Nb	
C12A	0.3	0.6	0.04	0.08	
10Cr1W1V	0.5	0.35	0.04	-	

Small-scale ingot of alloy 10Cr1W1V was casted in an induction furnace, using ferroalloys, tungsten, low carbon steel scrap, deoxidizer and slag coagulants. The alloy was poured in a no-bake sand mold at 1650°C. The chemical composition of the ingot was determined using optical emission spectrometry and TC400 Nitrogen/Oxygen Determinator. The ingot was submitted to the following heat treatment: austenitizing at 1075°C, oil quenching and tempering at 750°C, in accordance with ASTM A217 heat treatment parameters. The alloy microstructure was characterized by optical microscope analysis and the mechanical properties were measured with tensile tests and microhardness tests.

3. Results and Discussion

3.1. Phase fraction and heat treatment temperatures for the novel alloy

Thermochemical equilibrium data was calculated employing MatCalc software and it was used to evaluate the heat treatment parameters and to investigate the stability of detrimental phases for the creep resistant steel proposed.

According to the simulation (Fig. 1), the temperatures for austenitizing of C12A steel (1040 - 1080°C) can be applied in the heat treatment procedures for the new alloy composition.

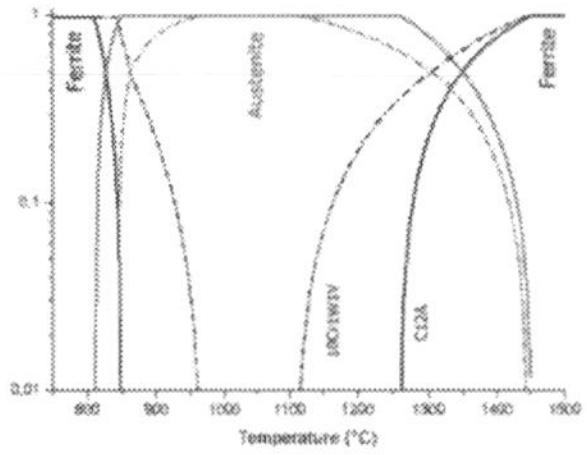

Fig. 1. Calculated temperature dependent phase fraction of the C12A and 10Cr1W1V

The simulated data suggests the extrapolation of the usual content of vanadium to increase the phase fraction of MX carbonitrides and to stabilize the detrimental Z-phase (Fig. 2).

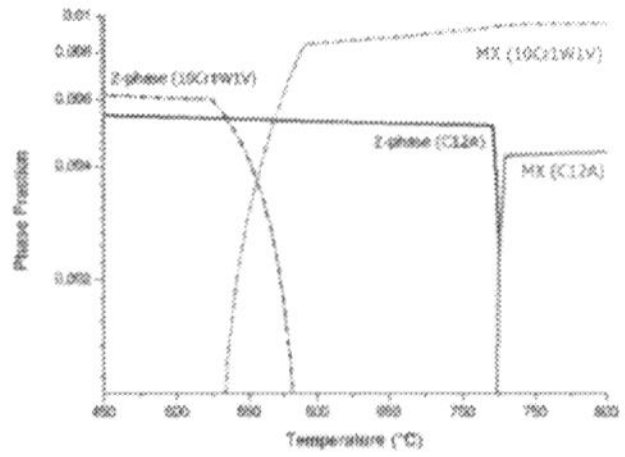

Fig. 2. Comparative of Z-phase and MX temperature dependent phase fraction between C12A and new alloy

The chemical composition 10Cr1W1V is more favorable thermodynamically in comparison with C12A steel to form stable nitrides, the phase fraction is around to 0.1% against 0.04% for conventional steel. Around the operating conditions (550°C – 650°C and 25MPa – 30MPa) and , the new composition the Z-phase content is inferior in comparison with C12A. Above 600°C the detrimental Z phase is not expected thermodynamically and for this reason does not reduce the long-term creep-resistance.

3.2. Chemical analysis and phase fraction calculated of the casted ingot

The final composition of the ingot (table 2) is the closest to the simulated composition but presents a variation on the vanadium content. This variation did not have a significant influence on the Z-phase stability, however it did influence the austenite matrix stability, according to the equilibrium data simulated ans presented in (Fig. 3).

Table 2. Chemical composition of cast ingot (%wt)

Steel	C	Cr	Mo	W	V
Ingot	0.12	10.08	-	1	1.1
	Si	Mn	N	Nb	
Ingot	0.47	0.34	0.035	-	

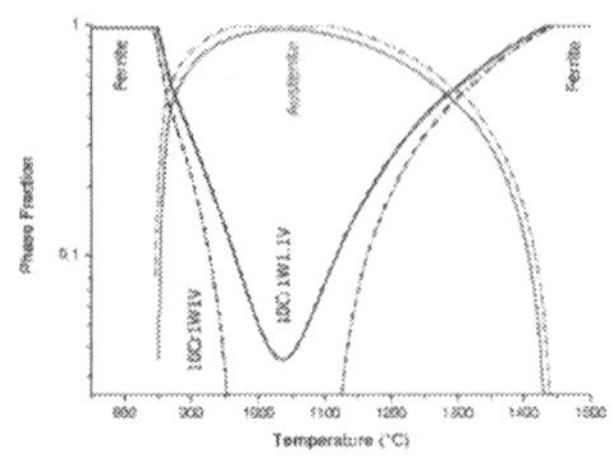

Fig. 3. Calculated temperature-dependent phase fraction of simulated 10Cr1W1V and new casted alloy

3.3. Mechanical properties and microstructure

The tensile specimens of 10Cr1W1V cast steel presented ~560 MPa for ultimate tensile strength (UTS) and ~20% elongation (EL). This results are not in accordance with ASTM A217, where the minimum tensile requirements of C12A cast steel is 585MPa and present 18% elongation. The metallographic analysis showed a dual-phase microstructure (etching with $HCl:HNO_3$ (3:1)) (Fig. 4). The different regions on the microstructure presented different values of hardness: the dark regions ~ 215 HV and light regions ~170 HV.

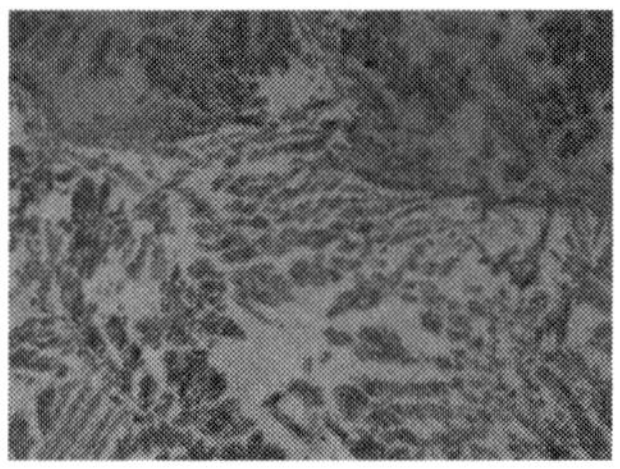

Fig. 4. Dual-phase microstrucuture of cast steel

The analysis suggests the ingot of the new composition alloy has not been fully martensitic transformed during the oil quenching resulting in a dual-phase microstructure, martensitic (dark region) and ferritic (light region), differing to the C12A which present a fully martensitic microstructure. The incomplete transformation may be a result of ferrite stabilizers in the alloy which do not permit a full austenitic transformation during the austenitizing.

4. Conclusion

In this research, a thermodynamic-designed alloy has been developed and casted. The new alloy, according to simulated data, shall not contain deleterious Z-phase at temperatures above 600°C. The excessive vanadium content in the alloy is responsible for not achieving a complete martensitic matrix after the heat treatment procedure.

References

[1] Abe,F. (2016). Progress in creep-resistant steels for high efficiency coal-fired power plants. J. Pressure Vessel Technol. 138, 040804-1 - 040804-21.

[2] Danielsen, H. K. (2016). Review of Z phase precipitation in 9–12 wt-%Cr steels. Materials Science and Technology. 32(2), 126-137.

THE DEVIATION FROM EUTECTIC COMPOSITION IN BOUNDARY LAYER FOR EUTECTIC GROWTH: A PHASE-FIELD STUDY

Zhixin Tu, Jianxin Zhou*, Yajun Yin*, Xiaoyuan Ji, Xu Shen

Huazhong University of Science and Technology, State Key Laboratory of Materials Processing and Die &Mould Technology, Wuhan 430074, Hubei Province, People's Republic of China
**Corresponding e-mail: zhoujianxin@hust.edu.cn, yinyajun436@hust.edu.cn*

Key words: Phase-Field; Eutectic Growth; Asymmetry; Lamella Morphology; Solidification

Abstract: In this paper, the deviation from eutectic composition in boundary layer for eutectic growth is studied by phase-field. According to a series of artificial phase diagram, the regular eutectic growth of these alloy is simulated during directional solidification. At steady-state, average growth velocity of eutectic lamella will equal to the pulling velocity. Due to the asymmetry of phase diagram, the composition in boundary layer deviate from eutectic composition, which is consistent with the results predicted by Magnin-Trivedi model. In addition, the S/L interface of β phase appear a depression because of the deviation.

1. Introduction

The Jackson and Hunt theory of eutectic growth give a perfect understanding of the physical mechanisms driving the eutectic solidification [1]. In the analysis by P. Magnin et.al. [2,3], the boundary layer composition determined by growth mechanism would deviate from eutectic composition, C_e. The origin lies in necessity to compensate the difference in the liquidus slopes and curvature undercooling of the phases. The value of deviation, δC, although it is very small, can drastically influence the growth behavior of some eutectic systems, such as oscillatory and divorced growth. In this paper, the multi-phase filed [4-6] is adopted to investigate the deviation, δC, in different asymmetrical phase diagram. The influence of this deviation on the eutectic morphology is also studied.

2. Simulation method

The multi-phase field model is chosen, which was originally established by Steinbach et.al and developed by Kim et.al and Eiken et.al [4-6]. In the model, the phase fields φ with constrain $\sum \varphi_i = 1$ denote the different phase in a system. In the case of a double obstacle potential, the corresponding governing equations are [5]:

$$\frac{\partial \varphi_i}{\partial t} = -\frac{2}{n}\sum_{j\neq i}^{n} s_i s_j M_{ij}\left[\begin{array}{l}\sum_{k\neq i}^{n}\left(\frac{\epsilon_{ik}^2}{2}\nabla^2\varphi_k + \omega_{ik}\varphi_k\right) - \\ \sum_{k\neq j}^{n}\left(\frac{\epsilon_{jk}^2}{2}\nabla^2\varphi_k + \omega_{jk}\varphi_k\right) + \Delta G_{ij}\end{array}\right] \quad (1)$$

$$\frac{\partial c}{\partial t} = \nabla\left(\sum_i \varphi_i D_i \nabla c_i\right) \quad (2)$$

where ϵ is gradient energy coefficient with $\epsilon_{ij} = \epsilon_{ji}$ and $\epsilon_{ij} = \frac{4}{\pi}\sqrt{\frac{\sigma_{ij}\eta}{2}}$. ω_{ik} is height of double obstacle potential with $\omega_{ij} = \omega_{ji}$, and $\omega_{ij} = 4\sigma_{ij}/\eta$. M_{ij} is phase-field mobility, calculated by thin-limit [5], $M_{ij} = 8\sigma_{ij}D\sqrt{2\omega_{ij}}/\left(\pi\Delta S \cdot m \cdot \Delta c \cdot \epsilon_{ij}^3\right)$. ΔG is driving-force for phase transformation with $\Delta G_{ij} = \Delta T \cdot \Delta S_{ij}$ and D is diffusion coefficient.

In order to investgate the influence of phase diagram symmetry on eutectic morphology, a series of phase diagram were established artificially. As the Fig.1 shows,all the phases have the same eutectic concenterations but the equilibrium liquidus slopes of β phase. The more detal physcial parameters are shown in Table.1.

Table 1. Physical properties used in the simulation

Symbol	Definition	Value
$\sigma_{\alpha L}$, $\sigma_{\beta L}$, $\sigma_{\alpha\beta}$	Interface energy	6×10^{-3} J/m^2
D_L	Liquid diffusivity	5×10^{-10} m^2/s
D_S	Solid diffusivity	5×10^{-14} m^2/s
$m_{L\alpha}$	Liquidus slope of α	-80 K/mol
$m_{L\beta}$	Liquidus slope of β	80—800 K/mol
ΔS_α, ΔS_β	Entropy of fusion	2×10^4 J/(m^3·K)

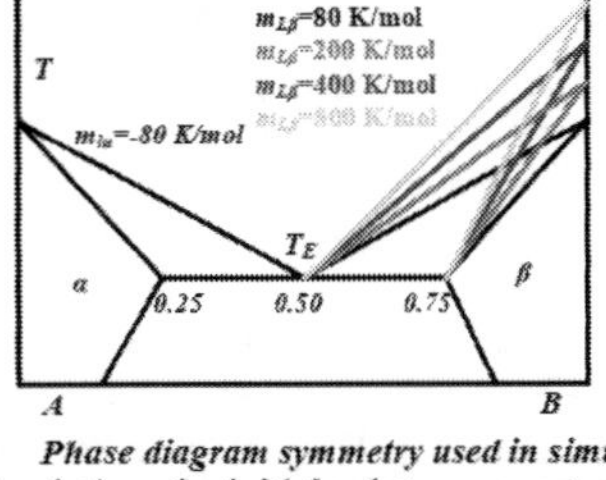

Fig. 1. Phase diagram symmetry used in simulation

For all simulation, the initial solute concenctration was set as eutectic concenctration. The lamellar spacing λ_{min} at minimum undercooling was calculated by JH model. In order to recover the Gibbs–Thomson effect with a good accuracy, $\lambda_{min} = 80\Delta x$ and the interface width $\eta = 7\Delta x$.The initial lamellar spacing is set as λ_{min}. The domain size is fixed as 160×4000, which contains two lamellar spacings. The growth velocity V is 2μm/s and the temperature gradient G is 10×10^{3}K/m. A periodic boundary condition was imposed on the system boundaries parallel to the thermal gradient direction, while adiabatic condition on the other two boundaries.

3. Results and discussion

3.1 The establishing of steady state

Because the concentration field and eutectic morphology deviate from the steady state at initial, there exists an adjustment before reaching steady state.

When reaching to steady, average growth velocity will equal to the pulling velocity and aveverge undercooling will equal to a constant, as the Fig 2 shows. The undercooling at λ_{min} increases as $m_{L\beta}$ increase.

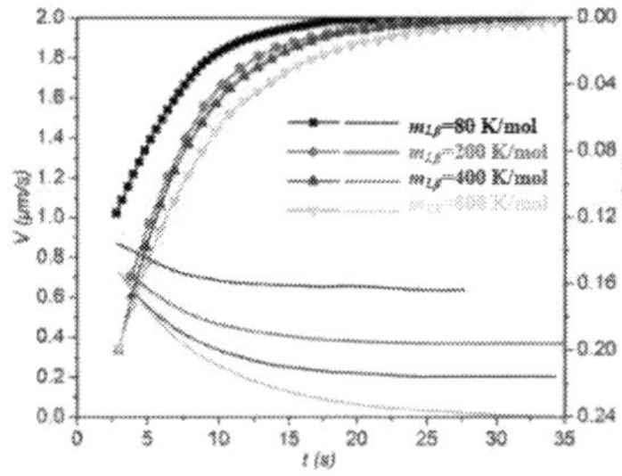

Fig. 2. Growth velocity and undercooling from initialization to steady-state

3.2 The concentration field at steady-stead

The distribution of the solute concentration along three horizontal lines, 0, $\lambda/8$, and $\lambda/4$, is shown in Fig. 3. The periodic alternation of the solute concentration behaved in a form of the sinusoidal function. And as the distance from the S/L interface increases, the amplitude of concentration curves decreases. The simulation result is in accordant with analytic solution.

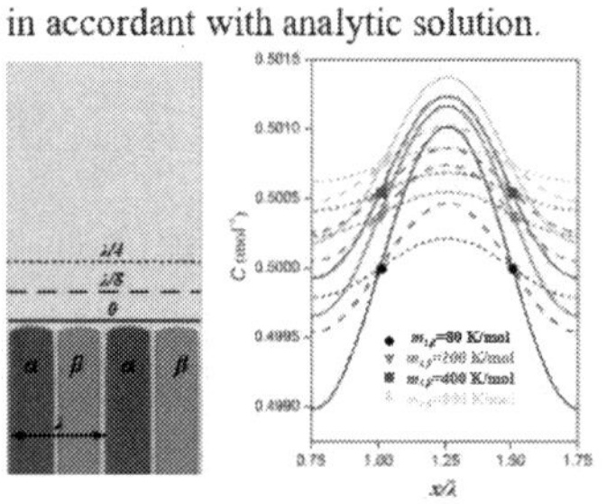

Fig. 3. Periodic composition field

For each value of $m_{L\beta}$, the intersections of periodic composition curves have the same composition, which is equal to boundary layer composition. Boundary layer composition increases with $m_{L\beta}$. As Fig.4 shows, boundary layer composition deviates from eutectic composition when $m_{L\beta}$ is not equal $m_{L\alpha}$. The value of deviation, δC, is small, usaully less than 1%, and is agree with predction by Magnin-Trivedi model.

3.3 The influences on lamellar morphology

Fig.5 shows the lamellar morphology at steady-state with different $m_{L\beta}$. The outline is defined by $\varphi_i = 0.5$. As $m_{L\beta}$ increases, the lamellar morphology of α phase change little. Nevertheless, it appear depression for β phase gradually. Due to the deviation of boundary layer composition, average constitutional undercooling of β phase becomes smaller. According to the hypothesis of isotherm for S/L interface, the variations of the constitutional undercooling must be compensated by an opposite change in the curvature undercooling. As a result, if the depression is larger enoughly, it wlould lead to a non-isothermal S/L interface.

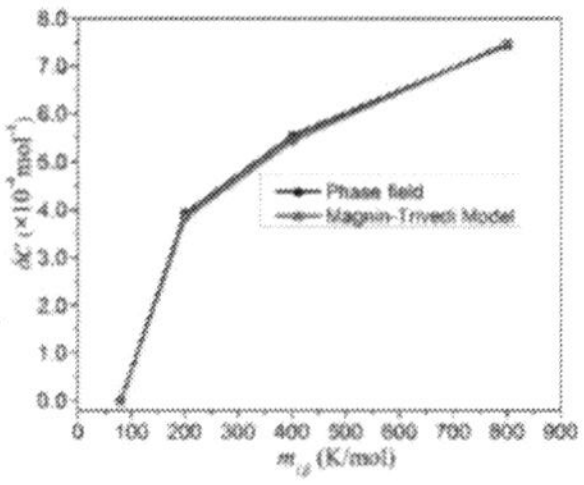

Fig. 4. Deviation of boundary layer composition

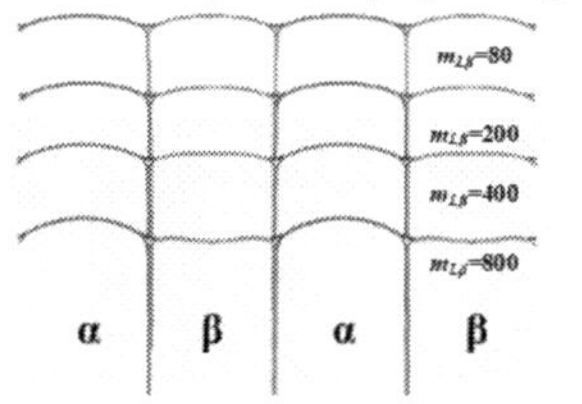

Fig. 5. Interface morphology at steady-state

4. Conclusion

1) At steady-state, average growth velocity will equal to the pulling velocity and average undercooling of S/L interface will equal to a constant.
2) Due to the asymmetry of phase diagram, the composition in boundary layer deviates from eutectic composition.
3) The eutectic phase will develop depression under influence of deviation.

Acknowledgments

This research was financially supported by a project aided by the National Nature Science Fund Projects of China (No. 51475181, No. 51605174, No. 51305149).

References

[1] Jackson K A, Hunt J D. Lamellar and rod eutectic growth[J]. AIME Met Soc Trans, 1966, 236: 1129-1142.

[2] Magnin P, Trivedi R. Eutectic growth: A modification of the Jackson and Hunt theory. Acta metallurgica et materialia, 1991, 39(4): 453-467.

[3] Magnin P, Mason J T, Trivedi R. Growth of irregular eutectics and the AlSi system. Acta metallurgica et materialia, 1991, 39(4): 469-480.

[4] Steinbach I, Pezzolla F. A generalized field method for multiphase transformations using interface fields. Physica D: Nonlinear Phenomena, 1999, 134(4): 385-393.

[5] Kim S G, Kim W T, Suzuki T, et al. Phase-field modeling of eutectic solidification. Journal of crystal growth, 2004, 261(1): 135-158.

[6] Eiken J, Böttger B, Steinbach I. Multiphase-field approach for multicomponent alloys with extrapolation scheme for numerical application. Physical Review E, 2006, 73(6): 066122.

The 73rd World Foundry Congress
„Creative Foundry"
23 - 27 September 2018, Krakow, Poland

EFFECT OF COOLING RATE ON CONSTITUENT PARTICLE FORMATION IN HIGH-SPEED TWIN-ROLL CAST Al-Mn BASED ALLOY STRIPS

Ram Song [1]*, Shinji Muraishi [2], Shinji Kumai [2]

[1]*Tokyo Institute of Technology, Department of Metallurgy and Ceramics Science, 2-12-1 Ookayama, Meguro-ku Tokyo 152-8552, Japan*
[2]*Tokyo Institute of Technology, Department of Materials Science and Engineering, 2-12-1 Ookayama, Meguro-ku Tokyo 152-8552, Japan*
**Corresponding address: e-mail: song.r.aa@m.titech.ac.jp*

Key words: high-speed twin-roll casting, cooling rate, constituent particle, aluminum-manganese-iron alloy, aluminum-manganese-silicon alloy

1. Introduction

Wrought A3xxx alloys contain Mn as a major alloying element, and Fe and/or Si as alloying elements or natural impurities. During solidification, Mn is not only solid-soluted in the Al matrix which results in supersaturated solid solution, but also formation of Mn-containing second-phase particles with other alloying elements, Fe and/or Si. The Mn supersaturation as well as size, distribution, type, and the amount of second-phase particles are quite important factors controlling various characteristics of the wrought A3xxx alloys [1,2].

For aluminum sheet production, DC-casting and subsequent hot- and cold- rolling are commonly used. On the other hand, high-speed twin-roll casting (HSTRC) has been paid attention as an aluminum alloys strip fabrication process because of its high cooling rate over 10^3 K/sec, in particular at the near-surface area of the strip [3]. For A3xxx alloys, it is expected that the HSTRC can lead to Mn supersaturation and refined constituent particles as well as their homogeneous distribution.

This study focused on the constituent particles in HSTRC Al-Mn-Fe and Al-Mn-Si alloy strips. By comparing to other casting methods with different cooling rates, the effect of cooling rate on constituent particles was investigated.

2. Experimental

Al-1.2Mn-1.0Fe (referred as AMF) and Al-1.2Mn-1.0Si (reffered as AMS) in wt.% ternary alloys were used in this study. In order to investigate the effect of cooling rate on constituent particle formation in high-speed twin-roll cast strip, three kinds of casting methods with different cooling rates were selected; DC-casting, HSTRC, and single-roll casting. The cooling rate of the DC-cast billets produced by a lab-scale machine was estimated as less than 10^1 K/sec.

A schematic of HSTRC is shown in Fig. 1. The diameter and width of the water-colled copper roll were 300 mm and 100 mm, respectively. One of the rolls was fixed by a series of springs to control initial roll seperation force as 20-60 kN. The roll rotating speed was 60 m/min. The contact length between the melt and the roll surface was about 100 mm. The initial roll gap was set as 1mm. For HSTRC, the DC-cast billets were re-melted in an electric furnace to produce the 2.5 kg molten alloy for one charge. Before the casting, a degassing was conducted using Ar gas for 20 min. The casting temperature was 30 K higher than the liquidus temperature of the melt. Cast strips of around 3 m-long and 100 mm-wide were obtained. The middle part of the resultant strip around 1 m-long showed a constant thickness of around 2.2-2.5 mm along casting direction. Cooling rate during HSTRC was estimated over 10^3 K/sec by the previous work [3].

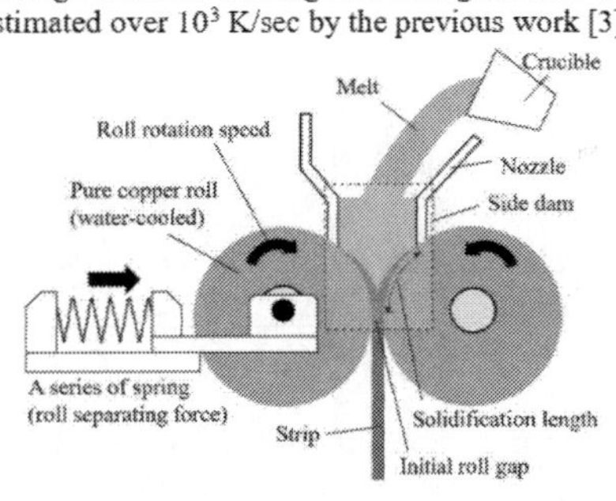

Fig. 1. Schematic of the high-speed twin-roll caster.

The DC-cast billets were also re-melted for fabrication of the ribbon specimen by single-roll casting (NEV-A05, NISSIN GIKEN). A small amount of the melt was ejected onto a copper wheel, 250 mm in diameter, rotating at 1800 rpm (23.5 m/sec) under Ar atmosphere. The gas pressure for ejecting the melt was 0.01 MPa. The 75 μm-thick and 5 mm-wide as-spun ribbons were obtained.

OM and FE-SEM observation were carried out to investigate solidification structure, constituent particle morphology and its distribution in the as-cast state. TEM observation including composition analysis was also applied for phase identification of the particles.

3. Results and Discussion

In order to examine the constituent particles morphology and distribution condition, deep-etched samples were observed as shown in Fig. 2. Relatively large primary particles as well as Al-Mn eutectic particles were observed in DC-cast AMF as shown in Fig. 2 (a). In contrast to that, large primary particles were hardly observed in HSTRC AMF. Instead, fine constituent particles were mainly located at cell or inter-dendritic areas. Basically, the constituent particles are found in the vicinity of cell/grain boundaries. Due to the high cooling rate of the HSTRC, α-Al dendrites can be refined compared to those of the DC-cast. Thus, fine constituent particles were also located in the inter-cellular areas as well as grain boundaries as shown in Fig. 2 (b). In as-spun AMF ribbon, much finer constituent particles were distributed homogeneously (Fig. 2 (c)). In DC-cast AMS, constituent particles were inter-connected at boundaries (Fig. 2 (d)). Most of them had dendritic morphology. The dendritic constituent particles along cell/grain boundaries were also observed in HSTRC AMS as marked in Fig. 2 (e). Inside of the grains, nano-size constituent particles with irregular morphologies were randomly distributed. These types of constituent particles were also observed in as-spun AMS ribbon as shown in Fig. 2 (f). Compared with other casting methods with different cooling rates, the constituent particles in DC-cast AMS showed mainly dendritic morphology, while constituent particles inside of grains in HSTRC AMS and constituent particles in as-spun AMS ribbon indicated irregular-type. From this observation results, it is considered that the cooling rates determine the initial morphology of the constituent in AMS.

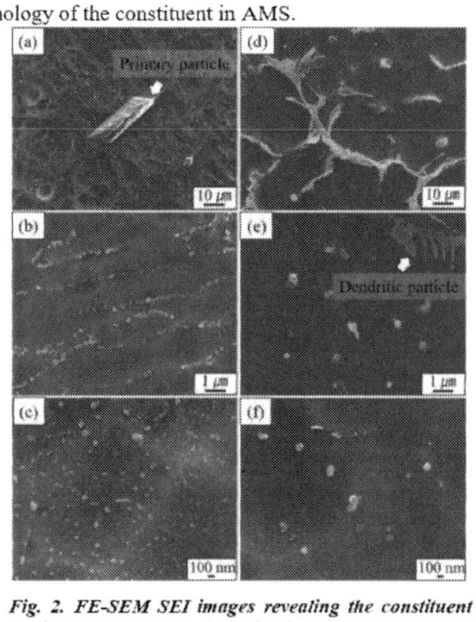

Fig. 2. FE-SEM SEI images revealing the constituent particle morphology and distribution. (a) DC-cast AMF, (b) HSTRC AMF, (c) AMF ribbon, (d) DC-cast AMS, (e) HSTRC AMS, (f) AMS ribbon.

The Mn concentration ratio of the constituent particles was examined by TEM-EDS point analysis for AMF and AMS alloy produced by different casting methods with different cooling rates. The result is shown in Fig. 3. In AMF alloy, Mn/(Mn+Fe) ratio decreased from 0.30 to 0.18 with increasing the cooling rate. Since the solid solubility of Fe in aluminum is quite low, most of Fe are expected to be segregated at the inter-dendritic region and boundaries, which corresponded to the growth front during solidification. The normal segregation effect leads to the formation of the constituent particles and increases Fe concentration in the constituent particles. In this region, on the other hand, Mn is relatively easy to solid-soluted in aluminum. Thus, it is considered to be reasonable that Mn concentration in the constituent particles decreases with increasing the cooling rate. It is also possible the formation of different phases as constituent particles by different cooling rate. In AMS alloy, Mn/(Mn+Si) ratio decreased from 0.61 to 0.16. The composition of the constituent particles in AMS alloy is difficult to interpret but the tendency also can be seen clearly that Mn concenration ratio decreased with increasing the cooling rate of casting method.

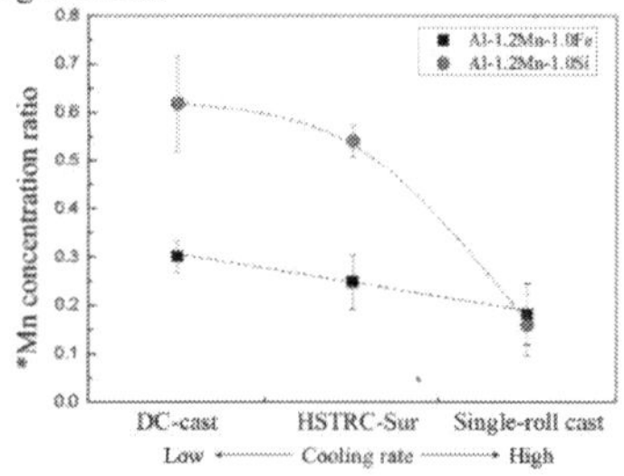

*Fig. 3. Mn composition ratio of constituent particles in Al-Mn-Fe and Al-Mn-Si alloy for each casting methods. *Note: Mn concentration ratio means Mn/(Mn+Fe) ratio in Al-Mn-Fe alloy, Mn/(Mn+Si) ratio in Al-Mn-Si alloy, respectively.*

4. Summary

High-speed twin-roll cast Al-Mn based alloy strips produced by high cooling rate contained fine nano-size constituent particles. Morphology and chemical composition ratio of the constituent particles changed depending on the cooling rate.

References

[1] A.M.F. Muggerud, E.A., Mørtsell, Y.J. Li, and R. Holmestad, (2013) Dispersoid strengthening in AA3xxx alloy with varying Mn and Si content during annealing at low temperatures, Materials Science & Engineering A, 567, 21-28.

[2] K. Hung, Y.J. Li, & K. Marthinsen, (2015) Effect of heterogeneously distributed pre-existing dispersoids on the recrystallization behavior of a cold-rolled Al-Mn-Fe-Si alloy, Materials Characterization, 102, 92-97.

[3] R. Song, Y. Harada & S. Kumai (2018). Influence of Cooling Rate on Primary Particle and Solute Distribution in High-Speed Twin-Roll Cast Al-Mn Based Alloy Strip, Materials Transactions, 59(1), 110-116.

The 73rd World Foundry Congress
„Creative Foundry"
23 - 27 September 2018, Krakow, Poland

SUPPRESSING OF ULTRA-HIGH TEMPERATURE WETTING BETWEEN MOLTEN Si AND SiC BY USING h-BN SPRAY COATINGS

Wojciech Polkowski [1]*, Natalia Sobczak [1,2], Donatella Giuranno[1,3], Artur Kudyba[1], Grzegorz Bruzda[1], Rafał Nowak[1], Adelajda Polkowska[1]

[1] *Foundry Research Institute, Zakopiańska 73 Str., 30-418 Krakow, Poland*
[2] *Institute of Precision Mechanics, Duchnicka 3 Str., 01-796 Warsaw, Poland*
[3] *CNR-Institute of Condensed Matter Chemistry and Technologies for Energy, Via E. De Marini, 6 16149 Genova, Italy*
**Corresponding e-mail: wojciech.polkowski@iod.krakow.pl*

Key words: AMADEUS Project, silicon, hexagonal boron nitride, wettability, interfaces

1. Introduction

An efficiency and performance of power&energy devices strongly depend on applied working parameters, especially on operating temperature. For this reason, high temperature systems for energy storage and conversion are of crucial importance for the most important industry branches. The positive effect of increased temperature is clearly observed in the case of gas turbine engines or concentrated power systems (CPS) combined with latent heat thermal energy storage (LHTES). For the latter devices, a possibility for rising of operational temperature is determined by performance of involved phase change materials (PCMs) and contacting refractories. Actually, operating temperature and corresponding energy density of state-of-the-art LHTES systems based on molten salts acting as PCMs, are limited to ~800°C and 150 kWh/m^3, respectively. On the other hand, it has been recently proposed in European Commission founded AMADEUS Project [1] that implementation of silicon as PCM should allow increase the operating temperature of LHTES system up to 2000°C and reach the energy density as high as 1000-2000 kWh/m^3, i.e. an order of magnitude greater than that of salt-based counterparts.

An introduction of silicon working at such harsh conditions (i.e at temperature much higher than its melting point) requires a selection of proper ceramics for building PCM container, that would ensure a long-life time of the silicon based LHTES system. The simplest approach to the selection assumes an inertness between PCM and a contacted refractory. In other words, the main criteria are a lack of wetting and negligible reactivity. Nevertheless, as it comes from the literature survey [2, 3] almost all oxide, carbide, nitride and boride ceramics are well wetted by molten Si, while the only one exception showing the non-wetting behavior at temperature <1500°C is hexagonal boron nitride (h-BN). In our previous work [4], by using sessile drop tests we demonstrated that this behavior is maintained up to 1650°C.

The h-BN may be easily accessed in the form of commercially available sinters, powders, oil lubricants or spray coatings (slurries). In this work, we focus on examining the ultra-high temperature performance of h-BN spray coatings in contact with molten Si. For this reason, we cover SiC polycrystalline substrates (SiC has been recognized to be well wetted by molten Si, with a corresponding contact angle $\theta<50°$ [5]) and then subjected them to contact heating with Si at temperatures up to 1750°C .

2. Materials and methods

Following materials were used in this work:

a) polycrystalline silicon (7N);
b) polycrystalline SiC sintered substrates (Frialit® SiC 198D);
c) two types of h-BN coatings (Henze HeBoCoat® 21E - polymer slurry and Henze HeBoCoat® 401E - ceramic slurry).

Before experiments, contacting surfaces of polycrystalline SiC were subjected to mechanical polishing in order to achieve a roughness of ~150 nm. After that, the surfaces were coated by one of h-BN slurries and then dried. The uncoated silicon carbide substrate was also used as the reference sample.The Si/h-BN coating/SiC sandwiches were subjected to sessile drop experiments at temperatures up to 1750°C (Fig. 1) combined with a contact heating procedure, by using experimental complex for examining high temperature capillary phenomena. After the test, the structure and chemistry of solidified couples were characterized by means of optical and scanning electron microscopy.

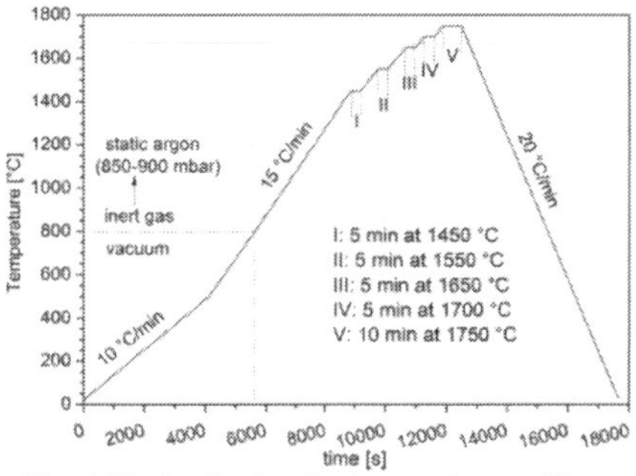

Fig. 1. The heating/cooling temperature profile used upon the wetting test of Si /h-BN coatings/SiC systems (more detailed description is shown in [4]).

3. Results and conclusions

The wetting kinetics curves (showing a change of θ vs. testing time) calculated for Si on SiC substrates coated and uncoated by h-BN slurries, are shown in Fig. 2. It is documented that application of boron nitride coatings allows suppressing wettability in the Si/SiC system – even at 1700°C the contact angle values remained in the non-wetting range (θ>90°). This behavior was in a strong contrast to Si/uncoated SiC couple that shows a contact angle θ~60° in a 1450 to 1750°C temperature range, while at 1750°C the Si drop was suddenly consumed due to the fast infiltration of molten metal into the polycrystalline SiC substrate.

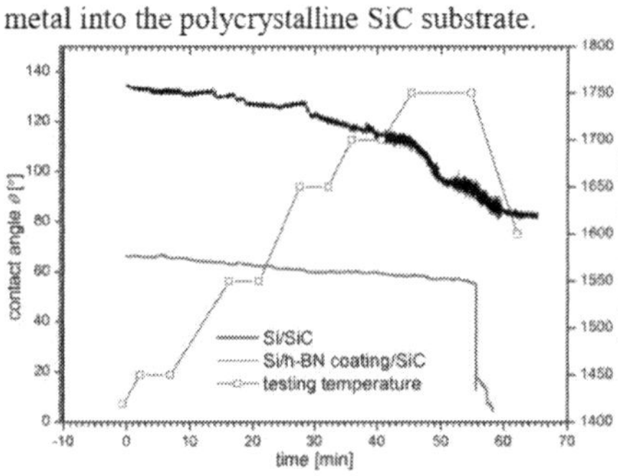

Fig. 2. The wetting kinetics curves obtained for Si on h-BN coated and uncoated SiC substrates at temperatures up to 1750°C.

Nevertheless, it should be noted that h-BN coatings exhibited a rather poor adhesion reflected by a severe spallation after cooling down to room temperature, what makes them unsuitable for operating under cyclic thermal loadings.

Additionally, the microscopic examinations of cross-sectioned couples (Fig. 3) revealed that the h-BN coating does not completely suppress the reactivity in the Si/SiC system at temperatures up to 1750°C – some effects of SiC dissolution followed by reprecipitation of SiC crystals at the interface, are clearly observed.

Finally, the most important findings regarding the performance of h-BN spray coatings are as follows:

(I) they allow substantially slow down the wetting kinetics even in a such reactive system as the Si/SiC;

(II) since a short term interaction (up to ~20 min, and temperature up to 1550°C) almost completely hinder the wettability in the investigated system, it is reasonable to assume that h-BN coatings may be successfully introduced into a metal casting practice, e.g. as releasing agents, or into measurements of thermophysical properties in high temeparature devices.

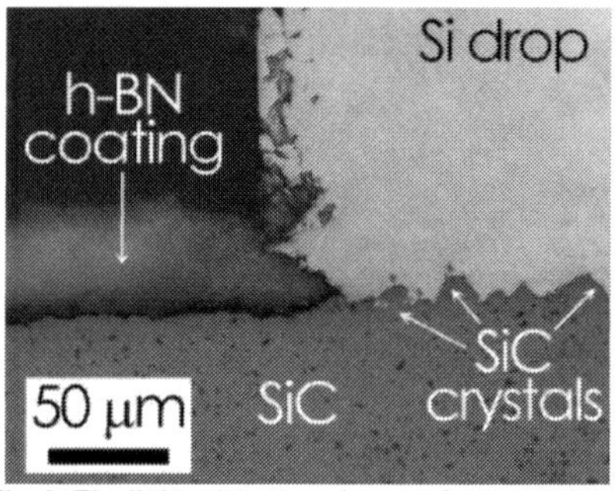

Fig. 3. The light microscopy image of cross sectioned Si/h-BN coating/SiC couple at triple point area.

Acknowledgments

The project AMADEUS has received funds from the European Union's Horizon 2020 research and innovation program, FET-OPEN action, under grant agreement 737054. The sole responsibility for the content of this publication lies with the authors. It does not necessarily reflect the opinion of the European Union. Neither the REA nor the European Commission are responsible for any use that may be made of the information contained therein.

References

[1] A. Datas, A.B. Cristobal, C. del Cañizo (2017). AMADEUS: Next Generation Materials and Solid State Devices for Ultra High Temperature Energy Storage and Conversion. SolarPaces 2017 Conference, Santiago, Chile, 2017. Article submitted to AIP Conference Proceedings.

[2] B. Drevet, N. Eustathopoulos (2012). Wetting of ceramics by molten silicon and silicon alloys: a review. Journal of Materials Science 47, 8247–8260.

[3] Z. Yuan, W.I. Huang, K. Mukai (2004). Wettability and reactivity of molten silicon with various substrates. Applied Physics 78, 617–622.

[4] W. Polkowski, N. Sobczak, R. Nowak, et al. (2017). Wetting behavior and reactivity of molten silicon with h-BN substrate at ultrahigh temperatures up to 1750°C. Journal of Materials Engineering and Performance. https://doi.org/10.1007/s11665-017-3114-8.

[5] M. W. Barsoum, P. Darrell Ownby (1981). The effect of oxygen partial pressure on the jetting of SiC, AlN, and Si_3N_4 by Si and a method for calculating the surface energies involved. In: J. Pask et al. (eds.), Surfaces and Interfaces in Ceramic and Ceramic — Metal Systems. Plenum Press, New York.

The 73rd World Foundry Congress
„Creative Foundry"
23 - 27 September 2018, Krakow, Poland

INDICATOR OF THE EFFICIENT PROTECTION AGAINST THE HUMIDITY Wso, AS A NEW PARAMETER CHARACTERISING PROTECTIVE COATINGS

Natalia Kaźnica [1*], Jerzy Zych [2]

[1]AGH University of Science and Technology, Faculty of Foundry Engineering, Al. Mickiewicza 30, 30-059 Krakow, Poland
*Corresponding address: e-mail: kaznica@agh.edu.pl

Key words: protective coating, moisture sorption, moulding sand, surface layer

1. Introduction

The hereby paper concerns the development of the indicator of the efficient protection against the humidity W_{so}, as a new parameter characterising protective coatings. This indicator's aim is a determination, to what degree covering of cores and sand moulds influences advantageously or disadvantageously the humidity absorption from surroundings into surface layers, in comparison with layers without covering, which at a high humid conditions of the atmospheric air absorb the moisture [1-4].

Simultaneously, this indicator will allow the efficient comparison of tendencies of individual coatings and their layers number for the moisture sorption. Thus, it can constitute a practical information in selecting the coating kind for the moulding sand, on which it will be deposited. Especially taking into account the time of waiting - of moulds and cores covered with coatings - for pouring with liquid metal and the atmospheric conditions to the influences of which they will be exposed.

2. Materials and methods

Investigations concerning the determination of the indicator of the efficient protection against the humidity W_{so} , were performed on cuboidal shaped elements (Figure 1) of dimensions 50x30x5 mm, made of moulding sands:

- alkyd,
- furan,
- with hydrated sodium silicate (water glass).

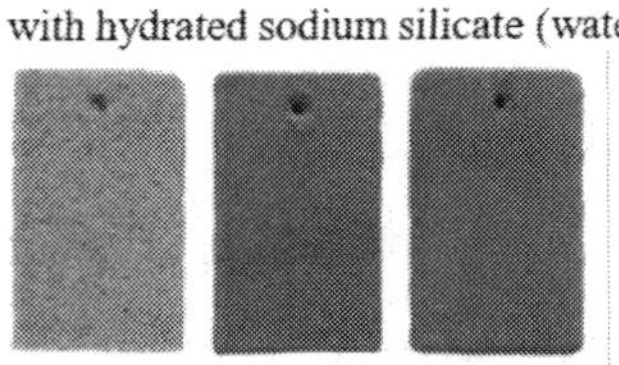

Fig. 1. Shaped elements for testing: without coating, with one layer of PCG4 coating, with two layers of PCG4 coatings

The shaped element thickness of 5 mm, at double-sided moisture exchanges (presented schematically in Figure 2), which occurs during tests, reflects conditions in surface layers of moulds and cores down to the depth of 2.5 mm.

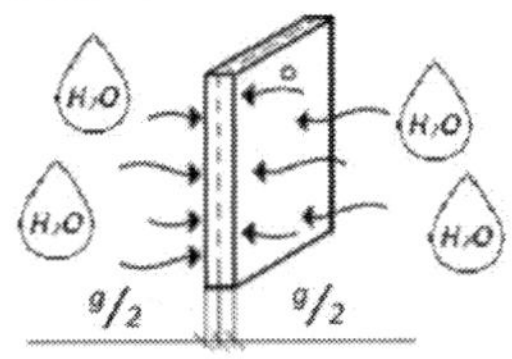

Fig. 2. Scheme of moisture exchange between surroundings and the shaped element

Alcoholic coatings and a water coating were applied as the protective coatings (see Table 1).

Table 1. Protective coatings deposited on cores

Alcoholic coatings	Water coating
zirconium PCM2-VK	graphite WT1-D
zirconium – graphite PCG 4	

For better understanding the occurring effects, investigations of the moisture sorption from surroundings, by protective coatings themselves, were performed. A glass plate, limiting the moisture sorption into the inside, functioned as the classic shaped element made of a moulding sand.

Quantitative measurements of the moisture sorption, based on the gravimetric method (described in details in [2]) taking into account the modification allowing to deposit the protective coating [3], was applied as the investigation method. The immersing time was 5 s at the relative viscosity of coatings, measured by the Ford's cup of the outflow hole diameter equal 4 mm, being approximately 20 s. The prepared samples were placed on the measuring stand for 12 hours, under the thermal-humid conditions being:

- 27 - 32°C – air temperature,
- above 95% of the relative humidity of the atmospheric air.

Changes of the sample weight, due to the moisture absorbed from surroundings, were continuously measured and recorded every 30 s, by means of the computer.

3. Determination of the indicator

The determination of the indicator of the efficient protection against the humidity, W_{so} , is done on the basis of the collected data concerning increases of the sample weight, due to the moisture sorption from the surroundings in time. The indicator W_{so} is expressed by equation (1):

$$W_{so} = \frac{I_{w_0} - I_{w_p}}{I_{w_0}} \cdot 100\% \qquad (1)$$

where:
W_{so}- indicator of the efficient protection against the humidity, %,
I_{w_0}- humidity amount of the shaped element without the protective coating, g/cm^2,
I_{w_p} - humidity amount of the shaped element covered by the protective coating, g/cm^2.

The positive indicator value shows that the amount of the moisture absorbed by surface layers covered by coatings was smaller (in a percentage formulation), than by layers not covered. In other words, it indicates that these coatings provide the efficient protection against the humidity. In a similar fashion, the negative value indicates that layers covered with coatings absorb more moisture than layers not covered, it means that the protective function is not provided.
The W_{so} indicator can be determined for the arbitrary period during which cores or sand moulds are exposed to the atmospheric conditions influence.

4. Analysis of the results

Only examples of the investigation results concerning the indicator of the efficient protection against the humidity W_{so} , related to the humidity absorption by surface layers of cores and moulds, are presented in this paper.

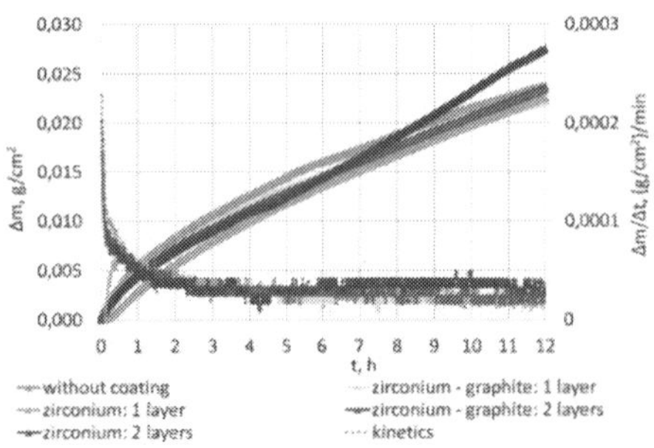

Fig. 3. Sorption process of surface layers of cores covered by alcoholic protective coatings

Figure 3 presents the process course of the moisture sorption in time (together with its kinetics) for surface layers of moulds and cores made of a moulding sand with water glass and covered by alcoholic protective coatings. Surface layers of this moulding sand absorb the highest amounts of the moisture from surroundings as compared to the remaining moulding sands (with alkyd or furan resins), regardless of being or not being covered by protective coatings. Nevertheless the application of coatings slightly reduces the surface layers ability to absorb the moisture from surroundings (with an exception of 2 layers of zirconium coating). It is especially important in the initial stage of the process, since in this period the surface layers are especially exposed to the humidity harmful influence.

The results obtained in such way, constitute a point of departure for the determination of the W_{so} indicator, which graphical interpretation is presented in Figure 4.

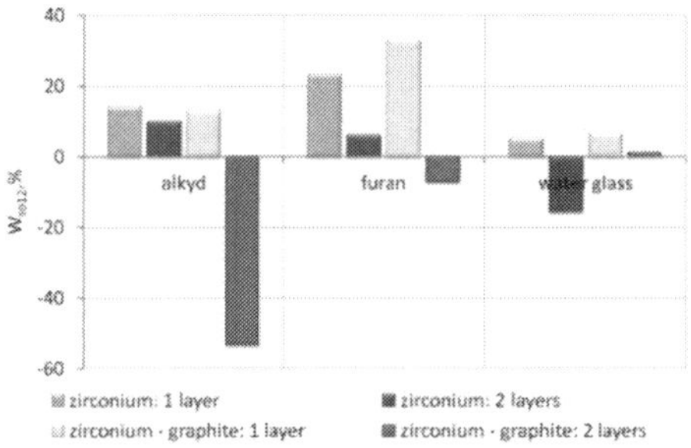

Fig. 4. Indicator of the efficient protection against the humidity W_{so}, for surface layers covered by alcoholic protective coatings

W_{so} presents the efficiency of applying alcoholic protective coatings deposited on shaped elements and exposed for 12 hours to the influence of atmospheric conditions. The significant difference in the efficiency of the protective coatings is seen between individual moulding sands as well as between the numbers of coating layers, where the protective function decreases when the number of coating layers increases.

Acknowledgments
This research received funding from NCBiR No. POIR 01.01.01-00-0042/17-00.

Bibliografia

[1] Kaźnica N., Zych J. (2015). Moisture sorption and desorption processes on the example of moulding sands'surface layers. Archives of Foundry Engineering. 15 (Spec. Iss. 4), 63 – 66.
[2] Kaźnica N., Zych J. (2015). Investigations of the sorption process' kinetics of sand moulds' surface layers under conditions of a high air humidity. Archives of Foundry Engineering. 15(Spec. Iss. 3), 29 – 32.
[3] Kaźnica N., Zych J. (2016). Effectiveness of protective action of coatings from moisture sorption into surface layer of sand moulds. Archives of Metallurgy and Materials. 61 (4), 2111 - 2116.
[4] Kaźnica N., Zych J. (2017). Role of sand grains in sorption processes by surface layers of components of sand moulds. Archives of Foundry Engineering. 17(1), 87 – 92.

The 73rd World Foundry Congress
„Creative Foundry"
23 - 27 September 2018, Krakow, Poland

THE INFLUENCE OF REMELTING ON THE PROPERTIES OF AlSi9Cu3 ALLOY WITH HIGHER IRON CONCENT

M. Matejka *, D. Bolibruchova

University of Žilina, Department of Technological Engineering, Univerzitná 1, 010 26 Žilina, Slovakia
**Corresponding author. E-mail address: marek.matejka@fstroj.uniza.sk*

Key words: Al-Si-Cu, remelting, mechanical properties, metallograph, heat treatment

1. Inrouction

The application of secondary Al-Si-Cu aluminium alloys is increasing in recent years. These alloys are suitable material in the production of castings for various industries. A key requirement is to reduce costs while maintaining high quality, thereby increasing overall competitiveness. Secondary alloys, in addition to economic benefits, are more energy efficient compared to primary aluminium. When using such alloys, it is necessary to find a compromise between the price and the resulting quality.

Recycled or multiple remelted alloys are common materials used in the foundry (gating systems, risers, etc.), which are used as a batch materials to reduce costs. When using a return material, it is problematic to determine the number of the remelts and thus its effect on the resulting casting quality. The reason for examining the impact of multiple remelting is how and to what extent it affects the material properties, which subsequently affect the mechanical and foundry properties [1, 2].

The main disadvantage of recycled alloys is the higher amount of iron. Iron with other elements situated in alloy creates intermetallic phases. The presence of these phases causes degradation of the mechanical properties of the alloys because are formed during solidification of the eutectic and can affect the fluidity and support porosity. Since the iron cannot be removed from the molten alloy, for neutralizing it is necessary to use neutralizing iron correctors, such as Cr, Be, Ni, and Mn. Another possibility to achieve an increase of mechanical properties is the use of heat treatment. In the presence of copper and magnesium, the intermetallic phases Al_2Cu and Mg_2Si are formed, which allow a significant increase in strength characteristics [3-5].

Our scientific team at technological engineering has been evaluating for several years the issue of increased iron content and its adverse effect.

The basis for initiating this work is insufficient theoretical knowledge in the field of use of Al-Si-Cu-type recycled alloys with higher iron content and its influence on the final quality of castings.

2. Materials and experiments procedure

As an experimental material was used secondary alloy AlSi9Cu3 (D1 - reference sample), in which iron amount was intentionally increased to approximately 1.4 wt. %, by master alloy AlFe10. Tab. 1 shows the chemical composition. The alloy was subsequently subjected to 6 times multiple remelting in an electric resistance furnace. Samples for structural analysis and for evaluation of mechanical properties were cast from the first melt (D1 as the reference sample) and from each second melt (D3, D5, D7) into a metal mold.

Table 1. Chemical composition of the AlSi9Cu3 alloy after addition of Fe

Elem.	Si	Fe	Cu	Mn
(wt. %)	9,347	1,416	1,741	0,178
Elem.	Mg	Ti	Cr	Sr
(wt. %)	0,427	0.034	0.025	<0.002

3. Evaluation of chemical composition

Impact multiple remelting, the chemical composition may be changed due to the burning process of certain elements. Changing the chemical composition has a significant effect on mechanical properties. In tab. 2 are referred to changes in wt. % of the selected elements depending on the number of remelting.

Table 2. Relationship between wt. % of selected elements and cast number (CN)

CN	Si	Fe	Cu	Mn	Mg	Cr
D1	9.347	1.416	1.741	0.178	0.427	0.025
D3	9.306	1.51	1.741	0.186	0.417	0.04
D5	9.302	1.738	1.667	0.181	0.402	0.065
D7	9,179	1.889	1.633	0.187	0.339	0.106

Occurrence of manganese and chromium may lead to formation of so called "sludge phases". Amount of these phases depend mainly on the concentration of iron, chromium, and manganese, expressed by sludge factor. In Tab. 3, referred sludge factor (SF) is dependent on the amount of remelting. Tab. 3 also shows the ratio of Mn and Fe. The Mn / Fe parameter indicates the probability of iron phase formation in negative needle form. The addition of manganese is most often recommended in an amount of at least half the iron content ($Mn/Fe \geq 0.5$).

Table 3. Relationship between SF and cast number and between ratio Mn/Fe and cast number

CN	D1	D3	D5	D7
SF	1.85	2.01	2.3	2.89
Mn/Fe	0.16	0.11	0.1	0.09

4. Structure

The microstructures photos of alloys (D1, D3, D5, D7) were analysed in casted state, after the heat treatment (T5) and after natural aging. In Fig. 1 to 3 shows the comparison of the alloys D1 and D7 in the casted, after T5, and after natural aging.

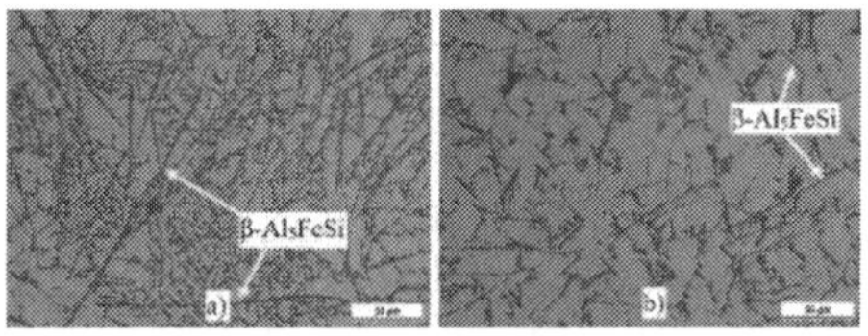

Fig. 1. Optical micrographs showing microstructure of a) D1 and b) D7 in cast state

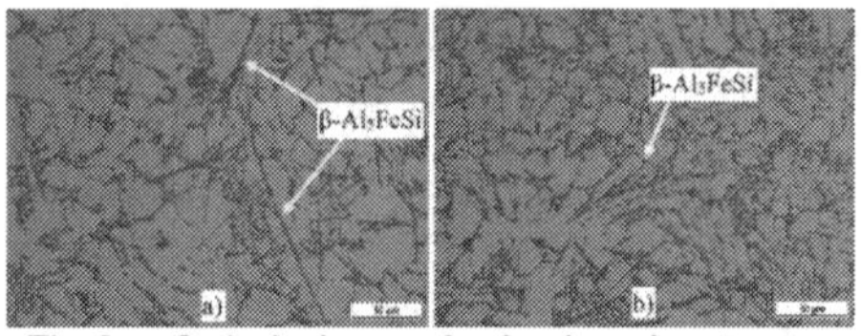

Fig. 2. Optical micrographs showing microstructure of a) D1 and b) D7 in after artificially aged

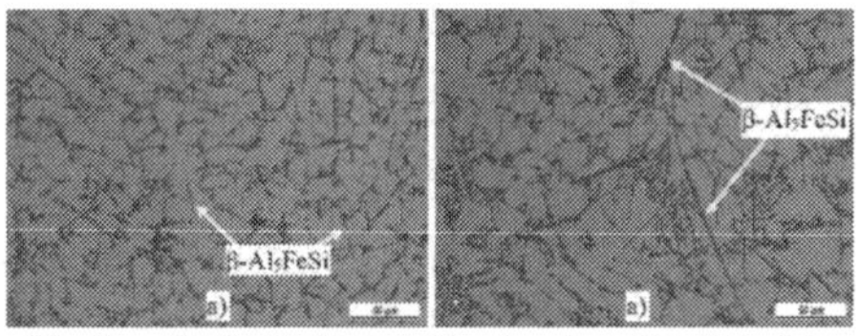

Fig. 3. Optical micrographs showing microstructure of a) D1 and b) D7 in after naturally aged

In structures of alloys after T5 and after natural aging were measured lengths of iron-based phases. The results are shown in Tab. 4.

Table 4. Results of measurements of length of iron based particles

Cast number		D1	D3	D5	D7
Length of iron phases (µm)	Artificial aged	37,83	26,28	55,11	72,56
	Natural aged	55,67	27,67	71,35	89,42

5. Gas content of AlSi9Cu3

The amount of gas content has a direct effect on the porosity of the castings. Tab. 5 show the values of the AlSi9Cu3 Alloy Density Index graded the amount of remelting.

Table 5. Density index (DI) of AlSi9Cu3

Cast number	D1	D3	D5	D7
DI (%)	5.97	10.86	9.36	10.9

6. Mechanical properties

In Tab. 6 to 8 are shown the results of ultimate tensile strength, elongation and Brinell hardness tests.

Table 6. Ultimate tensile strength of AlSi9Cu3 alloy with different cast number

Cast number		D1	D3	D5	D7
Ultimate tensile strength (MPa)	Cast state	164	193	147	144
	Artificial aged	186	197	174	165
	Natural aged	182	195	144	141

Table 7. Elongation of AlSi9Cu3 alloy with different cast number

Cast number		D1	D3	D5	D7
Elongation (%)	Cast state	0.6	1.4	0.5	0.5
	Artificial aged	0.5	0.8	0.3	0.3
	Natural aged	0.8	1.1	0.4	0.4

Table 8. Brinell hardness of AlSi9Cu3 alloy with different cast number

Cast number		D1	D3	D5	D7
Brinell hardness (HBW)	Cast state	93	84	86	94
	Artificial aged	97	90	111	108
	Natural aged	97	91	96	92

7. Conclusions

From the obtained results can be concluded, that multiple remelting leads to a change in chemical composition and changes in the resulting structure, increasing the formation of sludge phases. Due to the remelting, iron phase needles length has grown. The use of the heat treatment T5 led to a positive effect on the microstructure while the impact of natural aging was only beneficial to the third remelting.

The significant effect of multiple remelting on selected mechanical properties occurs at the fourth remelting, when significant decrease occured. With lower numbers of remelting, artificial aging and natural aging are approximately equally effective, but with the increasing number of remelting, the effectiveness of natural aging is declining and therefore, it is better to use artificial aging.

References

[1] Bolibruchová, D., Macko, J. & Brůna, M. (2014). Elimination of negative effect of Fe in secondary alloys AlSi6Cu4 (EN AC 45 000, A 319) by nickel. Archives of Metallurgy and Materials. 59(2), 717-721.

[2] Krivoš, E., Pastirčák, R. & Madaj, (2014). Effect of technological parameters on the quality and dimensional accuracy of castings manufactured by patternless process technology. Archives of Metallurgy and Materials. 59(3), 1069-1072. ISSN 1733-3490.

[3] Elsharkawi, et al. (2010). Effects of Mg, Fe, Be additions and solutions heat treatment on the π-AlMgFeSi iron intermetallic phase in Al-7Si-Mg alloys. Journal of Materials Science. 45, 1528-1539.

[4] Cao, X. & Campbell, J. (2006). Morphology of Al5FeSi Phase in Al-Si Cast Alloys. Materials Transactions. 47(5), 1303-1312.

[5] Taylor, J.A. (2012). Iron-containing intermetallic phases in Al-Si based casting alloys. Procedia Materials Science. 1, 19-33.

The 73rd World Foundry Congress
„Creative Foundry"
23 - 27 September 2018, Krakow, Poland

STUDY OF DAMPING PROPERTIES OF COMPSITE MATERIALS ON A METTALIC MATRIX

[1]Magdalena Poręba*, [2]Janusz Buraś, [3]Ewa Olejnik,
[1]*AGH University of Science and Technology. Faculty of Foundry Engineering.*
Department of Foundry Processes Engineering
Reymonta 23, 30-059 Krakow, Poland
**Corresponding address: e-mail: poreba.m03@mail.com*

Key words: in-situ composite, carbides, metal-based composites, TiC, vibration damping

1. Abstract

Composites are increasingly used as substitutes for other engineering materials due to their very good structural and operational properties. In particular, composite materials reinforced with fibers or particles are perceived as future-proof materials due to the favorable strength-to-density ratio, as well as satisfactory stability of mechanical properties at elevated temperatures and sufficient resistance to corrosion [1].

The reinforcing phase strengthens the composite and improves mechanical, chemical and thermal properties. The combination of the matrix of the composite with the reinforcing components is possible due to the existence of adhesion forces and chemical interactions [2].

MMCs (metal matrix composites) materials are used in areas where the choice of material is based on the useful function fulfilled and the price does not play a significant role, for example, the aerospace or cosmic industries and in industries where a compromise is sought between a usable function and the cost of the composite [3].

The ability to damp vibrations is one of the most important features of construction materials, the more attractive it can be determined based on non-destructive methods. In recent years, composite materials have aroused particular interest, which is associated with a number of their attractive properties, including their high ability to damp vibrations. This is the subject of many studies focusing on the mechanism of vibration damping depending on the type of reinforcement, temperature, deformation amplitude or the method of surface preparation of reinforcement molecules before their introduction into liquid metal or powder [4].

The aim of the work was to examine immunity for damping the vibrations of castings made of composite materials on a metallic matrix. The material tested was a composite made in-situ. The matrix component was aluminum, while the reinforcing phase consisted of TiC titanium carbide particles with a variable weight fraction. Pure aluminum was used as the base material, constituting the matrix of the composite. The research involved a composite material based on aluminum with a TiC titanium carbide weight ratio - 9.25% and 7.4% in relation to the batch mass. The echo method (pulse - echo method) was used to determine the attenuation intensity [5]. Ultrasonic wave generated by the head after passing through the tested material is dispersed and weakened. On the basis of the wave echo image, the damping coefficient was determined as a measure of the material's ability to damp vibrations. The tests were carried out for longitudinal wave attenuation, frequency 1 MHz. Presents an image of the program window showing Figure 1 (Fig.1.) the results of ultrasonic wave attenuation measurements for pure aluminum.

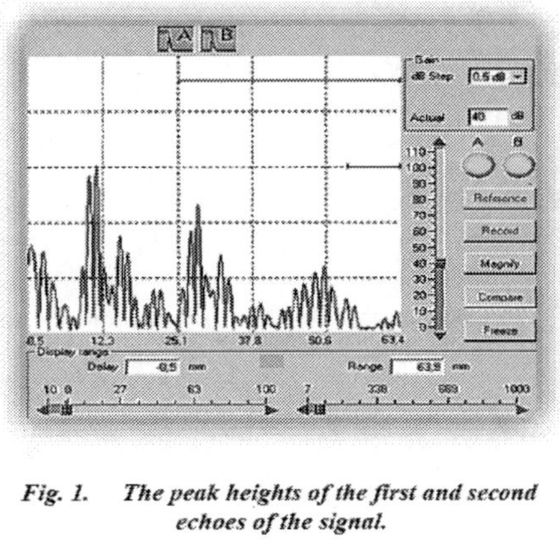

Fig. 1. The peak heights of the first and second echoes of the signal.

Analysis of the test results showed that the composite material with a higher weight fraction of TiC particles of 9.25% in relation to the mass of the load is characterized by a better vibration damping ability of approx. 150 dB/m, compared to a composite with a lower TiC content (Fig. 2.). The value of the damping coefficient for both composite materials is at the level above 100 dB/m, which places these composite materials in the group of materials with high damping properties.

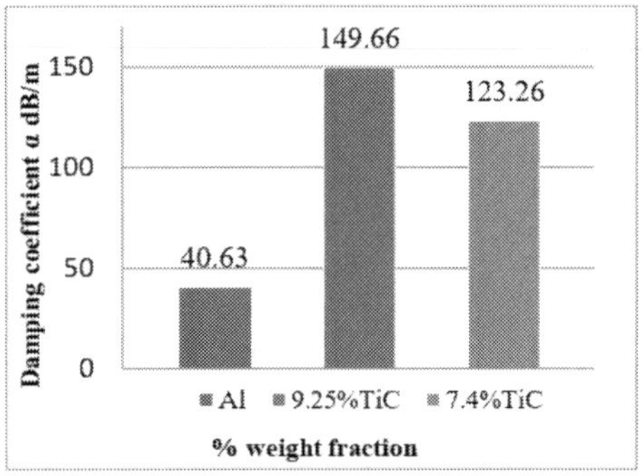

Fig. 2. Comparison of values of damping coefficients for the tested composite samples.

For the purpose of comparison, the value of the damping coefficient for the base alloy (pure aluminum) was also determined, which constituted the matrix of the composite, for which the damping coefficient value is low and amounts to approximately 40 dB / m.

The ultrasonic wave echo suppression method used to determine the ultrasound echo suppression properties is a very sensitive method and gives a large dispersion of the sample value, depending on the point where the probe contacts the surface of the tested sample. Probably this may be due to, among others the heterogeneity of the structure of the composite material, manifested by uneven distribution of TiC or intermetallic phases in the composite structure. Therefore, this method requires a large series of measurements to determine the reliable averaged values of the damping factor attenuation coefficient within the same.

Acknowledgments

The composite material that was tested was designed and manufactured as part of the project entitled In-situ molded composites reinforced with ceramic "phase nanoparticles", No. UMO -2016 / 21 / B / ST8 / 01181, financed by the NCN.

References

[1] St.Wierzbiński. 2003. Kraków. Właściwości Użytkowe Kompozytów Metalowych. Instytut Metalurgii i Inżynierii Materiałowej im. A. Krupkowskiego Polskiej Akademii Nauk, Instytut Techniki Akademii Pedagogicznej im. Komisji Edukacji Narodowej. Wydawnictwo Orekop.

[2] Z. Konopka. 2011. Metalowe Kompozyty Odlewane. Politechnika Częstochowska, Wydawnictwo Politechniki Częstochowskiej.

[3] J. Sobczak, S. Wojciechowski. 2002. Współczesne Tendencje Praktycznego Zastosowania Kompozytów Metalowych. Kompozyty (Composites). R. 2, nr 3. 24-37.

[4] K. Pietrzak, A. Klasik, K. Makowska, J. Sobczak, A. Wojciechowski, D. Rudnik. 2012. O Tłumieniu Drgań W Materiałach Kompozytowych. Prace Instytutu Odlewnictwa, T. 52, z. 4.53-96.

[5] J. Deputat. 1979. Badania Ultradźwiękowe. Instytut Metalurgii Żelaza im. St. Staszica. Gliwice.

The 73rd World Foundry Congress
„Creative Foundry"
23 - 27 September 2018, Krakow, Poland

TRIBOLOGICAL PROPERTIES OF COMPOSITE MATERIALS ON THE METALLIC MATRIX

Barbara Gospodarczyk[1]*, Janusz Buraś [1]

[2]Sławomir Zimowski, [1]Ewa Olejnik

[1] AGH University of Science and Technology. Faculty of Foundry Engineering. Department of Foundry Processes Engineering Reymonta 23, 30-059 Krakow, Poland,

[2] AGH University of Science and Technology. Faculty of Mechanical Engineering and Robotics, Department of Machine Design and Technology Mickiewicza 30, 30-059 Krakow, Poland,

**barbara.g@onet.com.pl*

Key words: tribological properties, composite, lubrication, wear, friction .

1. Abstract

Composites are modern construction materials with a wide range of applications. Composites with an aluminum alloy matrix reinforced with ceramic particles are of particular importance, as they have been used in aviation, in the military industry, but first and foremost in the automotive industry. These composites are used, among others, for sleeves, pistons, cylinders or bearings. These materials are characterized by high hardness, abrasion and corrosion resistance, show good strength and rigidity while maintaining low product density. [1]

The introduction of ceramic particles of the oxide type (Al_2O_3, ZrO_2, TiO_2), carbides (SiC, TiC) or graphite for aluminum alloys allows for the creation of sliding, abrasion resistant and higher strength composites. Reinforcement with ceramic particles also ensures higher working temperature [2].

Metal composite materials based on light alloys, in which particles with different morphology, improving mechanical properties and providing appropriate properties belong to the group of in situ composites [3]. The transformation of the classical cast material into a composite material has a significant impact on improving its mechanical properties [4]. Using casting techniques, the structure of composites can be obtained in various ways, including thanks to chemical reactions taking place inside a metal bath (so-called "in situ" composites). Composites produced in "in situ" methods are referred to as the new generation of composite materials [5]. In this kind of cast composites, the reinforcing phase is created as a result of chemical reactions between the components of the molten metal to form carbides, borides or nitrides.

The aim of the research was to determine the tribological properties of castings made of the composite produced in-situ. The subject of the research was a composite material based on aluminum reinforced with titanium carbide TiC particles with a variable weight fraction of TiC particles, i.e. 5.5%, 3.7% and 1.8% in relation to the mass of the charge. On the basis of tribological tests carried out using the ball-on-disc method (Fig. 1), the volume wear index was determined, which is a measure of wear resistance and coefficient of material friction when sliding on another material under dry friction conditions. In addition, the hardness of the tested composite was measured using the Vickers method and the Young's elastic modulus, which is a measure of the deformation properties of the material.

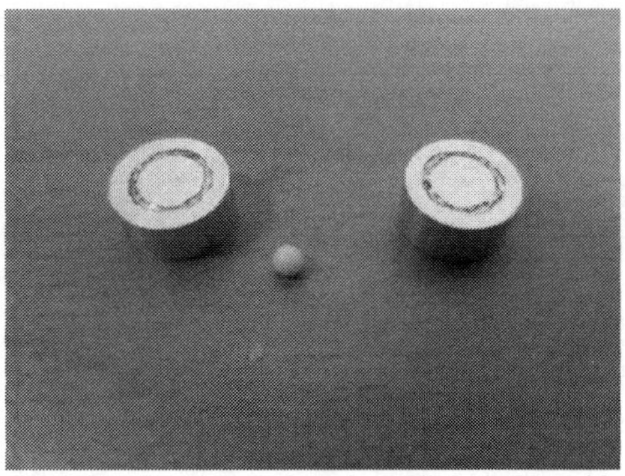

Fig. 1. Samples of composites after a tribological test with visible grooves and a ball (counter-sample) with Al_2O_3

Analyzing the results of tribological tests, it can be seen that the smallest abrasive wear determined by the volumetric wear index is characterized by a composite material with a weight fraction of titanium carbide TiC particles of 3.7% in relation to the mass of the charge (Fig. 2.), Which was also characterized by the lowest coefficient friction and the highest hardness.

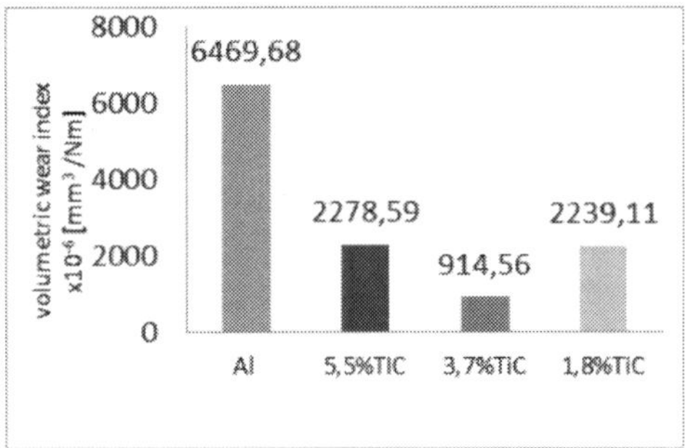

Fig. 2. Volumetric wear index of a composite with variable TiC content in relation to the mass of the charge

Table 1. predicts the individual results obtained for each of the composites. Four samples were made for the base and composites (IC7, IC8, IC9).

Table 1. Scoreboard – Volumetric wear index

	Base	IC7	IC8	IC9
Volumetric wear index	Wv 1x10^{-6} mm^3/Nm	Wv 1x10^{-6} mm^3/Nm	Wv 1x10^{-6} mm^3/Nm	Wv 1x10^{-6} mm^3/Nm
1	5680,21	1963,75	1116,25	2954,75
2	7011,34	1950,00	608,75	1121,25
3	6916,51	2446,25	588,74	1175,44
4	6266,37	2754,38	1344,5	3705,00
Average	6469,68	2278,59	914,56	2239,11

It should also be noted that with the weight fraction of titanium carbide TiC particles at 5.5% and 1.8% in relation to the weight of the charge, the resistance of the tested composite material to abrasion is comparable. The highest hardness is characterized by a composite with the highest weight fraction of titanium carbide particles at the level of 5.5% in relation to the mass of the charge. Based on the analysis of the results, it can be concluded that the presence of hard TiC particles clearly improves the tribological properties of the tested composite.

The tested composite material samples have a large dispersion of hardness value and abrasive wear indicators, which may be caused by, among others, inhomogeneity of the material structure, uneven distribution of the TiC reinforcing phase particles.

Apart from research on tribological properties, hardness tests were carried out.

The graph (Fig. 3.) below presents the results of hardness tests.

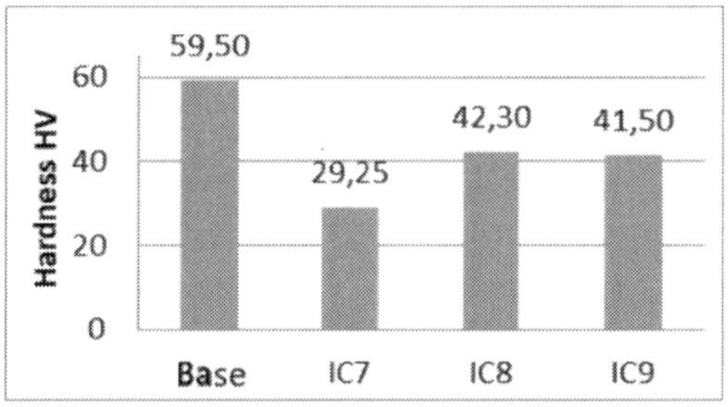

Fig. 3. Average values of HV hardness for the base alloy and samples IC7, IC8 and IC9

2. Acknowledgements

The composite material that was tested was designed and manufactured as part of the project entitled "In-situ cast composites reinforced with ceramic phase nanoparticles", No. UMO-2016/21 / B / ST8 / 01181, financed by NCN

References

[1] A. Dolata-Grosz, J. Śleziona, J. Wieczorek, K. Pietrzak (2003) „Wytwarzanie i kształtowanie struktur kompozytowych odlewów aluminiowych o złożonym rozmieszczeniu uzbrojenia, Politechnika Śląska, Wydział Inżynierii materiałowej i metalurgii,

[2] J. Bieniaś (2014) „Struktura i właściwości materiałów kompozytowych" (opracowanie) Politechnika Lubelska, Katedra Inżynierii Materiałowej,

[3]A. Olszówka – Myalska (2016) ,"Spiekane kompozyty magnezytowe typu In situ" Mechanik Nr 5-6/

[4] P. Mierzwa, E. Olejnik, A. Janas,(2012), Achives of Foundary Engineering, „Nowoczesne materiały kompozytowe zastępujące tradycyjne materiały odlewnicze", Vol. 12, Special Isuee 1/2012, 137-142

[5] A. Dolata-Grosz, A. Formanek, J. Śleziona, J. Wieczorek, (2004)," Aluminiowe kompozyty z hybrydowym mocnieniem faz międzymetalicznych i ceramicznych", Archiwum Odlewnictwa, Rocznik 4, nr 14

The 73rd World Foundry Congress
„Creative Foundry"
23 - 27 September 2018, Krakow, Poland

DYNAMIC MEASUREMENT AND FEM ANALYSIS OF RESTRAINT FORCE FROM MOLD ON GRAY CAST IRON CASTINGS AND CONTRACTION OF CASTINGS DURING COOLING IN GREEN SAND MOLD

Masayuki Kaneko [1], Yusuke Sano [1], Seigo Ueno [1], Hiroyasu Makino [2], and Makoto Yoshida [3]

[1] *Graduate School of Waseda University, Department of Modern Mechanical Engineering, 3-4-1 Shinjyuku-ku Okubo, Tokyo 169-8555, Japan,*
[2] *SINTOKOGIO LTD., Research and Develop Center, 3-1 Honohara, Toyokawa Aichi 442-8505, Japan,*
[3] *Waseda University, Kagami Memorial Laboratory for Materials Science and Technology, 2-8-26 Nishi-Waseda, Shinjyuku-ku, Tokyo, 169-0051, Japan*

Key words: sand mold, sand casting, gray cast iron, analysis, continuous comparison

Introduction

The presence of residual stress and distortion are significant problems when producing sand castings such as cylinder heads and machine tool beds made from cast iron. Parkins identified these defects as caused by the restraint force from sand mold to casting [1]. In recent years, thermal stress analysis by FEM has been done against these defects. Until now, some researchers performed FEM analysis that incorporated the behavior of sand mold, and compared with experimented measurements. Baghani et al. (2014) has measured the residual stress on casting in sand mold, and compared with analysis results that used elasto-plastic constitutive equation for sand mold after shakeout [2]. However there is no previous research that evaluate the precision of analysis by continuously comparing the measurement from the end of the solidification process until the final cooling stage. In this study, firstly, we conducted a dynamic measurement of the restraint force from mold and the contraction of the gray cast iron in green sand mold. Then, we performed thermal stress analysis by taking account of the continuous behavior of sand mold. By comparing the analysis results with the experimental measurements, we verified the prediction precision of the restraint force from mold and the contraction of casting.

Experimental

Fig. 1 demonstrates schematic illustration of the equipment. Cross section of the casting is 30×30 mm and it has 90×90 mm flange at one end. The casting was made from gray cast iron JIS FC 300(equivalent to ASTM 45). We squeezed sand mold at 0.8MPa. Thermocouples were set at the center, flange and sprue of the casting and at the points distance from the casting 10 mm, 20 mm and 30 mm in the sand mold to obtain the temperature. Obtained thermal histories were used for thermal analysis. Tie bar, load cell and rigid plate were assembled. The casting end without the flange was fixed by a clamp part after pouring. Longitudinal contraction of the casting was dynamically obtained by measuring the displacement of the quartz rod by using laser displacement gauge. Simultaneously, the restraint force reacted on the flange of the casting was dynamically measured by the load cell. Fig. 2 shows continuous measurement result of the restraint force and the contraction of the casting. Solid lines indicate the restraint force and dot-lines indicate the contraction. Horizontal axis indicates the temperature at the center of the casting. In both restraint force and contraction, influences of eutectic solidification and A1 transformation are confirmed at about 1150°C and 723°C, respectively. The maximum restraint force was approximately 2.5 kN at about 600°C, and the final contraction was roughly 5 mm.

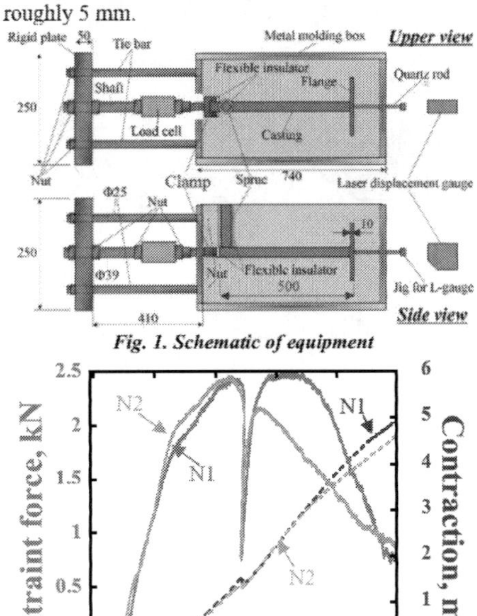

Fig. 1. Schematic of equipment

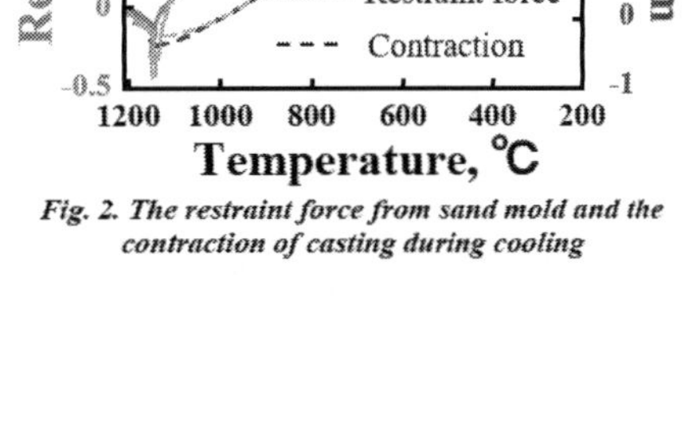

Fig. 2. The restraint force from sand mold and the contraction of casting during cooling

Analysis

Analysis model consisting of casting, sand mold, metal frame and clamp part is shown in **Fig. 3. Fig. 4** compares the experiment and analysis value of restraint force, while **Fig. 5** compares the contraction of gray cast iron by both experiment and analysis. In both graphs, solid line indicates the average of two experimental measurements, and dot-line indicates analysis results. Abaqus 6.14 was used to perform the analysis. Linear expansion coefficient of casting was calculated from the contraction of casting without flange, which was measured with temperature continuously during cooling. Elasto-plastic-creep model was quoted for constitutive equation of casting, and mechanical properties of casting were obtained from tensile test from 1000°C to room temperature [3]. Mechanical properties for over 1000°C were extrapolated from that of below 1000°C. For constitutive equation of the sand mold, elasto-plastic model (von-Mises model) was used. The assigned mechanical properties were estimated by inverse analysis in such a way as to recreate the experimental measurement of the restraint force in **Fig. 4**. Error of the final contraction between analysis result that calculated by the inverse analysis and measurement are minimized within +8% by fitting the restraint force from start of shrinking of the casting to A1 transformation.

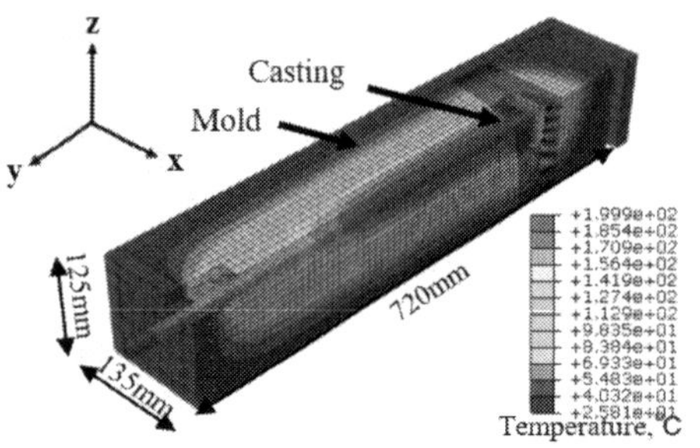

Fig. 3. Thermal stress analysis model

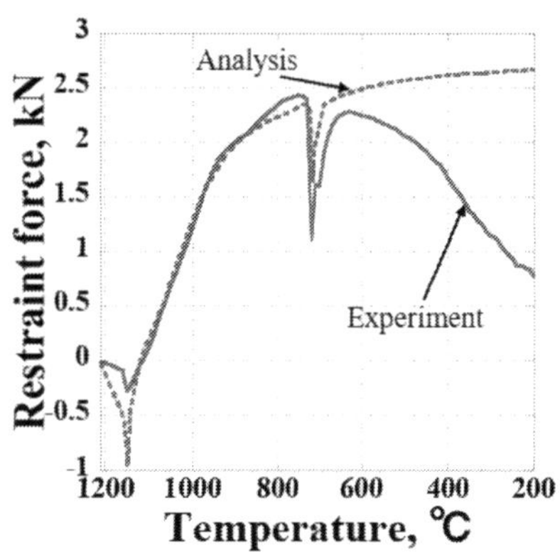

Fig. 4. The difference between the amount of the restraint force

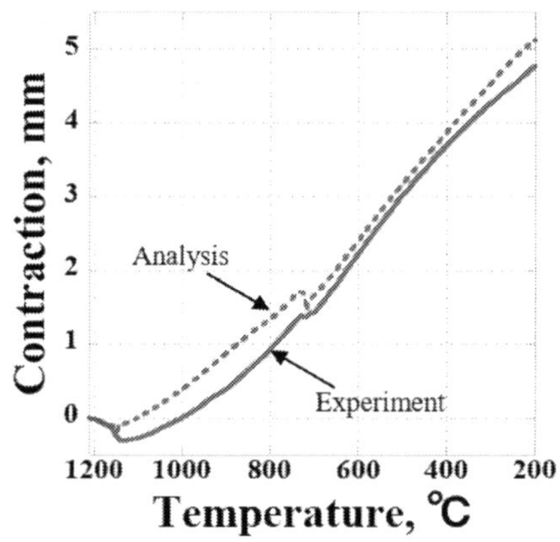

Fig. 5. The difference between the amount of the contraction of casting

Discussion

According to **Fig. 4** and **Fig. 5**, in the analysis, restraint from the sand mold to casting was overestimated at eutectic solidification expansion. A n error between analysis and experiment was shown below 700°C, though the restraint force from the start of shrinking of the casting to A1 transformation was fitted. Although, it is assumed that the water condensation layer of the sand mold affected the experimental result, it has not been clarified yet. In the contraction of the casting, influence of the error was not found. It is thought that the strength of the casting was far stronger than the sand mold at this temperature. By simulating behavior precisely at eutectic solidification expansion, subsequently the analysis of the contraction will get close to the measurement.

Conclusion

We measured the restraint force from sand mold and the contraction of casting in green sand mold continuously. In thermal stress analysis, elasto-plastic creep model was used for constitutive equation of casting and elasto-plastic model (von-Mises model) was used for constitutive equation of sand mold. Mechanical properties are estimated by the inverse analysis from measurement. Then, we evaluated the analysis precision by comparing measurement and analysis continuously that has not performed yet. Analysis results were successfully estimated at the final contraction within +8% error.

Reference

[1] Parkins, R.N., Cowan, A., (1953). Effects of Mould Resistance on Internal Stress in Sand Castings. Inst. Brit. Found. Paper No. 1062, A101-A109.

[2] A. Baghani, P. Davami, (2014). Metall. Mater. Trans. B 45, pp.1157-1169.

[3] Y.Motoyama, H.Shiga, T.Sato, H.Kambe, M.Yoshida, (2017). Metall. Mater. Trans. A, vol.48, issue 6, pp.2960-2970.

The 73rd World Foundry Congress
„Creative Foundry"
23 - 27 September 2018, Krakow, Poland

DEFORMATIONS IN CHEMICALLY BONDED MOULDING SANDS

Aleksandra Grabarczyk [1*], Stanisław M. Dobosz [1], Katarzyna Major-Gabryś[1], Jarosław Jakubski[1], Jan Morek[2]

[1]AGH University of Science and Technology, al. A. Mickiewicza 30, 30-059 Kraków, Poland
Faculty of Foundry Engineering
[2] Multiserw-Morek Company, Marcyporęba 36, 34-114 Brzeźnica, Poland
*Corresponding address: e-mail: agrab@agh.edu.pl

Key words: moulding sands, mechanical deformation, thermal deformation, hot-distortion, core fracture.

In modern foundries as well as in other branches of industry, one can observe a continuous striving to improve the quality of products and a growing emphasis on production efficiency and reducing the number of defects. More and more attention is also paid to the environmental performance and energy consumption of manufacturing processes. This leads to continuous work on the development of used materials, production processes and new measuring equipment that would guarantee the achievement of required quality.
Single-use moulds, made on the basis of quartz matrix, are one of the leading methods of casting production. Proper selection of moulding and core materials has a significant impact on the quality of the finished casting. There are a number of factors associated with moulding materials that affect the casting process, such as: solidification, and the process of knocking out the finished casting. In addition, a number of foundry defects are associated with moulding materials, the occurrence of which determines the quality of the cast element. The basic parameters of the casting include:
• mechanical properties,
• dimensional accuracy,
• smoothness of the surface.
In addition, the following are also important aspects of foundry production:
• economic,
• ecological,
• technological.
Analyzing these factors in terms of moulding materials, it can be noticed that they fit in a different degree in most of them. Of course, the usable properties of the casting depend mainly on the used alloy. However, the selection of an inadequate moulding sand may result in changes in the top layer of the casting, thus affecting its properties or even forcing the sanding of the layer, which may result in failure to maintain dimensional accuracy.
The foundry industry is constantly developing, and the broadly understood automation of production processes following this development creates both new working conditions and new requirements for previously used materials [1-2].
The high requirements described above create the need to develop elements of the casting process, which haven't been fully researched. These include two types of technological processes.
The first is the moment of setting the cores in the foundry moulds, where, especially in the case of highly mechanized foundries, core damage can occur, without the possibility of removing the defective form from the processing line. This process has been analyzed in this work.
The second process that can affect the accuracy of the casting is the initial pouring process in which the uneven heating of the cores occurs, leading to their thermal deformation. This is the second of the processes that has been studied in detail in this work.

1. Moulding sand's deformation

Moulding sand is exposed to various forces during preparation, transport and forming. The next stages, such as removing the core from the box, transferring the cores, storing them and assembling them in the mould, and further on to the casting process, are moments in which cores are exposed to damage. The nature of which can be both mechanical and thermal.

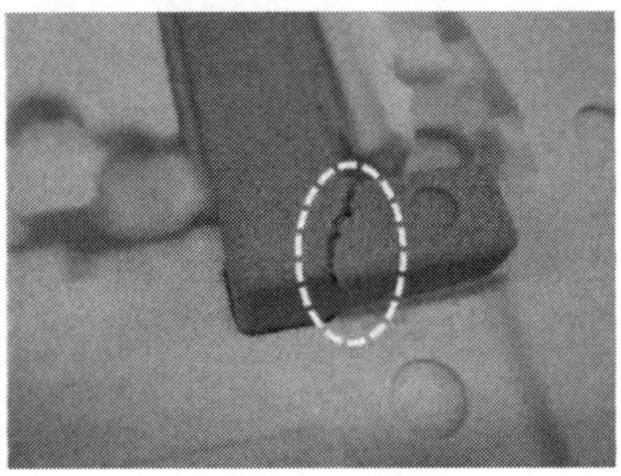

Fig. 1. Core fracture [1]

Even small cracks or dimensional changes of the cores may later become a source of defects arising in the castings (fig. 1). The importance of this problem has been increasingly recognized in foundries with a high degree of automation focused on mass production, with as few scrapings as possible, both

among ready casts as well as prepared moulds or cores.[1-2]
For the purpose of the research two groups of parameters were chosen. One that leads to mechanical deformation of the moulding samples, and the other connected to thermal deformation. The first one covers such parameters as: moulding sand bending strength, moulding sand mechanical deformation size/elasticity, the nature of the fracture. For the second: thermogravimetric analysis of binders and moulding sand samples, thermal deformation of the samples/hot distortion parameter.

2. Own research

To determine the behavior of moulding sand sample, during both bending and heating, a multipurpose device manufactured by Multiserw-Morek Company, was used (fig. 2).

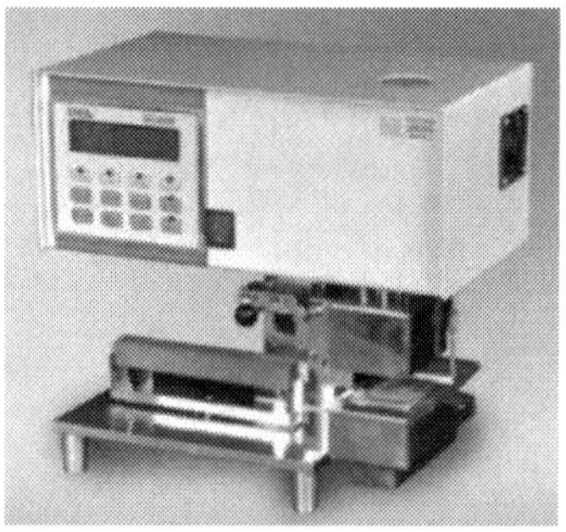

Fig. 2. Universal Hot-distortion and elasticity/bending strength tester [3]

Tests were conducted on both organic and inorganic bonded sands. The most popular moulding technologies were selected, such as: self-hardening moulding sands with furfuryl resin, alkyd resin and hydrated sodium silicate. Also technologies representing hot and cold-box processes, using both organic and inorganic binders were tested.
The elasticity/mechanical deformation tests were performed on standard longitudinal samples, which were also used to determine moulding sand's bending strength. While hot-distortion tests were performed on dedicated thin samples.

Detailed information on both, the deformation tests and used measuring equipment, can be found in previous works of authors [4-7].
The research proved that the thermal deformation analysis (D_{HD}) and the size of mechanical deformation (D_E) of tested moulding sands, can significantly affect the formation of casting defects, through deterioration of dimensional accuracy, or even defects in casting, caused by the breakage of the core. The values of these parameters obtained in this work are used to compare the moulding sands in terms of these properties. The reported values should be related to the dimensions of the test sample. The value of these parameters depends on the size of the sample, its mass and shape and of course the binder used, thus giving an overview on moulding sands behavior in different parts of the casting process.

References

[1] Gröning P., Schreckenberg S., Jenrich K. (2015). Herstellung von hoch-komplexen Zylinderkurbel-gehäusen. GIESSEREI Vol. 102, Iss. 01, p.42-47 (part I)

[2] Gröning P., Schreckenberg S., Jenrich K. (2015). Herstellung von hoch-komplexen Zylinderkurbel-gehäusen. GIESSEREI Vol. 102, Iss. 01, p.48-53 (part II)

[3] Multiserw-Morek Company (2014) Retrieved: January 2018, from: http://multiserw-morek.pl/

[4] Dobosz St. M., Grabarczyk A., Major-Gabryś K., Morek J. (2016) Quality assessment of the moulding mixtures from the elasticity point of view. Technolog, 2, p. 19-23.

[5] Grabarczyk A., Dobosz St. M., Major-Gabryś K., Jakubski J., Morek J. (2015) Elasticity - a New Quality Assessment Criterion for Moulding Sands. Archives of Foundry Engineering vol. 15 spec. Iss. 4, p. 39–42.

[6] Dobosz St. M., Grabarczyk A., Major-Gabryś K., Bolibruchova D. (2017) Elasticity of cores manufactured in cold box technology. Archives of Metallurgy and Materials Vol. 62, Iss. 1, p. 351–354.

[7] Grabarczyk A., Dobosz St. M., Kusiński J., Major-Gabryś K. (2018) The tendency of moulding sands to generate core cracks. Archives of Foundry Engineering, vol. 18, Iss. 1, p. 157-159.

The 73rd World Foundry Congress
„Creative Foundry"
23 - 27 September 2018, Krakow, Poland

AUSTEMPERING IN ZAMAK BATH: INFLUENCE OF AUSTEMPERING TIME ON THE PROPERTIES OF DUCTILE CAST IRON

Leonardo Pereira [1*], Matheus Roberto Bellé [1], Luiz Francisco Seibel Júnior [1], Willian Martins Pasini [1], Regis Fabiano do Amaral [1], Vinicius Karlinski de Barcellos [1]

[1] Federal University of Rio Grande do Sul - UFRGS, School of Engineering, Foundry Laboratory, 9500 Bento Gonçalves Ave. Porto Alegre City, Rio Grande do Sul State, Brazil.
**Corresponding address: e-mail: leonardo.pereira@ufrgs.br*

Key words: Austempering, Ductile Iron, ADI, Zamak

1. Introduction

Austempered Ductile Iron (ADI) exhibits a good combination of mechanical resistance and tenacity, with elevated wearing resistance. ADI has mechanical properties similar or superior to forged steel, with a fabrication cost by resistance unit much inferior. The material properties are a result of its microstructure, ausferrite which is composed by high carbon austenite, acicular ferrite and also graphite nodes [1]. ADI is a low cost alternative for welded, forged or cast steel and aluminum components, creating opportunities for the cast iron industry [2].

ADI properties are related to the production variables and the heat treatment cycle. Regarding the fabrication, the main variable is the alloy composition which may receive Mn and Mo additions to increase hardness and austemperability, and Cu and Ni to increase ductility and also resistance. The main heat treatment cycle parameters are the austenitizing temperature ($T\gamma$), austenizing time ($t\gamma$), austempering temperature (TA) and austempering time (tA) [3].

Austempering process is tradicionally performed in cast salt bath which is a mix of potassium nitrate, sodium nitrate and sodium nitrite [4]. The great disadvantage of the salt bath is its pollution potential [5]. An alternative to this process is metallic bath. Zn-Al and Zn-Sn eutectic alloys were successfully utilized in a laboratorial level [6]. Zamak (Zn-Al-Mg-Cu), according to the former study, can be used as austempering bath for ADI obtention, at austemper temperatures superior to 375°C [7].

The austempering reaction occurs in two stages, at the first stage the nucleation and growth of acicular ferrite begins, which provides high carbon austenite formation. At the end of the first stage the material shows its best properties. At the second stage, undesirable, the high carbon austenite transforms into ferrite and carbides [8].

2. Materials and Methods

From as-cast parts destined to the obtention of high mechanical resistance ADI, blocks intended to the different austempering cycles were built. To each cycle was employed a three block set (measuring 12x12x105 mm each block) aimed at the obtention of tension specimens, and four block set (measuring 12x12x65 mm each block) aimed at the designing of impact test specimens (Charpy).

Ductile cast iron, of perlite matrix, has 150 nodes/mm^2 and 95% of the nodes are type I and II. Fe-C-Si ternary alloy received additions of Mn, Cu, Ni and Mo, the composition of ductile iron is shown at Table 1. Ductile cast iron agrees with ASTM A897M requirements and recommendations.

Table 1. Ductile cast iron chemical composition (%wt).

C	Si	Mn	Cu	Ni	Mo
3.56	2.30	0.32	0.66	0.53	0.18
Mg	**Sn**	**S**	**P**	**Cr**	**Fe**
0.034	0.02	0.07	0.021	0.03	Bal.

The blocks sets were submitted to five different austempering cycles, where only the austempering time was changed. The following parameters were employed on the austemper heat treatment cycles:

- T_γ = 900°C
- t_γ = 90 minutes
- T_A = 380°C (±5°C)
- t_A = 15, 30, 60, 90 and 120 minutes

As austempering environment, a Zamak alloy was used, possessing the following chemical composition:

Table 2. Zamak alloy chemical composition (%wt).

Al	Cu	Mg	Fe	Si	Zn
4.25	1.90	0.12	0.30	0.02	93.4

After the heat treatment cycle implementation, the blocks were air-chilled. The thin layer of Zamak that covered the blocks was manually removed. Then the blocks were forwarded to machining operations, in order to manufacture the test specimens and execute the tensile test (ASTM E8M) and Charpy impact test (ASTM E.23). In regions without deformation on the test specimens it was performed hardness measurements and microstructure analysis.

3. Results and Discussion

The average results of the tensile tests as a function of austempering time are presented in figure 1. The austempering time of 120 min is the one that contributes most significantly to the increase of ultimate tensile strength (UTS), yield strength (YS) and elongation (ε).

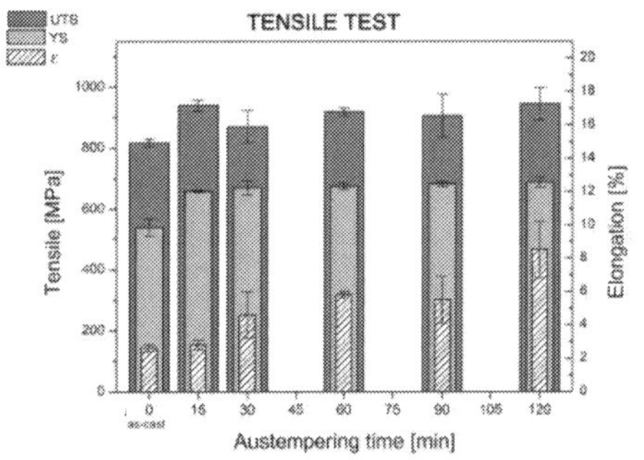

Fig. 1. Mechanical properties for as-cast and austempered. Ultimate Tensile Strength(UTS), Yield Strength (YS) and Elongation in 36mm (ε).

In the impact test it was observed a tenacity reduction at 15 min and 30 min, as it is shown in figure 2. A significantly increase in hardness at 15 min has also occurred.

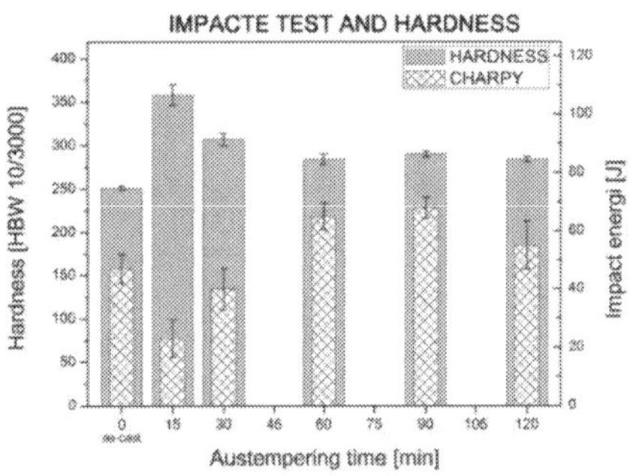

Fig. 2. Brinell hardness and unnotched charpy test. The values are an average of the highest three test values of the four tested samples.

Acicular ferrite growth and carbon diffusion to remaining austenite are phenomenons that require time. Short austempering times are not enough to turn remaining austenite thermal and mechanically stable, so that the material, when requested or chilled, leads to martensite formation, causing tenacity reduction and hardness increase.

The microstructural analysis of test specimens from the different cycles enabled the verification of ausferrita formation. It was observed martensite formation in significant regions of the material from the 15 and 30 min cycles. Martensite presented an average micro hardness of 527 HV, while ausferrite of 382 HV. After 60 min of austempering, as shown in figure 3, relevant microstructural modification no longer occurred.

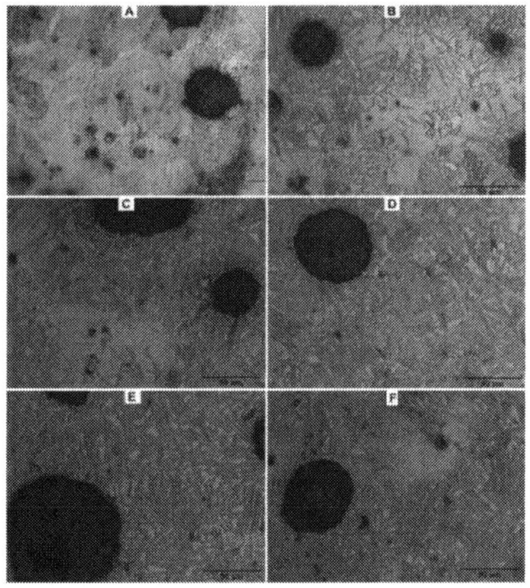

Fig. 3. Microstructural evolution with austemper time. As-cast material (A), austempered with 15 min (B), 30 min (C), 60 min (D), 90 min (E) and 120 min (F).

4. Conclusion

The best combination of mechanical properties happens on the 60 min austemper cycle. The 15 and 30 min cycles were not adequate for the evaluated ductile iron alloy. Zamak demonstrated to be a satisfactory austemper environment.

5. References

[1] Keough, J. R., & Hayrynen, K. L. (2000). Automotive applications of austempered ductile iron (ADI): a critical review(No. 2000-01-0764). SAE Technical Paper.

[2] Keough, J. R., Hayrynen, K. L., & Pioszak, G. L. (2010). Designing with austempered ductile iron (ADI). AFS proceedings, 1-15.

[3] Trudel, A., & Gagne, M. (1997). Effect of composition and heat treatment parameters on the characteristics of austempered ductile irons. Canadian Metallurgical Quarterly, 36(5), 289-298.

[4] Handbook, A. S. M. (1991). Heat treating. vol, 4, 744.

[5] Yang, C., Wei, X., Wang, W., Lin, Z., Ding, J., Wang, Y., & Yang, J. (2016). NOx emissions and the component changes of ternary molten nitrate salts in thermal energy storage process. Applied Energy, 184, 346-352.

[6] de Souza, B. V., Ribeiro, T. M., Francois, A., & dos Santos, C. A. (2018). Austempering heat treatments of ductile iron using molten metal baths. Materials and Manufacturing Processes, 1-7.

[7] Pereira, L et al. (2017) Obtainment of ADI by austempering treactment using Zamak 5 bath , p. 757-764. In: 72nd ABM Annual Congress, São Paulo.

[8] Meier, L., Hofmann, M., Saal, P., Volk, W., & Hoffmann, H. (2013). In-situ measurement of phase transformation kinetics in austempered ductile iron. Materials Characterization, 85, 124-133.

The 73rd World Foundry Congress
„Creative Foundry"
23 - 27 September 2018, Krakow, Poland

Influence of AlP in solidification sequence of Al-10%Si-0.3%Mg

Yiyang Zhao [1]*, Susumu IKENO[2], Seiji Saikawa [3]

[1]Unibersity of Toyama, Graduate School of Science and Engineering for Education,3190, Gofuku, Toyama, 930-8555, Japan,
[2] Professor emeritus at University of Toyama,3190, Gofuku, Toyama 930-8555, Japan,
[3] University of Toyama, Science and Engineering for Research, 3190,Gofuku.Toyama 930-8555,Japan,
*Corresponding address: e-mail: zhaoyiyang0101@gmail.com

Key words: quench, heterogeneous nucleation, microstructure, Al-Si-Mg alloy, AlP

1. Introduction

In order to reduce CO_2 emissions from the viewpoint of protecting the global environment, in the automobile industry, Improve fuel economy by reducing the weight of the car body. For this purpose, the automobile industry is attempted conversion from iron-based materials of automobile parts to aluminum, aluminum to magnesium, resin. with a low specific gravity. Aluminum alloy is applied to automobile parts because it is easier to recycle, cheap and environmentally friendly than resin. Most of these alloys are based on Al-Si system alloy, and Al-Si system alloy added with a small amount of Cu, Mg, Ni, are used as the casting alloy. On account of the Al-Si alloys are low-density, good castability, and excellent corrosion resistance. Especially the hypo-eutectic Al-Si alloy which containing 5 to 11 mass% Si (hereinafter mass% is abbreviated as "%") and a small Mg is the important material that has been widely used for automobile road wheels, suspension arms.

We investigated the effect of AlP on the microstructure of Al-10mass%Si-Mg system alloy by using defferent purity aluminum (to reduce the nucleation frequency of AlP caused by P content). In DTA cooling curve of using low purity Al-10%Si-0.8%Mg alloy, observed unknown peak. The peak was researched with quench at any temperature during solidification . In the result, it was found that the peak occured by growth of plate-like eutectic Si phase formed with AlP as the heterogenius nucleation site. At the above investigation, plate-like eutectic Si phase was observed even in a temperature range higher than the eutectic reaction temperature. In this study, has investigated influence of P content on Al-10%Si-0.3%Mg system alloy by using different purity aluminum. In addition, has investigated the solidification sequence of Al-10%Si-0.3%Mg alloy.

2.Experimental procedure

The target composition were Al(2N)-10%Si-0.3%Mg and A(5N)-10%Si-0.3%Mg alloys. The experimental composition and alloy codes are shown
in Table 1. 2N alloy and 5N alloy that different of P content due to investigate effect of AlP was prepared. Aluminum of commercial purity (99.7%) or high purity (99.999%) was melted with crucible made by alumina in electric resistant furnace. High purity Si (99.999999999%) and magnesium (99.999%) were added into molten metal. The molten metal was poured into Y-block shaped mold at 720°C. The Y-block shaped mold that had the sampling portion with 60 mmH, 100mmW and 3.5mmT in lower side and the riser with inverted trapezoidal shape in the upper side was used for casting.
For investigation of the solidification sequence, measured crystallization temperature and quenched microstructure.

1) Measurement of crystallization temperature and crystallization reaction.

About 3.5g of specimens were cut from the sampling portion of the Y-block shaped castings. Specimens were enclosed into capsule made by SUS304 with K-type thermocouple. The capsule was hung in cylinder shaped furnace by wire. Specimens were melted at 700 °C and held for 30min, the molten metal was slowly cooled (CR≈0.05 °C /s) and obtain cooling curve. Schematic drawing of electric resistance furnace is shown in Fig. 1.

2) Quenched microstructure

Quenched into water during solidification as same as 1 by cutting wire and thermocouple.
The quenching specimens were polished with sandpaper, buffed with 1μm diamond paste and etched in a 0.5%HF solution. The microstructures were observed by optical microscope.

Table 1. The chemical compositions of the alloys

Code	Alloy	Si	Mg	Fe	Al
2N alloy	Al(2N)-10%Si-0.3%Mg	9.92-10.01	0.31-0.32	0.08-0.09	Bal.
5N alloy	Al(5N)-10%Si-0.3%Mg	10.00	0.31	0.00	Bal.

Black character: mass%

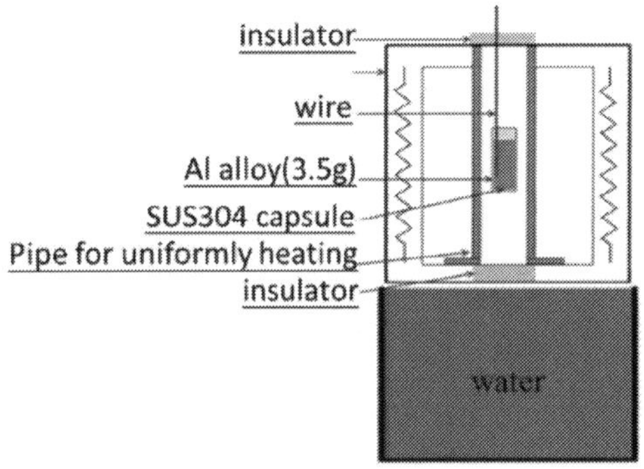

Fig. 1. Schematic drawing of electric resistance furnace.

3.Results and Discussions

Fig. 2 shows the eutectic growth temperatures taken from cooling curves obtained during the eutectic solidification of the low purity 2N alloy and high purity 5N alloy. The inflection points were observed at about 595℃ and about 575℃ for both 2N alloy and 5N alloy. It seems to the crystallization of primary α-Al phase at about 595℃ and the crystallization of Al-Si eutectic phase at about 575℃. In addition, supercooling at about 2 ° C was observed only in 5N alloy.

The microstructure of 2N and 5N alloys were quenched at 450°C (full solidification) are shown in Fig 3. In the 2N alloy shown in Fig.3(a), as shown by the arrows there are many plate-like eutectic Si phases formed with AlP as the heterogenius nucleation site. On the other hand, in the 5N alloy shown in Fig. 3(b), the plat-like eutectic Si phase was not observed but observed the eutectic Si phase was finer than 2N alloy.

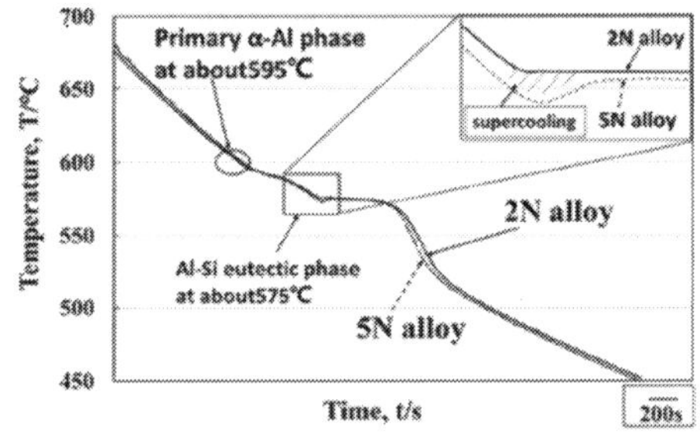

Fig. 2. The eutectic growth temperatures taken from cooling curves obtained during the eutectic solidification of the low purity 2N alloy and high purity 5N alloy.

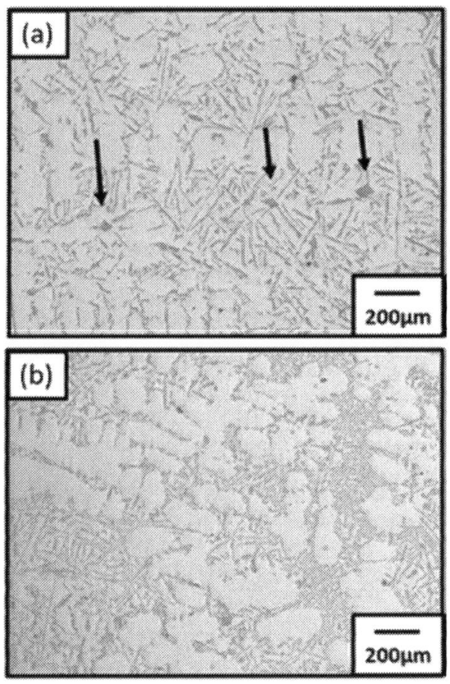

Fig. 3. Quenched microstructure at 450°C (full solidification) in 2N alloy and 5N alloy.

References

Effects of boron on eutectic modification of hypoeutectic Al–Si alloys

Scripta Materialia, Volume 48, Issue 3, February 2003, Pages 307-313

K Nogita, A. K Dahle

The 73rd World Foundry Congress
„Creative Foundry"
23 - 27 September 2018, Krakow, Poland

EFFECT OF AlP IN SOLIDIFICATION SEQUENCE OF Al-6%Mg-3%Si

Arisa Osugi [1]*, Susumu IKENO [2], Seiji Saikawa [3]

[1] *University of Toyama, Graduate School of Science and Engineering for Research, 3190, Gofuku, Toyama, 930-8555, Japan,*
[2] *Professor emeritus at University of Toyama,3190,Gofuku, Toyama 930-8555, Japan,*
[3] *University of Toyama, Science and Engineering for Research, 3190,Gofuku,Toyama 930-8555,Japan,*
**Corresponding address: e-mail:* g.arisa0516@gmail.com

Key words: Al-Mg-Si alloy, microstructure, quench, heterogeneous nucleation, AlP

1. Introduction

In recent years, weight reduction of vehicle body has been required in the automobile industry. Aluminium alloy have characteristic of strength/weight ratio and excellent corrosion resistance. Therefore, the application of aluminium alloy have been expanded in the auto parts. Al-Mg-Si alloy has both of good strength and high ductility without heat treatment, which is attracting attention as new die-cast alloy. However, this alloy is hard to cast, hot-tearing is easily occured during solidification. Therefore, it becomes a problem in the manufacture of parts.

We investigation the castability of Al-2~6%Mg-1~3%Si system alloy. As a result, the composition that hot-tear decresed was Al-6%Mg-3%Si alloy, because the composition of this alloy was close to quasi-binary eutectic of Al-Mg_2Si [1]. In addition, it is found that hot-tearing was not occurred in the case of 0.04%Sr addition to Al-6mass%Mg-3mass%Si alloy. Eutectic Mg_2Si phase of Sr-free alloy was observed plate-like Mg_2Si having AlP nuclei. On the other hands, Eutectic Mg_2Si phase of 0.04%Sr addition alloy was the remarkable refinement and not present plate-like Mg_2Si [2].

In this study, Al-6%Mg-3%Si alloy was water-quenching at any temperature, and the crystallization behavior of plate-like Mg_2Si phase having AlP nuclei was investigated in detail.

2. Experiment Procedures

The target composition was Al-6%Mg-3%Si. Two kinds of alloys that different of P content due to investigate effect of AlP was prepared. Chemical composition of the two alloys are shown in Table1.Aluminum of commercial purity (2N,99.7%) and Al-20%Si master alloy were melted with crucible made by alumina in electric resistant furnace. Al-10%Mg master alloy was added into molten metal. Further, 0.003%Be was added using Al-2.5%Be master alloy to prevent oxidation depletion of Mg at the molten metal state. On the other hand, high purity Aluminum (5N) was melted with crucible made by alumina in electric resistant furnace. High purity Si (11N) and magnesium (5N) were added into molten metal. The molten metal was poured into Y-block shaped mold at 730°C. The Y-block shaped mold have the sampling portion with 60mmH, 100mmW and 3.5 mmT in lower side and riser with inverted trapezoidal shape in the upper side. Measured crystallization temperature and quenched microstructure for investigation of solidification sequence.

1) Measurement of crystallization temperature and crystallization reaction.

About 3.5g of specimens were cut from the sampling portion of the Y-block shaped castings. Specimens were enclosed into capsule made by SUS304 with K-type thermocouple. The capsule was hung in cylinder shaped furnace by wire. Specimens were melted at 700°C and held for 30min, the molten metal was slowly cooled and obtain cooling curve. Schematic drawing of electric resistance furnace is shown in Fig. 1.

2) Quenched microstructure

Two quenching method were adopted. In the quenched temperature is less than 650°C, quenched into water during same solidification as 1) by cutting wire and thermocouple. In the quenching temperature is over 650°C, iron diffusion into the molten metal was concerned. Therefore, another method was used due to obtain freezing structure as same as mention above. The method is consist of three process: firstly chopped metals is into crucible, secondly heat up to molten state, thirdly quenched into water.

The quenching specimens were polished with sandpaper, buffed with 1μm diamond paste and etched in a 0.5%HF solution. The microstructures were observed by optical microscope.

Table1. The chemical composition of the alloys (black is mass%, red is ppm)

Alloy	Mg	Si	Fe	P	Al
Al(2N)-6%Mg-3%Si	5.41-5.80	3.22-3.32	0.14	7-11	Bal.
Al(5N)-6%Mg-3%Si	5.42-5.81	3.21-3.65	0.0016	0.5	Bal.

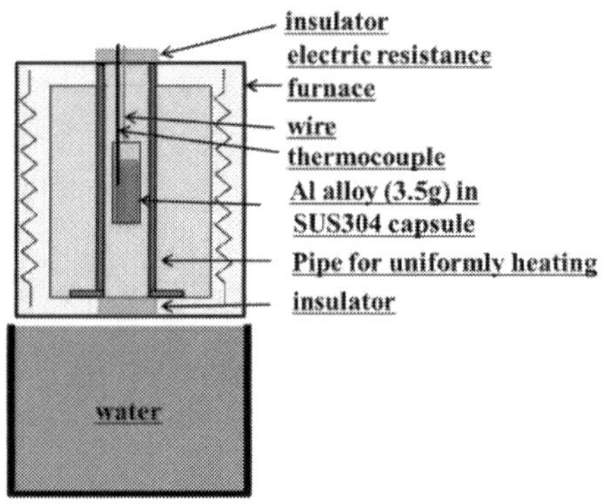

Fig.1. Schematic drawing of electric resistance furnace.

3. Results and Discussions

The cooling curves is shown in Fig.2. Al(2N)-6%Mg-3%Si alloy and Al(5N)-6%Mg-3%Si alloy was observed inflection points at about 610°C and 590°C. It seems to the crystallization of primary α-Al phase at about 610°C and the crystallization of eutectic phase at about 590°C. In addition, Al (2N) - 6% Mg - 3% Si alloy showed longer eutectic reaction time than Al (5N) - 6% Mg - 3% Si alloy. The quenched microstructure at 560°C of Al(2N)-6%Mg-3%Si and Al(5N)-6%Mg-3%Si alloy were shown in Fig.3. In the Al(2N)-6%Mg-3%Si alloy, there was plate-like Mg_2Si phase having AlP nuclei (shown by arrow in Fig.3(a)). Such heterogeneous nucleation with plate-like Mg_2Si as nuclei was occurred. Therefore, it is considered that the eutectic reaction time is prolonged by coarse growth of the eutectic phase. In the case of Al(5N)-6%Mg-3%Si, no such nuclei existed, there were very fine lamellar eutectic phase.

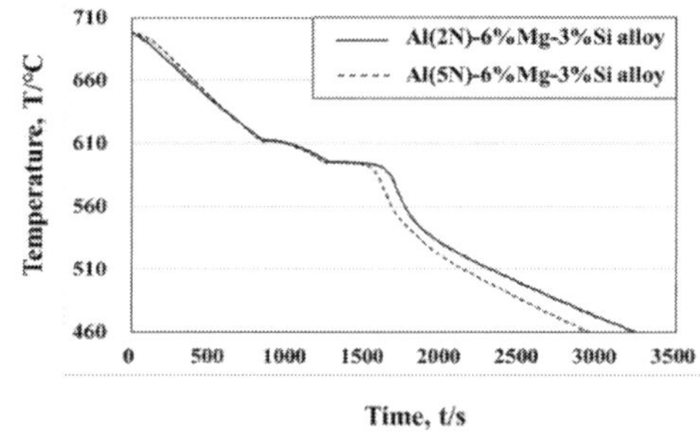

Fig.2. Cooling curves in Al(2N)-6%Mg-3%Si and Al(5N)-6%Mg-3%Si alloys.

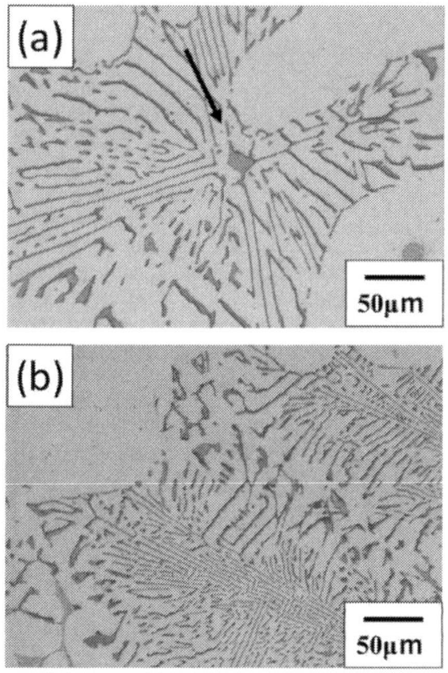

Fig.3. Quenched microstructure at 560°C in Al-6%Mg-3%Si alloy.
(a) Al(2N)-6%Mg-3%Si (b)Al(5N)-6%Mg-3%Si

References

[1] S. Saikawa, Y. Maeda, S. Ikeno, K. Ota, S. Orii, (2015) Journal of Japan Foundry Engineering Society, Vol. 87, 39-45.

[2] S. Saikawa, G. Aoshima, N. Hattori, S. Ikeno, E. Yanagihara, (2015) Journal of Japan Foundry Engineering Society, Vol. 87, 561-568.

The 73rd World Foundry Congress
„Creative Foundry"
23 - 27 September 2018, Krakow, Poland

Influence of P content in eutectic crystallization of Al-10%Si-0.8%Mg

Hisao Kazuta [1*], Susumu IKENO[2], Seiji Saikawa [3]

[1] *University of Toyama, Graduate School of Science and Engineering for Education, 3190, Gofuku, Toyama 930-8555, Japan,*
[2] *Professor emeritus at University of Toyama, 3190, Gofuku, Toyama 930-8555, Japan,*
[3] *University of Toyama, Science and Engineering for Research, 3190, Gofuku, Toyama 930-8555, Japan,*
**Corresponding address: e-mail: kzthso@gmail.com*

Keywords: Al-Si system alloy, microstructure, eutectic Si phase, heterogenous nucleation, AlP

1. Introduction

Hypoeutectic Al-Si system alloy is the one of widely used aluminum alloys. Al-6~10mass% alloy containing small amount of Mg have excellent castability, good strength and corrosion resistance by limiting impurity element such as Cu and Fe. AA356 and AA365 alloy are applied for automobile compo-nent such as road wheels, suspension and rear member. Morphology of eutectic Si phase and Mg_2Si phases affect the characterisic of these parts. It was reported that eutectic Si phase becomes coarse flake-like when AlP act as heterogeneous nucleation sites for eutectic Si phase in Al-Si alloy[1)].

We investigated influence of AlP in microstructure of Al-10mass%Si-Mg system alloy by using different purity aluminum as matrix ingot for reducing the amount of P due to decrease nucleation frequency of AlP. The low purity Al-10mass%Si-0.8mas%Mg alloy have unknown peak in DTA cooling curve. The peak was researched with quench at any temperature during solidification. It was found that the peak occured by growth of plate-like eutectic Si phase formed with AlP as the heterogeneous nucleation site. At the above investigation, plate-like eutectic Si phase was observed even in a temperature range higher than the eutectic reaction temperature.

In this study, influence of AlP in solidification sequence of Al-10%Si-0.8%Mg alloy investigated by quench at liquid state or during solidification.

2. Experiment procedure

2N alloys and 5N alloys that different of P content due to investigate effect of AlP was prepared. The target compositions were Al-10%massSi-0.8mass%Mg. Chemical composition of the two alloys are shown in Table1. Aluminum of commercial purity (99.7%) or high purity (99.999%) was melted with crucible made by alumina in electric resistant furnace. High purity Si (99.999999999%) and magnesium (99.999%) were added into molten metal. The molten metal was poured into Y-block shaped mold at 720°C. The Y-block shaped mold have the sampling portion with 60 mmH, 100 mmW and 3.5 mmT in lower side and riser with inverted trapezoidal shape in the upper side.

Measured crystallization temperature and quenched microstructure for investigation of solidification sequence.

1) Measurement of crystallization temperature and crystallization reaction.

About 3.5g of specimens were cut from the sampling portion of the Y-block shaped castings. Specimens were enclosed into capsule made by SUS304 with K-type thermocouple. The capsule was hung in cylinder shaped furnace by wire. Specimens were melted at 700°C and held for 30 min, the molten metal was slowly cooled (CR≈0.05°C/s) and obtain cooling curve.

2) Quenched microstructure

Two quenching method were adopted. In the quenched temperature is less than 700°C, quenched into water during same solidification as 1) by cutting wire and thermocouple. In the quenching temperature is over 700°C, iron diffusion into the molten metal was concerned. Therefore, another method was used due to obtain freezing structure as same as mention above. The method is consisted of three processes, firstly chopped metals is into crucible, secondly heat up to molten metal state, thirdly quenched into water.

The quenched specimens were polished with sandpaper, buffed with 1μm diamond paste and etched in a 0.5%HF solution. The microstructures were observed by optical microscope.

Table 1. The chemical composition of the alloys (black is mass%, red is ppm)

Alloy	Targetcom-position (mass%)	Chemicalcompositon (Black:mass%, Red:ppm)				
		Si	Mg	Fe	P	Al
2N	Al-10%Si-0.8%Mg	9.63-1.036	0.72-0.90	0.08-0.12	7.50	Bal.
5N	Al-10%Si-0.8%Mg	9.68-10.40	0.75-0.85	0.00-0.03	0.64	Bal.

3. Result and discussion

The cooling curves obtain in 2N alloy or 5N alloy when slowly cooled are shown in Fig.1. The primary α-Al phase and α-Al+Si eutectic crystalized in both alloys at about 595°C and about 575°C, respectively. The cooling curves of zoomed at eutectic reaction are shown in Fig.2. In cooling curve of 5N alloy, undercooling of 2°C was observed. On the othe hand, under cooling was hardly observed in 2N alloy of cooling curve. The microstrucuture of 2N and 5N alloy quenched at 567°C are shown in Fig.3. The plate-like eutectic Si phase formed with AlP as the heterogeneous nucleation site was observed only in 2N alloy. From above result, it was considered that undercooling was obeserved because heterogenous nucleation was not occured in 5N alloy.

The microstructure of 2N alloy and 5N alloy quenched at 700°C are shown in Fig.4. The plate-like eutectic Si phase was observed at liquid state only in 2N alloys. It was suggested that the AlP nuclei present even in liquid state.

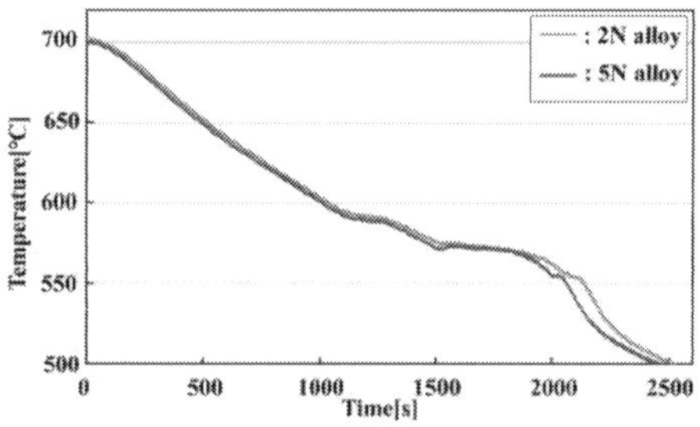

Fig.1. Cooling curve of 2N and 5N alloy

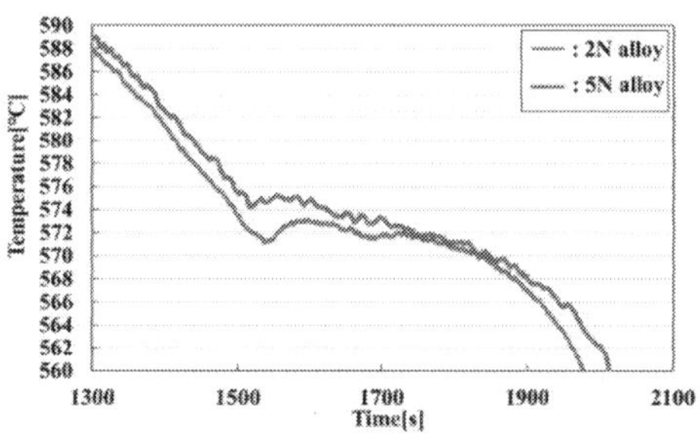

Fig.2. The cooling curves of zoomed at eutectic reaction in 2N and 5N alloy

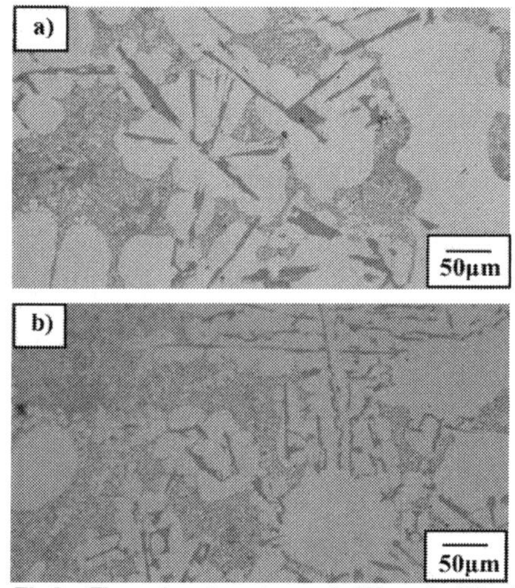

Fig.3. The microstrucuture of 2N and 5N alloy quenched at 567°C. a)2N alloy and b)5N alloy

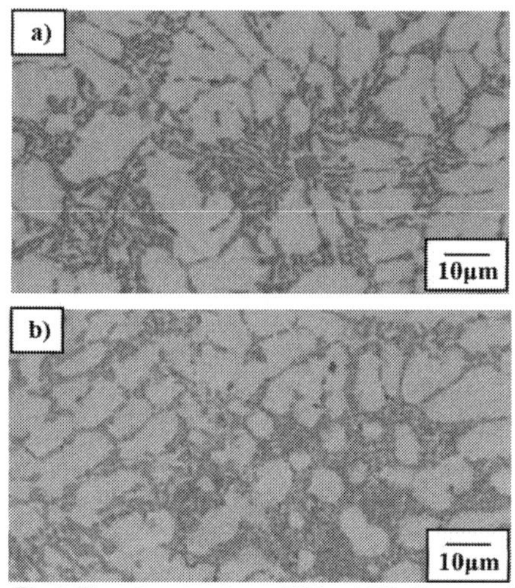

Fig.4. The microstructure of 2N and 5N alloy quenched at 700°C. a)2N alloy and b)5N alloy

Reference

[1] K. Nogita, S.D. McDonald, K. Tsujimoto, K. Yasuda, A.K. Dahle, (2004) J. Electron. Microsc. 53 361–369.

The 73rd World Foundry Congress
„Creative Foundry"
23 - 27 September 2018, Krakow, Poland

INFRUENCE OF TELLURIUM ADDITION TO SPHEROIDAL GRAPHITE CAST IRON ON THE NUMBER OF GRAPHITE PARTICLES

Ryuta Saito[1]*, Toru Maruyama[2], Toshiki Nakamura[1],
Hitoshi Yanagitani[3], Takahiro Sakai[3], Kouji Nakamoto[3]

[1]*Graduate School of Kansai Univ., Dept. of Chemistry and Mater. Engineering, 3-3-35Yamate-cho, Suita, Japan*
[2]*Kansai Univ., Dept. of Chemistry and Mater. Engineering, 3-3-35 Yamate-cho, Suita, Japan*
[3]*Kurimoto Ltd., Kagaya Factory, 2-1-64 Izumi, Suminoe-ku, Osaka, Japan*
**Corresponding address: ryubsk5@gmail.com*

Key words: Spheroidal graphite cast iron, graphite particles, Tellurium, graphitization, spheroidization

Abstract

Spheroidal graphite cast iron is superior in mechanical properties such as strength and ductility to flaky graphite cast iron. When the cooling rate during solidification is too large, it is a reduction in the mechanical properties due to the formation of chill structure. It is reported that the increase of spheroidal graphite particle number is effective for the reduction of chill structure. There are also reports that spherical graphite can form a sulfide of Mg or RE as nuclei. From these reports as reference, it is considered Te belonging to the same family as S might be a graphite core. In the experiment, two kinds of Mg amount of 0.020 mass% and 0.035 mass% were selected, and Te amount of 10 ppm, 20 ppm, 30 ppm and 40 ppm were performed for both samples. As a result, the addition of a small amount of Te in both samples was effective for increasing the number of graphite particles. However, when Te was added excessively, the number of graphite particles decreased.

1. Introduction

Spheroidal graphite cast iron is better in strength and ductility than flake graphite cast iron. So they are mainly used as cast iron tube and auto parts. Chill structure is formed in cast iron when cooling rate during solidification is too large. It is known that the mechanical properties are deteriorated due to chill structure. Especially thin castings of rapidly cooling rate often form chill structure. The phenomenon is serious problems. An increase in the number of graphite particles is known as an effective means for preventing chilling of spheroidal graphite cast iron [1]. In general, inoculation method is used for increasing graphite nucleus. One of the reports to increase the number of graphite particles is the addition of a trace amount of Bi. Sulfides of Mg and RE are known as graphite nucleus [2]. However, since excessive S consumes Mg and RE, graphitization and spheroidization are hindered. Te belonging to the same group as S is also an element that consumes Mg. Therefore, as in the case of S, an appropriate amount of Te may increase graphite nucleus. In the report of Nakamura et al. [3], The influence of Te on the number of graphite particles of spheroidal graphite cast iron produced by permanent mold casting was investigated. As a result, an increase in the number of graphite particles was reported. In the case of Te addition of about 10 to 30 ppm spheroidization was inhibited. On the other hand, the influence of the number of graphite particles by addition of Te in sand mold castings is unknown. Therefore, a sand mold casting sample in which the amount of Te and the amount of Mg were varied was prepared and the influence on the increase and decrease in the number of graphite particles was investigated.

2. Experimental Procedure

The target composition of the sample was Fe - 3.6mass% C - 1.7mass% Si - 0.3mass% Mn, and the target amount of Mg was 0.035 mass% (0.035 Mg) and 0.020 mass% (0.020 Mg). The target addition amount of Te was set to 10 ppm, 20 ppm, 30 ppm, and 40 ppm. As the raw materials, electrolytic iron, Fe - Si alloy, pure manganese, and a carburizing agent were used. Pure tellurium was used as the sample to which Te was added. As the inoculant, Fe - 70.6% Si - 1.6% Ba - 1.4 Ca% alloy was used. For melting, a 30 kVA high frequency induction furnace was used, casting was carried out at a total molten metal amount of 4.0 kg and a tapping temperature of 1550 ° C. The spheroidizing treatment was carried out by a sandwich method and poured into a sand mold (shell mold) for obtaining a casting of φ25 mm. In the specimens, the microstructure observation in the cross section and the measurement of the number of graphite particles were carried out. In the measurement of the number of graphite particles, particles under 5 μm in diameter were not included.

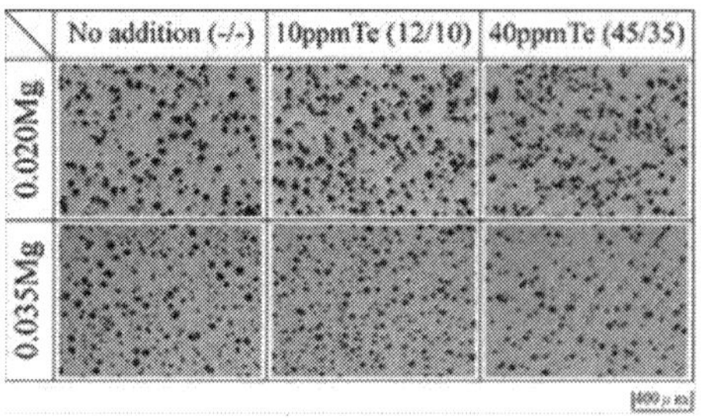

Fig.1 Graphite structure of spheroidal graphite cast iron with tellurium added

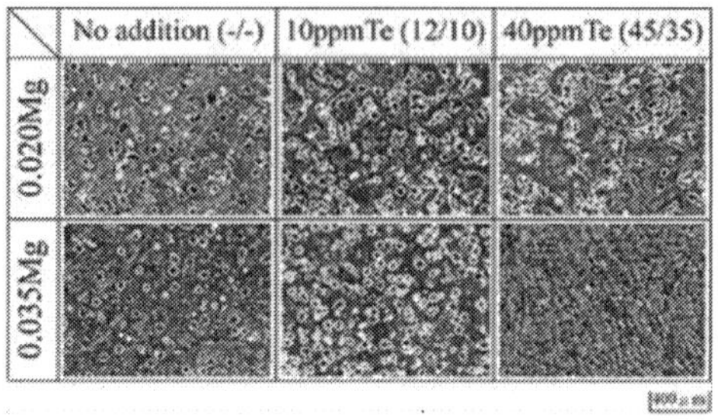

Fig.2 Optical microscopic texture of spherical graphite cast iron with tellurium added (Nital corrosion)

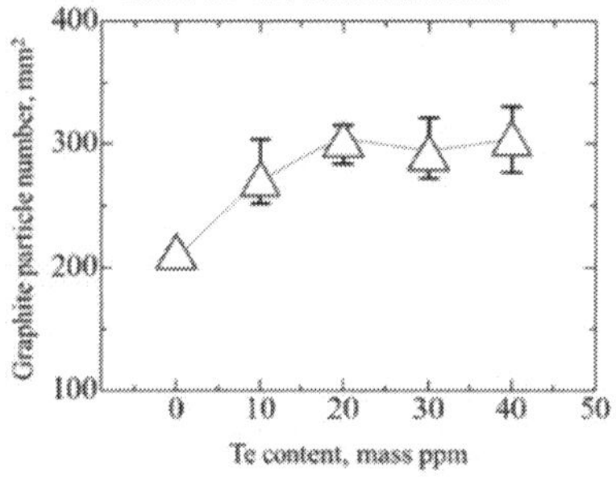

Fig.3 Relationship between the number of graphite particles and the amount of tellurium in spheroidal graphite cast iron containing about 0.020% Mg

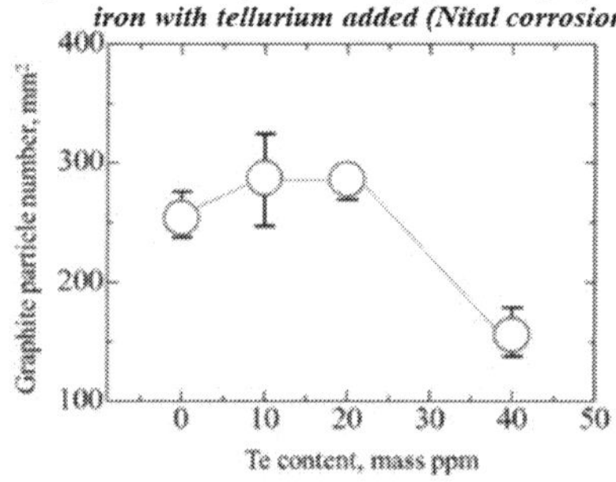

Fig.4 Relationship between the number of graphite particles and the amount of tellurium in spheroidal graphite cast iron containing about 0.035% Mg

3. Result and Discussion

Figure 1 shows the structure of various samples including Te and Mg before the etching. In the figure, the target addition amount of Te is shown, and the inside of parentheses shows the analysis value of Te. The right side value in parentheses is of 0.020Mg sample. The left side value in parentheses is of 0.035Mg sample. An increase in the number of graphite particles was shown in both the upper 0.020 Mg sample and the lower Te sample of the 0.035 Mg sample. On the other hand, it was recognized that the 0.020 Mg sample collapsed from the spherical shape as the Te addition amount increased. Figure 2 shows the microstructure of the specimens with different amounts of Te and Mg after etching for showing pearlite. A large amount of chill structure was observed in the as-cast samples after the ethings. In the 0.035 Mg sample, the amount of chill tended to increase due to an increase in Te content. The measurement results of the number of graphite particles in a Te-added sample containing 0.020 mass% of Mg are shown in figure 3. From figure 3, it was confirmed that the number of graphite particles increased due to the addition of Te in the sample of 0.020 Mg. In the sample added with 10 ppm or more of Te, there was no significant change in the number of graphite particles over 20ppm of Te. Figure 4 shows the measurement results of the number of graphite particles in a Te-added sample containing 0.035 mass% of Mg. From figure 4, it was confirmed that the number of graphite particles increased due to the addition of Te up to the amount of about 20 ppm. It was confirmed that the sample in which the addition amount of Te was further increased was sharply decreased in the number of graphite particles. Considering that Mg also forms a compound with Te similarly to S, the increase in the number of graphite particles due to the addition of Te is thought to be due to the fact that Te forms MgTe and the number of graphite particles increases as it becomes graphite nucleus. When the amount of Te added was excessive, the spheroidization ratio decreased in the 0.020 Mg sample and the chilling in the 0.035 Mg sample was promoted, so it is considered that there is an appropriate addition amount of Te corresponding to the Mg amount.

References

[1] Hiroshi Horie, Toshio Miyate, Minoru Saito and Toshinori Kowata, (1984), Effect of Carbon, Silicon and Nodule Count on the Chill Formation in Spheroidal Graphite Cast Iron with Thin Sections, IMONO, 56(8), 491-496.

[2] Hiroshi Horie, Toshinori Kowata, (1988), Effect of Bismuth on Nodule Count in Spheroidal Craphite Iron Castings with Thin Section, IMONO, 60(3), 173-178.

[3] Toshiki Nakamura, Toru Maruyama, Ryuta Saito, Hitoshi Yanagitani, Takahiro Sakai, Kouji Nakamoto, Reports of the 168th JFS Meeting, (2016), 55

The 73rd World Foundry Congress
„Creative Foundry"
23 - 27 September 2018, Krakow, Poland

INFLUENCE OF TI AND RE ON PRIMARY CRYSTALLIZATION AND WEAR RESISTANCE OF CHROMIUM CAST IRON

Malwina Dojka*, Rafał Dojka, Andrzej Studnicki, Marcin Stawarz

Silesian University of Technology, Department of Foundry Engineering, Towarowa 7, 44-100 Gliwice, Poland
**Corresponding address: e-mail: malwina.dojka@polsl.pl*

Key words: chromium cast iron, inoculation, titanium, mischmetal, wear resistance

1. Introduction

Castings made of chromium cast irons are generally working in hostile conditions, where intense wear is the main problem. Ball and ring mill parts, rollers, transporting elements, abrasion shields, excavator teeth etc. are parts which work conditions are very hard, that is why in order to avoid often replacements of those parts, selection of proper material is vital in order to achieve as long lifespan of the casting as possible [1,2]. One of the reasons why chromium cast iron is so attractive material is its excellent wear resistance to the production cost ratio.

But what if wear resistance of selected chromium cast iron is insufficient to our needs? One of the possibilities is to use different alloying elements than chromium which will form different carbides resulting in enhancement of wear resistance, but such action will highly increase the cost of production. However if the amount of the element would be small enough, rather than with alloying we would deal with inoculation, the cost will stay comparatively low and wear resistance will increase.

Inoculation is the perfect way to change the kinetics of the crystallization process which will affect formation of microstructure and characteristic crystallization temperatures which will result in improvement of casting properties.

Usually a fraction of percent by weight is enough to clearly see inoculants effect on the microstructure. However in this work higher amounts of inoculants were used so that not only the grain would be refined, but also the wear resistance would highly increase.

This work treats about two selected inoculants which were ferrotitanium (FeTi67) and mischmetal (about 67% of Ce, 33% La). The reason behind choosing titanium is high similarity of titanium carbide TiC lattice parameters to the Cr_7C_3 carbide lattice parameters, because of its high melting point and hardness up to 9.5 in Mohs scale [3,4]. In hypoeutectic chromium cast iron the TiC occurrence influent the austenite crystallization, which also causes M_7C_3 refining. Research conducted by Yinhu Qu showed that using Ce as inoculant results in refinement of the carbides due to formation of Ce_2O_2S [5,6]. Both mischmetal and ferrotitanium are often used as inoculants for Cr_7C_3 carbides [1-7]. In this work both inoculant were added to the melt simultaneously.

2. Range of the studies

Figure 1 presents the scheme of conducted research. Initial W0 melt was eutectic chromium cast iron melted using electric induction furnace. Charge consisted of material commonly used in industrial conditions: steel scrap, ferrochromium and carburizer. Inoculated melts were made by remelting W0 melt and adding inoculant into the ladle. Inoculants were added for sample TiMM004 in an amount of: mischmetal - 0.06% by weight of the charge, titanium - 0.06% by weight of the charge and for TiMM1 sample in amount of: mischmetal - 2% by weight of the charge, titanium - 2% by weight of the charge.

In this case TDA method was used to examine the kinetics of primary crystallization. Test stand consisted of computer linked to Crystaldigraph devices linked to two tester reflecting two different cooling speeds in order to examine crystallization with two different solidification times [8]:

ATD-C tester reflecting shorter solidification time,
ATD-Is tester reflecting longer solidification time.

Solidification time of the melt in ATD-Is tester is approximately two times longer than in ATD-C tester.

The EDS analysis was performed on PhenomPro-X scanning electron microscope. This part of studies can show the influence of inoculants on crystallization process and microstructure of analyzed chromium cast iron.

Wear resistance tests base on modified pin-on-disc method were conducted with the use of Tribotester 3-POD device. In this method the wear of examined sample was compared to the wear of reference sample made of Creusabro 8000 (C8) wear resistant steel. Tribotester 3-POD device work parameters were:

- abrasive disc: abrasive paper C120, SiC,
- rotational speed of the abrasive disc: 155 rpm,
- rotational speed of the sample holder: 400 rpm,
- load of the single sample: 220 G,
- time of wear resistance test cycle time: 10 minutes,
- number of cycles: 6,
- dry grinding.

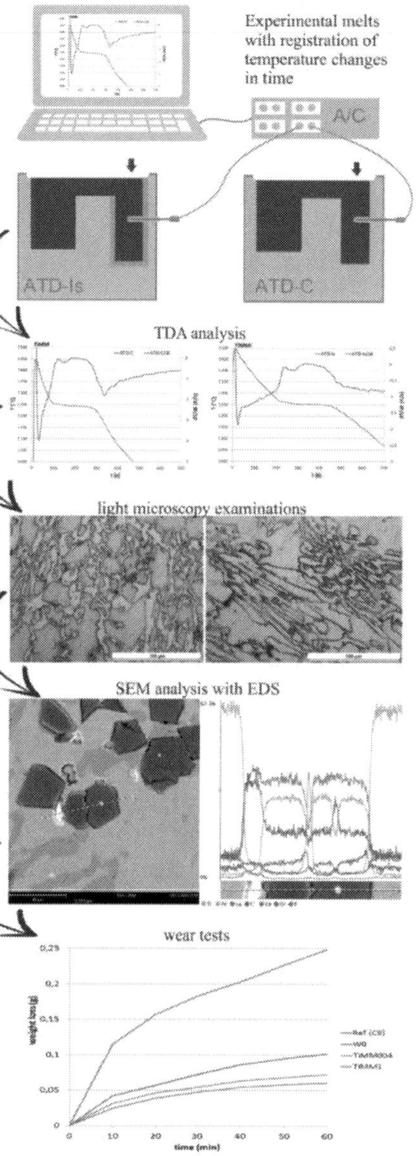

Fig. 1. Scheme of conducted research

3. Discussion of results

Studies allowed to state the impact of inoculation of chromium cast iron with Ti and RE. TDA analysis shows that in isolated testers each of characteristic temperatures of inoculated melts is higher than characteristic temperatures in not inoculated melt. It proves inoculation took place, because smaller supercooling was needed to start and finish the crystallization process.

Carbides in inoculated samples are rounded in comparison to not inoculated samples, so their impact strength should increase. From map analysis of selected area of TiMM1 sample it can be noticed that four elements determined the microstructure of alloy – Fe, Cr, Ti, C, but there can be seen some light particles in microstructure. Particles with higher amount of Ti and C were identified as TiC particles. On linear analysis it can be noticed, that on the light points there is high amount of cerium and lanthanum and close to those points we have not only M_7C_3 eutectic precipitations but also TiC, which means that the Ce and La compound could also be an underlay for titanium carbide particles crystallization. However, it can be assumed that the compounds of Ce and La as well as TiC serve as the underlays for the crystallization of chromium carbides.

Inoculated samples demonstrated better wear resistance than not inoculated sample. The highest improvement of wear resistance was clearly demonstrated by the sample with 2% addition of Ti and RE but it was only 17% better than for sample with 0,06% of additions. The important question is if achieving 17% less wear is worth spending over 3 times more expensive alloying elements. For some applications maybe it is, but the decision depends on many factors which shall be considered while planning to use this material.

References

[1] Studnicki A, Dojka R, Gromczyk M, Kondracki M. (2016) Influence of Titanium on Crystallization and Wear Resistance of High Chromium Cast Iron. Archives of Foundry Engineering. 16(1), 117 – 123.

[2] Kopyciński D., Guzik E., Siekaniec D. & Szczęsny A. (2015) The Effect of Addition of Titanium on The Structure and Properties of High Chromium Cast Iron. Archives of Foundry Engineering. 15(3), 35 – 38.

[3] Ma, Y., Li, X., Liu, Y., Zhou, S., Dang, X. (2012). Effect of Ti-V-Nb-Mo addition on microstructure of high chromium cast iron. China Foundry, 9(2), 148-153.

[4] Ma Y., Li X. & Liu Y. (2012). Effect of Ti-V-Nb-Mo addition on microstructure of high chromium cast iron. Bulletin of Materials Science, 36(5), 839-844.

[5] Zhi X., Liu J. & Xing J. (2014) Effect of cerium modification on microstructure and properties of hypereutectic high chromium cast iron. Materials Science and Engineering: A. 603, 98-103.

[6] Bedolla-Jacuinde A., Correa R. & Mejia I. (2007) The effect of titanium on the wear behaviour of a 16% Cr white cast iron under pure sliding. Wear. 263(1-6), 808-820.

[7] Wu X., Xing J., Fu H. & Zhi X. (2006) Effect of titanium on the morphology of primary M7C3 carbides in hypereutectic high chromium white iron. Materials Science and Engineering: A. 457(1-2), 180-185.

[8] Dojka M., Dojka R. & Studnicki A. (2017) Development of a New ATD-P Tester for Hard Wear Resistant Materials. Archives of Foundry Engineering. 17(1), 37-40.

The 73rd World Foundry Congress
„Creative Foundry"
23 - 27 September 2018, Krakow, Poland

CORROSION RESISTANCE OF Ti30Nb ALLOY IN HANK'S SOLUTION

Jagoda Ryba[1*], Halina Krawiec[1]

[1] *AGH-University of Science and Technology, Faculty of Foundry Engineering, Reymonta 23 Str., 30-059 Krakow*

Key words: corrosion, endoprostheses , titanium alloy, Hank solution, Young's modulus

Nowadays, a lot of joint and bone implants are made of titanium alloys. The chemical composition of titanium endoprostheses should be homogeneous both in terms of the elements present in it and various types of dispersion separations that may be present in the alloy. Titanium materials used for implants should also meet the requirements: very good corrosion resistance in body fluids, high biocompatibility, biofunctionality, as well as the Young's modulus of the material closer to the Young's bone modulus [1-5].

The most popular alloy on the market of endoprotic is currently Ti-6Al-4V. This alloy is classified as two-phase $\alpha + \beta$ alloys, characterized by an increase in melt strength both at 25 ° C and at elevated temperatures. Although it meets the requirements of biocompatibility, bofunctionality and corrosion resistance, it also has many unfavorable properties, for example a very high longitudinal elasticity coefficient (about 115 GPa for this alloy), which is much higher than the bone elasticity modulus (10-30 GPa). Biomedical Ti-6Al-4V titanium alloy contains toxic elements of aluminum (Al) and vanadium (V), which have an adverse effect on the body, causing, among others, tissue edema at the implant site or the presence of cancer cells. In the case when implants from the Ti-6Al-4V alloy are used, one should remember about the presence of bone tissue with rigid endoprosthesis, during which there are large amounts of frictional interactions. During the period of this implant in the human body, the formation of fibrous tissue and insufficiency of the entire joint occurs. This phenomenon usually leads to osteoporosis of bones surrounding the implant (made of Ti-6Al-4V) or loosening of the endoprosthesis, which leads to discomfort in the work of the joint. For these reasons, the service life of an endoprosthesis made of the above mentioned titanium alloy is usually limited to around 15 years. Therefore, more and more often scientific research is directed to titanium, monophasic β-stable alloys, which are characterized by better physical and mechanical properties compared to two-phase titanium alloys [1,3-8,10].

An example of such an alloy is Ti-30Nb alloy, considered as shape memory alloys. It has a very good biocompatibility and a relatively low modulus of elasticity similar to the Young's modulus of the human bone (about 50 GPa). This material is also characterized by very good bioactivity and biofunctionality. An important aspect that also speaks for the use of implants made of Ti-30Nb is their very good corrosion resistance [9, 11-12].

The aim of these research was was to study the corrosion mechanism of Ti-30Nb alloy in solutions simulating human body fluids. The corrosion tests were carried out in a Hank solution and its chemical composition is given in Table 1. The temperature of the solution was 37 ° C during the corrosion measurements.

Table 1. Chemical composition of the Hank's solution

Chemical compound	g/dm^3
NaCl	8
KCl	0.400
$CaCl_2$	0.140
$NaHCO_3$	0.350
NaH_2PO_4	0.217
Na_2HPO_4	0.060
$MgCl_2*6H_2O$	0.406
$MgSO_4$	0.029
G-glucose	1

Electrochemical measurements were performed in a classical 3-electrodes electrochemical cell, where the following electrodes were used: working electrode (WE) was tested alloy Ti-30Nb or pure metals like Ti, Nb, platinum grid was a counter electrode and the silver silver chloride electrode was used as reference electrode (Ag/AgCl electrode in 3M KCl). All corrosion tests were performed in Hank's solution at 37°C. In order to maintain a constant temperature of electrolyte (37 ° C), the electrochemical vessel was placed in a water bath. Corrosion tests were performed by means of the Potentiostat Autolab PGSTAT302N. Electrochemical impedance spectroscopy (EIS), potentiodynamic polarization curves (LSV) and long-term (48 h) measurement of open circuit potential (OCP) were carried out in Hank's solution for pure Ti, Nb and Ti-30Nb alloy. The potentiodynamic polarization curves (LSV – linear sweep volatamperommetry) have been preformed with a potential scan rate of 1 mV /s in the range of potentials from -1.0 to 2.0 V versus Ag/AgCl (3M KCl).

Fig.1. shows the evolution of open circuit potential (OCP) versus time measured in Hank solution for Ti30Nb alloy and for pure alloying elements. OCP after one hour obtained steady state of

-330 mV vs. Ag/AgCl electrode. This stable value of OCP is maintained for 24 hours. The values of OCP for Ti and Nb after 24 hours of immersion were -100 mV and -120 mV, respectively.

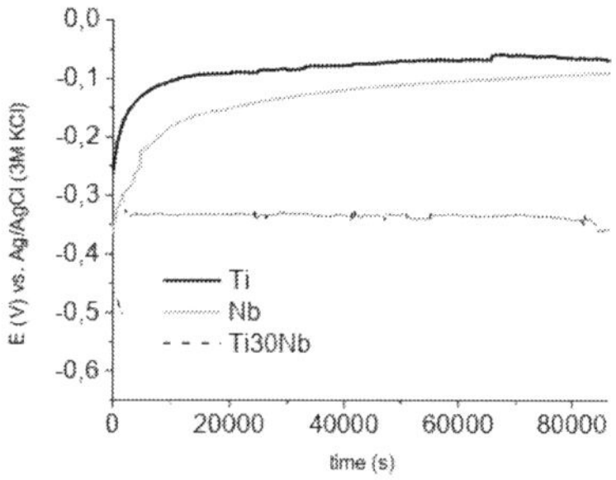

Fig. 1. Evolution of OCP for Ti-30Nb alloy, Ti, Nb in Hank's solution.

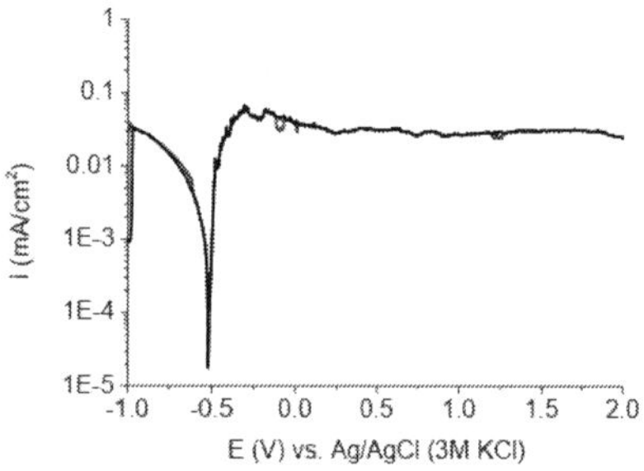

Fig. 2. Linear sweep voltamperometry curves (LSV) determined for Ti-30Nb alloy in Hank's solution.

In order to study the corrosion behaviour of Ti30Nb alloy, the polarisation curves (LSV) were performed in Hank's solution. As it can be seen in Fig.2, Ti30Nb titanium alloy is passivated in a wide range of potentials from -0.5 V up to 2 V. The current density in the passive range is 28 $\mu A/cm^2$. It indicates that this alloy exhibit very good corrosion resistance in Hank's solution. Formation of the passive film at the surface of Ti30Nb alloy ensures it very good corrosion resistance in Hank's solution. The passive film consists mainly of titanium dioxide TiO_2 and niobium oxides. The resistance of the passive film has been determined by electrochemical impedance spectroscopy (EIS).

Conclusions

1. The tested Ti-30Nb alloy exhibits very good corrosion resistance in Hank's solution, which results from the formation of a passive film on its surface consisting of titanium and niobium oxides.
3. Ti-30Nb alloy belongs to alloys which have very good biocompatibility and relatively low modulus of elasticity similar to Young's modulus of human bone.
4. The tested titanium alloy can replace the popular Ti-6Al-4V alloy in medicine.

References

[1] Nałęcz M. (2003). Biocybernetyka i Inżynieria Biomedyczna 2000: Biomateriały. Akademicka Oficyna Wydawnicza Exit, Warszawa
[2] Smith W.F. (1993). Structure and properties of engineering alloys. McGraw-Hill, Inc. USA
[3] Brunette D.M., Tengvall P., Textor M., Thomsen P. (2001). Titanium in medicine. Springer Germany
[4] Niinomi M. (2008). Biologically and Mechanically Biocompatible Titanium Alloys, Department of Biomaterials Science. Institute for Materials Research. Tohoku University Japan, Materials Transactions, Vol. 49, No. 10
[5] Choubey A., Balasubramaniam R., Basu B. (2004). Effect of replacement of V by Nb and Fe on the electrochemical and corrosion behavior of Ti-6Al-4V in simulated physiological environment. Journal of Alloys and Compounds 381. Pages 288-294.
[6] Niinomi M. (2018) Tiitanium Alloys, Reference Module in Biomedical Sciences.
[7] Sittig C.,Textor M., Spencer N.D., Wieland M., Vallotton P.H. (1999). Surface characterization of implant materials c.p. Ti, Ti-6Al-7Nb and Ti-6Al-4V with different pretreatments. Journal of Materials Science: Materials in Medicine, 10. Pages 35-46
[8] Wierzchoń T. (2003). Surface engineering of titanium alloys: new prospective applications. Materials Science Forum, 426-432. Pages 2563-2568
[9] Liu J., Chang L., Liu H., Li Y., Yang H., Ruan J. (2016). Microstructure, mechanical behavior and biocompatibility of powder metallurgy Nb-Ti-Ta alloys as biomedical material. Materials Science and Engineering.
[10] Song X., Wang L., Niinomi M. Nakai M., Liu Y., Zhu M. (2014). Microstructure and fatigue behaviors of a biomedical Ti–Nb–Ta–Zr alloy with trace CeO_2 additions; Materials Science and Engineering: A. Volume 619. Pages 112-118
[11] Aguilar C., Guerra C.,. Lascano S., Guzman D., Rojas P.A., Thirumurugan M., Bejar L., Medina A. (2016). Synthesis and characterization of Ti–Ta–Nb–Mn foams; Materials Science and Engineering: C, Volume 58. Pages 420-431
[12] Taddei E.B., Henriques V.A.R., Silva C.R.M., Cairo C.A.A. (2004). Production of new titanium alloy for orthopédie implants, Materials Science and Engineering C, 24. Pages 683-687

The 73rd World Foundry Congress
„Creative Foundry"
23 - 27 September 2018, Krakow, Poland

MAGNESIUM ALLOYS FOR BIOMEDICAL APPLICATIONS OF IMPLANTS

Andrzej Fijołek [1]*, Jagoda Ryba [1], Halina Krawiec [1], Łukasz Rogal [2], Janusz Lelito [1]

[1] AGH University of Science and Technology, Faculty of Foundry Engineering, 30-059 Cracow, 23 Reymonta Street, Poland,

[2] Institute of Metallurgy and Materials Science of Polish Academy of Sciences in Cracow, 30-059 Cracow, 25 Reymonta Street, Poland

*Corresponding address: e-mail: and.fijolek@gmail.com

Key words: *Biodegrability, Biomedical, Corrosion, Mg-Zn-Ca alloy, Ringer solution*

1. Introduction

Mg-Zn-Ca alloys are used in medicine because of their very favorable biodegradable and biocompatible properties compared to other alloys. Magnesium has many characteristic features as a biomaterial such as low weight and density, non-toxicity, excellent biocompatibility, not carcinogenic. Zinc has a positive effect on the corrosion resistance and mechanical properties of magnesium. The addition of calcium increases the metallurgical quality of magnesium alloys and reduces oxidation in both the melted alloy and during heat treatment. Adding calcium improves durability and creep resistance [1, 2].

The human body has an inherent tolerance of Mg, Zn and Ca [4]. The daily recommended dose for an adult human is 1000 mg / day for Ca [5], 420 mg / day for Mg [6] and 10 mg / day for Zn [7]. In addition, Zn and Mg ions have antibacterial properties, Zn inhibits catabolism of bacteria and Mg through pH alkalinizes the environment as a result of Mg dissolution [8, 9]. Therefore, it can be expected that these alloys also have antibacterial effects that can contribute to reducing the risk of bacterial infection around the implants, while maintaining good biocompatibility during implantation [10, 11].

Corrosion of magnesium alloys with a crystal structure is very fast, which in turn leads to the formation of gas bubbles under the skin of the living organism [12]. This phenomenon leads to acidification of the living organism and is unfavorable. Therefore, the kinetics of corrosion in biodegradable materials must be adjusted and monitored accordingly. Biomedical magnesium alloys are used as: pins, screws and bone implants [1, 12]. In addition to corrosion resistance, magnesium alloys for biomedical applications should have adequate mechanical and plastic properties. This is extremely important when stent application in the human body, which is made of magnesium alloy. Mechanical properties affect proper blood vessel support and ductile properties that facilitate the safe placement of this stent [1].

Corrosion is usually an undesirable phenomenon in engineering applications. However, in the field of biomedical applications, implants which are yielding "biocorod" arouse considerable interest. Placing them not only eliminates the need for implant surgery, but also bypasses the long-term negative effects of permanent implants. Mg-based glasses offer extended solubility for alloying elements plus a homogeneous single-phase structure, both of which can significantly change the corrosion behavior [3].

Currently, many scientists are working on this field, trying to obtain new magnesium alloys for medical applications, and research mechanical properties and their corrosion resistance [1-3].

The aim of the work was to determine the electrochemical behavior of alloys: Mg72Zn24Ca4 before and after rapid crystallization in the Ringer solution which simulating the human body environment. For this purpose, LSV and OCP tests were carried out from the above materials in the Ringer solution.

2. Materials and characterization

For the preparation of Mg72Zn24Ca4 alloy used high-purity alloy components with a purity of 99,99%. The melt was carried out in a resistance furnace under argon as an inert gas.

The Mg72Zn24Ca4 alloy was cast into the steel metal mould to obtain a cylinder sample with dimensions of: diameter = 20 mm and height = 50 mm. Final dimensions of samples for further testing after machining: diameter = 10 mm and height = 50 mm. In order to prepare a ribbons the following procedure was performed: the melt Mg72Zn24Ca4 alloy was cast on a melt spinner machine in an argon shield. This resulted in a thin ribbon of amorphous composite. Then the corrosion resistance of bulk and ribbons of Mg72Zn24Ca4 alloy has been determined. The corrosion tests have been performed in the Ringer solution at 37 C. The chemical composition of the Ringer solution is given in the table 1.

Table 1. The ingredients used to make 1 dm^3 of Ringer's solution

Component	Weight [g/dm^3]
NaCl	8.6
KCl	0.3
$CaCl_2$	0.25

The bulk material of Mg72Zn24Ca4 alloy used for corrosion tests was ground on abrasive papers with gradations from 1200 to 4000. After each polishing step, the samples were cleaned in ethanol using ultrasound. In order to determine the corrosion resistance of Mg alloy evolution of corrosion potential versus time (OCP – open circuit potential) and polarization curves (LSV – linear sweep voltamperometry) have been performed. The potential scan rate for LSV was 1 mV/s.

3. Results and discussion

Fig.1. shows the evolution of open circuit potential for Mg72Zn24Ca4 and ribbons of Mg72Zn24Ca4 in the Ringer solution.

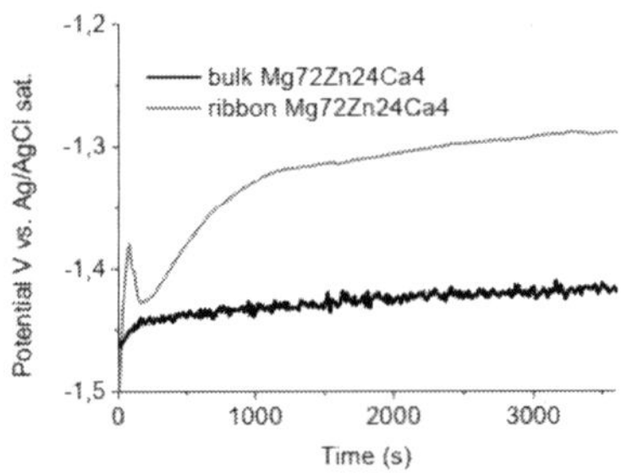

Fig. 1. Evolution of OCP for Mg72Zn24Ca4 alloy (bulk and ribbon) in Ringier's solution at 37°C, pH=7.2

The significant difference of OCP potential was detected. After one hour immersion of specimens in the Ringer solution the following values of OCP were measured: -1.417 V for bulk Mg72Zn24Ca4 and -1.293 V for ribbon, respectively. The higher value of corrosion potential detected for ribbon indicate that the surface undergoes the activation.

Fig. 2 depicts the polarisation curves determined for Mg72Zn24Ca4 and ribbons of Mg72Zn24Ca4 in the Ringer solution.

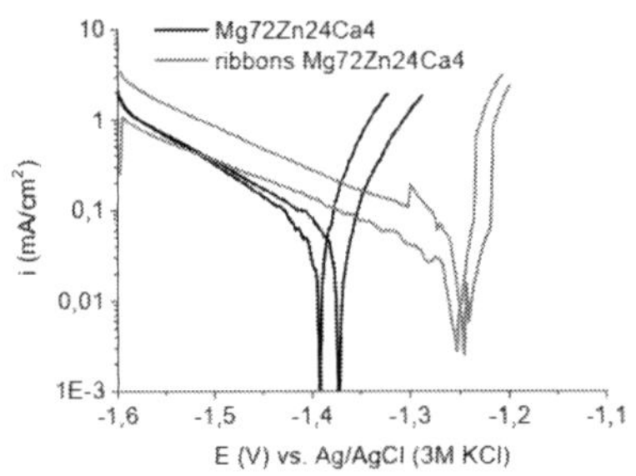

Fig. 2. Polarisation curves performed for Mg72Zn24Ca4 alloy (bulk and ribbon) in Ringier's solution at 37°C, pH=7.2

The both specimens (bulk and ribbon) exhibit the active corrosion. In the anodic branch the sharp increase of the current density is observed. Moreover, in the cathodic branch the higher current density is observed for ribbon specimen. It means, that the cathodic reaction (oxygen reduction) is promoted at the surface of ribbon specimen. Alkalization of the solution case the significant shift of the polarization curves to the higher potential values.

The corrosion tests have revealed that the ribbons specimens have higher corrosion resistance than the bulk material Mg72Zn24Ca4.

4. Conclusions

The corrosion tests have revealed, that the both specimens (bulk and ribbon) undergo the active corrosion. In the anodic branch a sharp increase in current density is observed for both specimens. Moreover, higher current density for the ribbon sample is observed in the cathodic branch than in the bulk sample. This results indicate that the cathodic reaction (oxygen reduction) proceeds faster on the surface of ribbon.

References

[1] Y. Zheng. (2015). Magnesium alloys as degradable biomaterials. 6000 Broken Sound Parkway NW, Suite 300: CRC Press Taylor & Francis Group.

[2] Y. Lu. (2014). Microstructure and degradation behaviour of Mg-Zn (-Ca) alloys. University of Birmingham.

[3] B. Zberg, P. J. Uggowitzer & J. F. Löffler. (2009). MgZnCa glasses without clinically observable hydrogen evolution for biodegradable implants. NATURE MATERIALS. 8, 887–891.

[4] J.D. Cao, N.T. Kirkland, K.J. Laws, N. Birbilis & M. Ferry. (2012). Ca–Mg–Zn bulk metallic glasses as bioresorbable metals. Elsevier. 8, 2375–2383.

[5] Food and Nutrition Board IoM. (1997). Dietary reference intakes for calcium, phosphorus, magnesium, vitamin D, and fluoride. Washington, DC: National Academies Press, 190–249.

[6] J. Vormann. (2003). Magnesium: nutrition and metabolism. Elsevier. 24 (1–3), 27–37.

[7] Food and Nutrition Board IoM. (2001). Dietary reference intakes for vitamin A, vitamin K, arsenic, boron, chromium, copper, iodine, iron, manganese, molybdenum, nickel, silicon, vanadium, and zinc. Washington, DC: National Academies Press, 442–501.

[8] R.C. Zeng, J. Chen, W. Dietzel, N. Hort & K.U. Kainer. (2007). Electrochemical behavior of magnesium alloys in simulated body fluids. 17, 166–170.

[9] W.C. Oliver & G.M. Pharr. (1992). An improved technique for determining hardness and elastic modulus using load and displacement sensing indentation experiments. 7, 1564–1583.

[10] D.A. Robinson, R.W. Griffith, D. Shechtman, R.B. Evans & M.G. Conzemius. (2010). In vitro antibacterial properties of magnesium metal against Escherichia coli, Pseudomonas aeruginosa and Staphylococcus aureus. Acta Biomater. 6, 1869–1877.

[11] J. Sheng, P.T.M Nguyen. & R.E. Marquis. (2005). Multi-target antimicrobial actions of zinc against oral anaerobes. Arch Oral Biol. 50, 747–757.

[12] Ch. Pil-Ryung, H. Hyung-Seop, Y. Gui-Fu, K. Yu-Chan, H. Ki-Ha, L. Seung-Cheol, J. Jae-Young, A. Jae-Pyeong, K. Young-Yul, Ch. Sung-Youn, B. Ji Young, L. Kang-Sik, Y. Seok-Jo & S. Hyun-Kwang. (2013). Biodegradability engineering of biodegradable Mg alloys: Tailoring the electrochemical properties and microstructure of constituent phases. SCIENTIFIC REPORTS. 3, 1–6.

Acknowledgment

This work is supported by H2020 MSCA -ITN-2017 – mCBEEs project.

The 73rd World Foundry Congress
„Creative Foundry"
23 - 27 September 2018, Krakow, Poland

MODELING OF THE KINETICS OF CARBONITRIDES PRECIPITATION PROCESS IN HSLA STEELS USING CELLULAR AUTOMATA METHOD

Przemysław Marynowski[1]*, Henryk Adrian[1], Mirosław Glowacki[2]

[1]*AGH University of Science and Technology, al. Mickiewicza 30,30-059 Kraków, Poland, EU,*
[2] *The Jan Kochanowski University (JKU), Kielce, Poland, EU,*
**Corresponding address: e-mail: pmarynow@agh.edu.pl*

Key words: Cellular Automata, precipitation, carbonitrides, microalloying elements, microalloyed steel.

1. Introduction

Carbonitrides precipitations of microalloying elements V, Nb and Ti, strongly influence the microstructure and mechanical properties of high strength low alloy steels. To predict the mechanical properties of these steels parameters of precipitates such as size (mean radius) and content (volume fraction) are required. Mathematical models describing the carbonitrides precipitation process enable to calculate these parameters [1,2]. The development of computer science enables simulations of precipitation process. The models are based on the algebraic equations describing the nucleation and growth kinetics of precipitates. One of the available tools to simulate the precipitation processes is the method of Cellular Automata. Cellular Automata model proves to be very efficient in modelling various phenomena in materials science [3]. Cellular Automata model of carbonitride precipitation in microalloyed steels containing such elements as Ti, Nb or V is presented in the paper. The model predicts kinetics of precipitation as well as stereological parameters of precipitates.

2. Microalloyed steels

Microalloyed steels with ferrite-pearlite microstructure are important group of structural steels, where high mechanical properties are achieved through small additions of elements such as Ti, Nb, V, introduced separately or comprehensively. High chemical affinity of these elements for interstitials (N, C) results in precipitation of binary compound, nitrides and carbides and products of their mutual solubility – carbonitrides. In steel containing one of the microalloying elements (Ti, Nb, V) the carbonitride described by chemical formula MC_yN_{1-y} is formed, where y means atomic fraction of carbon in carbonitride.

3. Model of the kinetics of carbonitride precipitation process

To calculate the kinetics of the carbonitrides precipitation process in the low alloy steel a model based on the classical theory of nucleation and growth (CNGT) was developed [4,5]. CNGT is based on the change in free energy, ΔG, associated with the formation of an embryo in a supersaturated solid solution. In the process of carbonitrides precipitation there are three stages: nucleation, growth and coalescence, which can occur simultaneously. The nucleation rate V_n, is described by the equation (1):

$$V_n = \frac{dN}{dt} = N_0 Z \beta^* exp\left(-\frac{\Delta G^*}{kt}\right) exp\left(-\frac{\tau}{t}\right) \quad (1)$$

where: β^*- the condensation rate of solute atoms in cluster of critical size, Z – Zeldovich parameter, N_0– number of nucleation site per unit volume, ΔG^*- critical Gibbs free energy for nucleus formation, k- Boltzmann constant, T – temperature, τ - incubation time, t – time.

Embryo critical radius r^* and parameters $\beta^*$and Z represent the following equations:

$$r^* = -\frac{2\gamma}{\Delta G_v} \quad (2)$$

$$\beta^* = \frac{4\pi r^{*2} DX}{a^4} \quad (3)$$

$$Z = \frac{v_{at}{}^P}{2\pi r^{*2}}\sqrt{\frac{\gamma}{kt}} \quad (4)$$

The incubation time, τ, is given by equation:

$$\tau = \frac{4}{2\pi\beta^* Z^2} \quad (5)$$

where: γ - interphase boundaries energy, ΔG_v - the driving force for precipitation per unit volume, D - diffusion coefficient of the metallic element, X - a fraction of the atomic metallic element dissolved in matrix, a- lattice parameter, $v_{at}{}^P$ - the average volume of an atom in precipitation.

In the case of the formation of precipitates of carbonitrides, described by formula MC_yN_{1-y} driving force of nucleation is equal to:

$$\Delta G_v = -\frac{RT}{V_{MCN}}\left[ln\left(\frac{X_M^S}{X_M^e}\right) + y ln\left(\frac{X_C^S}{X_C^e}\right) + (1 - y) ln\left(\frac{X_N^S}{X_N^e}\right)\right] \quad (6)$$

where X_i^s – the atomic fraction of the component X in the solution i X_i^e – equilibrium atomic fraction of component X in solution.

4. Cellular Automata (CA) model

The main idea of the CA technique is to divide a specific part of the material into one-, two-, or three-dimensional lattices of finite cells. Each cell in this CA space is called a cellular automaton, while the lattice of the cells is known as cellular automata space. Each cell is surrounded by neighbours, which affect one another. Neighbourhoods can be specified in one-, two-, and three-dimensional spaces. The most popular examples are the von Neumann and the Moore neighbourhoods [3], where in the 2D case each cell is surrounded by either four or eight neighbouring cells, respectively.

Each cell in the CA space is characterised by its state and by values of internal variables. The cells interactions within the CA space are based on the knowledge defined while studying a particular phenomenon. In every time step, the state of each cell in the lattice is determined by the previous states of its neighbours and the cell itself on a basis of a set of precisely defined transition rules:

$$Y_{i,j}^{t+1} = \begin{Bmatrix} if\ (\Lambda) \Longrightarrow newstate \\ else \Longrightarrow Y_{i,j}^{t} \end{Bmatrix} \quad (7)$$

where: $Y_{i,j}^{t+1}, Y_{i,j}^{t}$ - state of the cell i,j in the current and previous time step, respectively, Λ - logical function, which describes the condition when the state of the cell changes. Function Λ depends on:

$$\Lambda = \Lambda\left(Y_{i,j}^{t}, Y_{k,l}^{t}, \boldsymbol{p}, \boldsymbol{q}\right) \quad (8)$$

where: $Y_{k,l}^{t}$ - state of the cell k,l, which is a neighbour to the cell i,j, in the previous time step, **p** – vector containing external variables, e.g. temperature, **q** – vector containing internal variables, e.g. carbon concentration.

Since the transition rules control the cells behaviour during calculations, the proper definition of these rules in designing a CA model critically affects the accuracy of this approach. The transition rules of the developed Cellular Automata model of precipitation are based on the knowledge of experts, scientists, experimental observations, and available literature knowledge.

Acknowledgements

Work within the AGH project 11.11.110.300

References

[1] Adrian H., Augustyn-Pieniążek J., Marynowski P., Matusiewicz P., (2014). Model wydzielania węglikoazotków w stalach mikrostopowych. Hutnik – Wiadomości Hutnicze. 81/4. 208-214.

[2] Adrian H. (2011). Numeryczne modelowanie procesów obróbki cieplnej. Kraków: Wydawnictwo AGH.

[3] Wolfram S. (1983). Statistical mechanics of cellular automata. Reviews of Modern Physics. 55. 601-644.

[4] Dutta S., Sellars C.M. (1987). Effect of composition and process variables on Nb(C,N) precipitation in niobium microalloyed austenite. Materials Science and Technology. 3. 197-206.

[5] Dutta B., Valdes E., Sellars C.M. (1991). Mechanism and kinetics of strain induced precipitation of Nb (C,N) in austenite. Acta Metallurgica et Materialia. 40. 653-662.

The 73rd World Foundry Congress
„Creative Foundry"
23 - 27 September 2018, Krakow, Poland

GRAIN REFINEMENT OF Al-2%Cu ALLOY USING BY VIBRATING MOLD

Yasuo Yoshitake [1]*, Kaoru Yamamoto [2], Nobuya Sasaguri [2], Hidenori Era [3]

[1] *National Institute of Technology, Kurume College, Support Center of Education and Research, 1-1-1 Komorino, Kurume, Fukuoka, 830-8555, Japan,*
[2] *National Institute of Technology, Kurume College, Dept. of Materials System Engineering, 1-1-1 Komorino, Kurume, Fukuoka, 830-8555, Japan,*
[2] *Kyushu Institute of Technology, Faculty of Engineering, 1-1 Sensui-cho, Tobata-ku, Kitatakyushu-shi, Fukuoka, 804-8550, Japan,*
**Corresponding address: e-mail:yoshitake@kurume-nct.ac.jp*

Key words: aluminum alloy, grain refinement, vibration, frequency, amplitude

1. Introduction

The smaller the grain size, the better the mechanical properties of the metallic material. In the case of casting material, its grains have a tendency to become coarse in conventional casting, and it has poor mechanical properties. Therefore, it is greatly desired to produce the conventional castings with fine grains.

Up to now, many studies on grain refinement of cast alloy have been conducted by means of the ultrasonic vibration[1], magnetic stirring[2], magnetic vibration[3] and etc. In this study, the vibration method that gives simple mechanical vibration to a mold was introduced, and the effect of vibration conditions on the grain refinement of Al-alloy was investigated.

2. Experimental procedures

A schematic drawing of an apparatus giving vibration to mold is shown in Fig.1. The metal mold on the stage is reciprocated mechanically by an eccentric cam and a motor. A cavity size of the metal mold is 50mm(L)×50mm(W)×65mm(H).

Hypoeutectic Al-2%Cu alloy of 500g was melted in an electric furnace, and then was poured into the vibrating mold at 1023K. The mold can be vibrated with various frequency and amplitude. The frequency was changed from 5Hz to 50Hz, and the half amplitude was changed from 0.1mm to 5mm. The mold was vibrated before pouring and it was stopped after the molten metal had solidified. The ingot was cut off in the horizontal direction to vibration direction, and the macrostructure was observed using a stereo microscope. The grain size of specimen was measured in the range of 10mm to 20mm from the bottom of ingot.

In order to observe the convection direction of liquid and formation of microstructure, the water model experiment was carried out using 38% ammonium chloride solution. The solution was poured into a mold made by an acrylic plate at 363K, and its convectional direction and formation of grain were observed directly about with and without vibration.

3. Experimental results and Discussions

Fig.2 shows the macrostructures of casting under the with and without vibration conditions. In the case of the mold without vibration, columnar crystals grew from the mold wall to the center of mold. In the case of the mold with vibration, the columnar crystals became shorten and the equiaxed grains were observed in the center of mold. Based on these results, the grain refinement could be obtained by giving vibrating to mold. The relationship between frequency of vibration and an average grain size under the constant half amplitude is shown in Fig.3. In case of the specimen without vibration(0Hz), the average grain size is 6.4mm and it decreases to 0.76mm as the frequency increases to 15Hz. Fig.4 shows the relationship between a half amplitude of vibration and

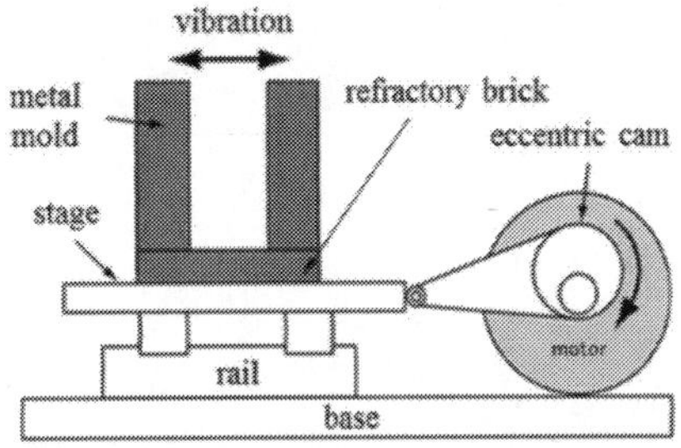

Fig. 1. Schematic drawing of mold vibration system.

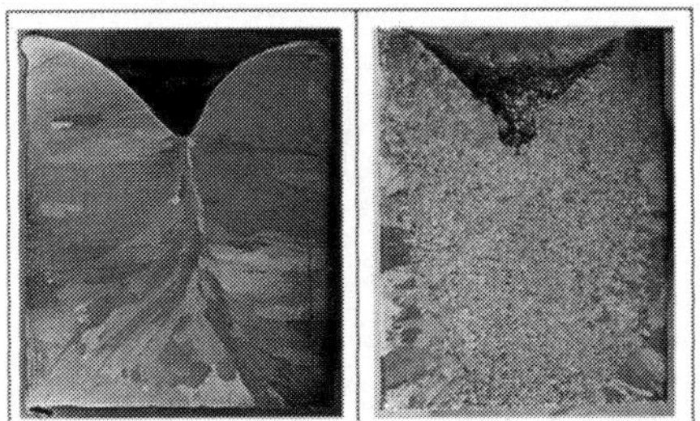

Without vibration	With vibration (40Hz, 0.1mm)

Fig.2. Macrostructures of specimens solidified under the with and without vibration.

an average grain size under the constant frequency. In case of that the half amplitude increases to 1.0mm, the average grain size sharply decreases from 6.4mm to 2.0mm. And the average grain size becomes 1.0mm at the half amplitude of 5.0mm.

It is clear that the average grain size is affected by both frequency and half amplitude of vibration. Therefore, it is thought that average grain size is summarized by excitation force given by equation (1) expressed by both frequency and amplitude.

$$F = ma(2\pi f)^2 \qquad (1)$$

where, F is excitation force(N), m is mass of specimen(kg), a is half amplitude(m) and f is frequency (Hz).

Fig.5 shows the relationship between excitation force and the average grain size. As the excitation force increases to 4.0N, the average grain size decreases to 1.0mm from 6.4mm. The influence of excitation force is little when the excitation force increases over 4N. As a result, it is said that the average grain size can be estimated by excitation force.

The water model experiment using ammonium chloride solution was carried out to clarify the effect of vibration on the molten metal.

Fig.6 shows the state of the solution in the mold with and without vibration at 15s after pouring. In the case of without vibration (Fig.6(a)), The crystals grew from the wall surface into the molten metal as dendrite morphology.

On the other hand, the vibration was given (Fig.6(b)), many crystals formed on the wall surface, then floated in liquid. And the crystals filled up in the mold at only 15s after pouring.

Therefore, also in the case of Al-2%Cu alloy, one explanation for grain refinement may say that many grains are crystallized on the wall of mold, and moved continuously into molten metal by vibration.

4. Conclusions

1. The grain of Al-2%Cu alloy was refined when the vibration was given to molten metal.
2. The grain size becomes small with an increase in amplitude and frequency of vibration.
3. The average grain size was can be estimated by the excitation force.
4. When the vibration was given to the mold, the convection was occurred and the number of crystal nucleation increased on the mold wall.

References

[1]Y.Osawa, G.Arakane, S.Takamori, A.Sato and O.Ohashi. (1999). Effect of Ultrasonic Vibrtion on Refining of Crystal Structures of Al-Si Alloys During Solidification. J.JSF, 71(2), 98-103.

[2]T.Momono and K.Ikawa. (1979). Structure change of hypereutectic Al-Si alloys by convective flow in solidification process. J.Jpn.Inst.Light Met. 29(6), 240-245.

[3]K.Miwa. (2002). New microstructural refinement process of metallic materials by electromagnetic vibrations. J.Jpn Inst Light Met. 52(12), 611-616.

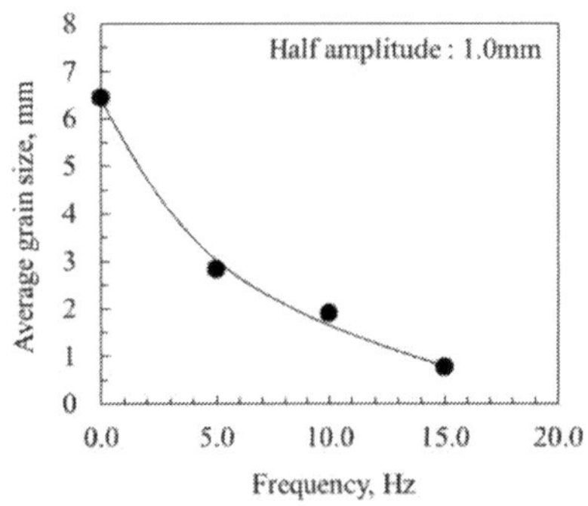

Fig.3. Influence of frequency on average grain size.

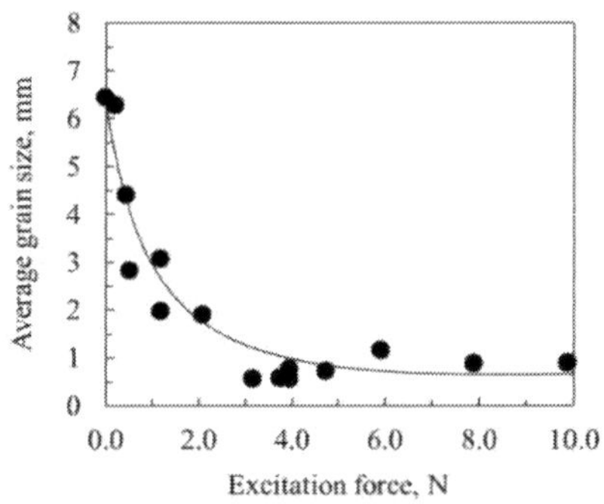

Fig.5. Relationship between excitation force and average grain size.

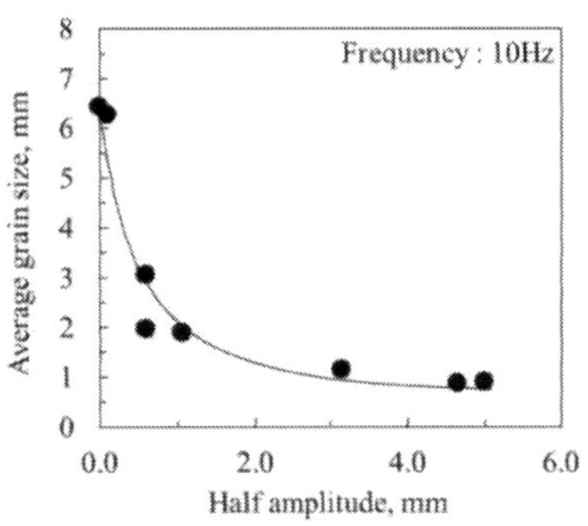

Fig.4. Influence of half amplitude on average grain size.

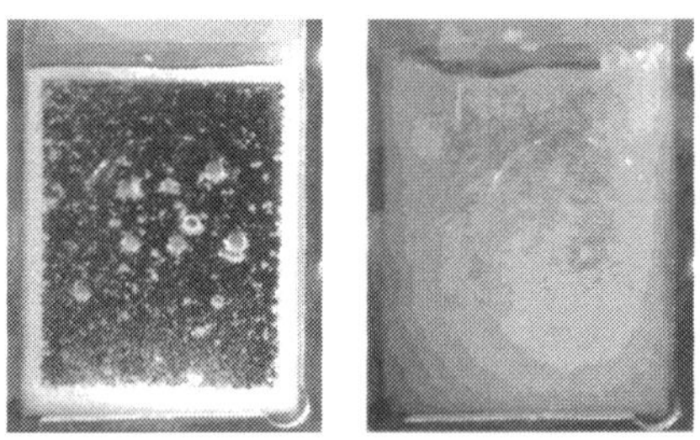

(a)Without vibration (b)With vibration (30Hz, 0.6mm)

Fig.6. States of crystallization at 15s after pouring in water model experiment.

The 73rd World Foundry Congress
„Creative Foundry"
23 - 27 September 2018, Krakow, Poland

SETTLING EFFECT OF Ti ON MELT QUALITY AND MECHANICAL PROPERTIES IN 356 CAST ALLOY

Özen Gürsoy *, Eray Erzi, Derya Dispinar

Istanbul University, Faculty of Engineering, Avcilar Campus, 34320 Avcilar-Istanbul, Turkey
**Corresponding address: e-mail: ozen.gursoy@gmail.com*

Key words: 356, Ti, settling, bifilm, fading, tensile

1. Introduction

Grain refinement with Al-Ti-B master alloy is the most commonly used method in aluminum casting processes. With the addition of grain refiner, TiB_2 and $TiAl_3$ particles heterogeneously nucleate α-Al during solidification [1]. Thus, higher mechanical properties are obtained by means of fine equiaxed grains. However, due the density differences of grain refiners, the particles show settling behavior when they are added to liquid aluminum. Therefore, the efficiency of Al-Ti-B master alloys has been a long going discussion by many researchers [2,3]. Bifilms whose density is almost equal to the molten metal are suspended in liquid aluminum [4,5]. Time dependently, settling of TiB_2 and $TiAl_3$ particles sweep down the bifilms to bottom of the crucible after grain refiner addition. This phenomenon directly influences the melt quality, microstructure, and mechanical properties.

2. Experimental Study

In this study, the influence of addition of Al-Ti-B grain refiner master alloy on melt quality, microstructure and mechanical properties was investigated. Two secondary 356 melts were prepared from the same scrap and melted at 725 °C. The ratio of 0.1 wt% Ti was achieved by Al-5Ti-1B master alloy addition into one of the melts. After 10 minutes of duration, samples were collected until the melt was completely consumed. Samples were poured into both die and sand molds. RPT samples were analyzed for bifilm index (BI) using SigmaScan image analysis software and the melt quality was quantified throughout the course of this study. SDAS analysis was performed from step-mold samples which have three different thicknesses (20, 10 and 5 mm). The microstructural analyses were carried out via Kameram image analysis software. The tensile test was performed to characterize the mechanical properties and Weibull analysis was used. Ti ratio of both melts was measured via spectral analysis during the study.

3. Results

According to spectral analyses (Fig. 1.), there is a slight decreasing of Ti ratio in the melt for the as-received alloy. However, in the Al-5Ti-1B grain refiner added melt, noticeable increasing of Ti content toward the bottom zone of the molten metal was observed.

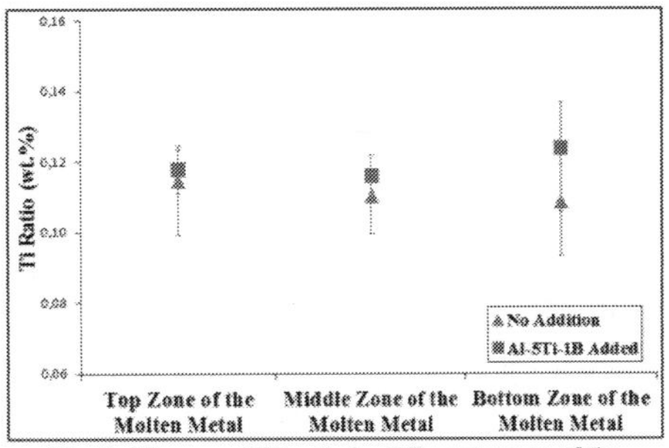

Fig. 1. Change of Ti ratio in different zones of the crucible.

BI analyses show clear time depending trend which is decreasing in both conditions. Interestingly, although the molten alloy had the same initial BI, Al-5Ti-1B master alloy added melt had lower quality. Master alloy addition decreases the melt quality because of oxide film entrapment during addition. Also, in both melts, BIs increase in the bottom zone of the molten metal (Fig. 2.). This is a good evidence of bifilm accumulation in the bottom of the crucible accompanied by the Ti settlement.

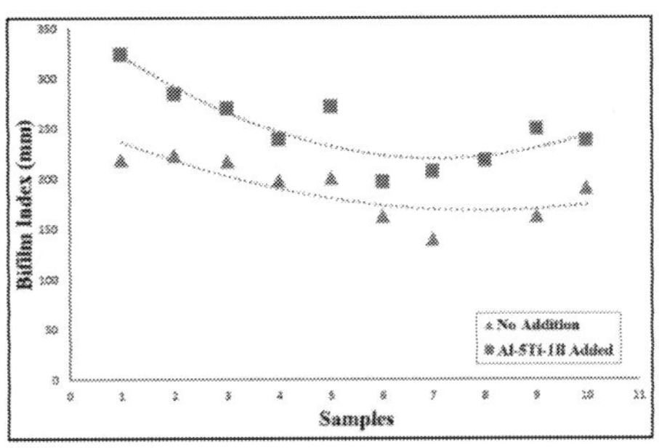

Fig. 2. BI changes of molten metals.

Conventional results cannot be observed in SDAS analyses except for few data. Despite the grain refiner addition, there is no successful refinement. Fig. 3 also shows the importance of cooling rate on grain size during solidification. SDAS data of die-cast samples are higher than sand cast samples'.

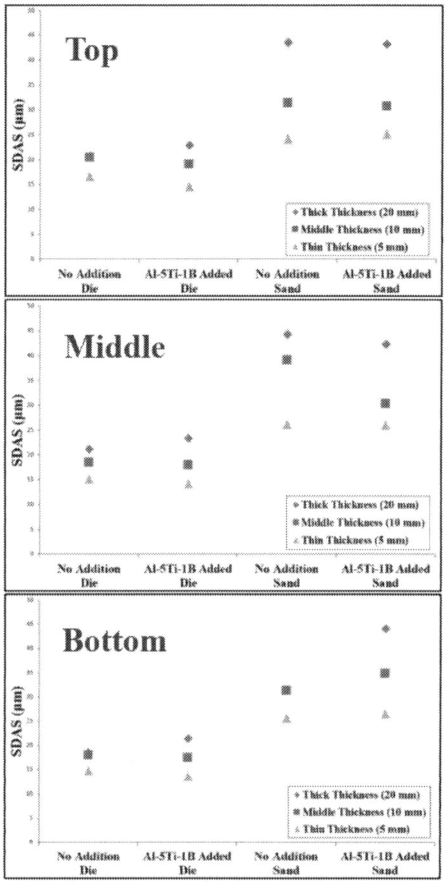

Fig. 3. Microstructural analyses of all conditions.

Despite the finer grains, the worst and inconsistent UTS and elongation values are observed in die casting (Fig. 3-5).

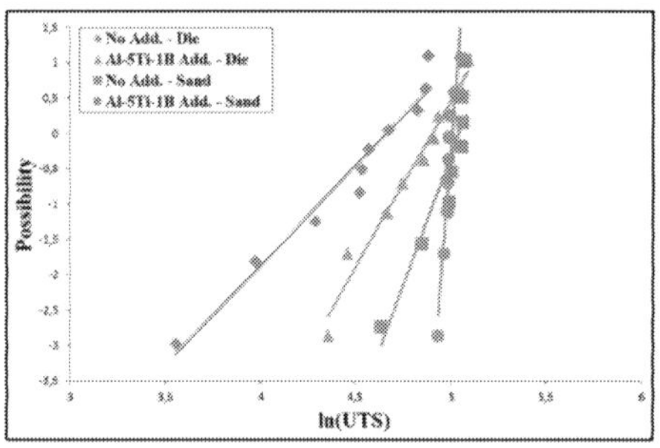

Fig. 4. Weibull analysis of UTS results of tensile tests.

According to Weibull analyses, grain refiner addition cannot rely a better result for UTS values and it has an adverse effect on elongation values in die cast samples. The best tensile test results are obtained distinctly in sand casting.

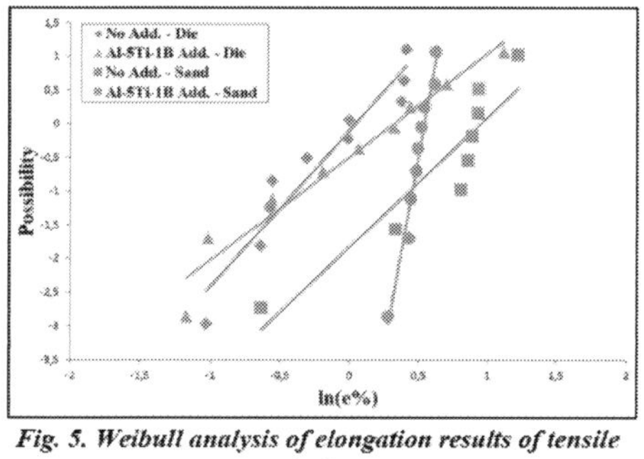

Fig. 5. Weibull analysis of elongation results of tensile tests.

Results of two different conditions (no add (i.e.: as-received) and Al-5Ti-1B add.) are close to each other. Al-5Ti-1B master alloy addition in sand casting has increased stability of UTS and elongation values. Low cooling rate results in unravelling o bifilms during solidification. So that, the notch effect of the failures is decreased in tensile testing compare to the compacted-scrambled bifilms.

4. Conclusions

- Addition of master alloy increases the bifilm index and decreases the melt quality.
- In these conditions, no clear relation between grain refiner addition and microstructure was detected.
- Duration (actually gravity) and particles from grain refiner, such as TiB_2 and $TiAl_3$, settle the bifilms.
- The shape factor of bifilm effects considerably on mechanical properties.

References

[1] P. S. Mohanty & J. E. Gruzleski. (1996). Grain refinement mechanisms of hypoeutectic Al-Si alloys. Acta Materialia. 44 (9), 3749-3760.

[2] P. L. Shaffer & A. K. Dahle. (2005). Settling behavior of different grain refiners in aluminium. Materials Science & Engineering A. 413, 373-378.

[3] C. Limmaneevichitr & W. Eidhed. (2003). Fading mechanism of grain refinement of aluminum-silicon alloy with Al-Ti-B grain refiners. Materials Science & Engineering A. 349, 197-206.

[4] D. Dispinar & J. Campbell. (2011). Porosity, hydrogen and bifilm content in Al alloy castings. Materials Science & Engineering A. 528, 3860-3865.

[5] D. Dispinar, S. Akhtar, A. Nordmark, M. Di Sabatino, L. Arnberg. (2010). Degassing, hydrogen and porosity phenomena in A356. Materials Science & Engineering A. 527, 3719-3725.

The 73rd World Foundry Congress
„Creative Foundry"
23 - 27 September 2018, Krakow, Poland

HOT TEARING TENDENCY AND TENSILE PROPERTY CHANGE OF Ni AND Cu ADDED A356 CASTING ALLOY

Kerim Yildirim [1]*, Johannes Brachmann [1], Derya Dispinar [2], Andreas Bührig-Polaczek [1], Uwe Vroomen [1]

[1]*RWTH, Gießerei-Institut, Aachen, Germany*
[2] *Istanbul University, Engineering Faculty, Metallurgical and Materials Eng, Avcilar, Istanbul, Turkey*
**Corresponding address: e-mail: kerim.yildirim@rwth-aachen.de*

Key words: hot tearing, tensile properties, degassing, porosity, hydrogen

1. Introduction

Aluminium-Silicon alloys are the most preferred choice of casting alloys due to their excellent properties mainly owing its features to the presence of silicon. A356 has 7 wt% Si and 0.2 wt% Mg. This alloy suffers high temperature mechanical properties. Therefore, the addition of minor ratios of different alloying elements has been a long going investigation by many researchers [1].

In this work, mechanical property change and hot tearing characteristics of Ni and Cu added A356 alloy was investigated. Ni was added up to 0.2 wt% while Cu addition was increased up to 2 wt% [2].

Moreover, the relation of the amount of dissolved hydrogen in the liquid aluminium with respect temperature was evaluated by means of porosity formation. It has also been determined by Dispinar and Campbell [3] that the amount of bifilms determine the mechanical properties of cast aluminium alloys. The quantification of bifilms were carried out by using Reduced Pressure Test. The size and distribution of pores on the cross section of vacuum solidified samples were measured. Bifilm index change with alloying element additions were recorded [4].

Several different casting experiments were carried out with the given alloy concentrations. The melt was degassed and upgassed slowly. Micro100 alloy was added to the melt to investigate relation between distribution of porosity and its correlation with mechanical properties. A ring permanent mould was used for measuring the hot tearing tendency of the alloys. Tensile test bars were poured into the standard permanent mould.

2. Results

The results of this research clarified that increasing the nickel and copper increased the mechanical properties of the material. In addition to this, the influence of mechanical properties on the change of hydrogen content and the cleaning of melt by degassing method was investigated and the effect of porosity on mechanical properties was investigated by adding micro100 alloy.

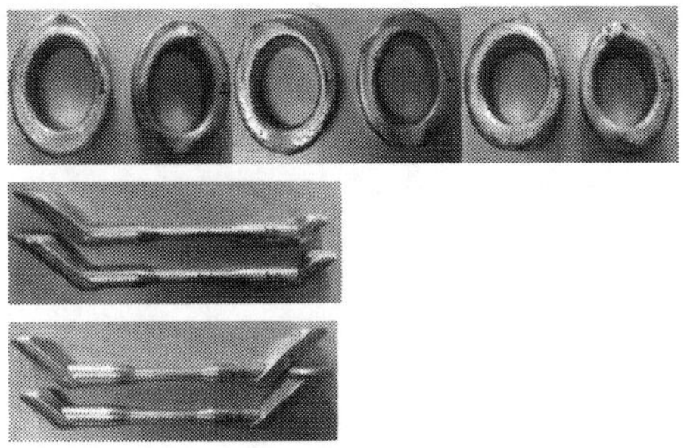

Fig. 1. Hot tearing and tensile test samples

Both hot and cold molds were used in each experiment to determine the hot tearing behavior of the alloys. It was observed that the copper addition increased the hot tearing, but the nickel addition did not show much effect on the hot tearing.

Since the effect of hot tearing was less noticeable in pre-heated molds, even when the alloy element was added to the composition, less hot tearing then cold mold was observed.

References

[1] Maria Teresa Di Giovanni (2014).The Influence of Nİ and V Trace Elements on the High Temperature Tensile Properties of A356 Aluminum Foundry. Norwegian University of Science and Technology.

[2] Daniele Casari, Thomas H. Ludwig, Mattia Merlin. (2014). The effect of Ni and V trace elements on the mechanical properties of A356 aluminum foundry alloy in as-cast and T6 heat treated conditions. Materials Science and Engineering A. Volume 610. 414-426.

[3] D. Dispinar, S. Akhtar, A. Nordmark. (2010). Degassing, hydrogen and porosity phenomena in A356. Materials Science and Engineering A. Volume 527. 3719-3725.

[4] D. Dispinar, J. Campbell. (2011). Porosity, hydrogen and bifilm content in Al alloy casting. Materials Science and Engineering A. Volume 528. 3860-3865.

The 73rd World Foundry Congress
„Creative Foundry"
23 - 27 September 2018, Krakow, Poland

EFFICIENCY OF TI ADDITION IN A356 ALLOY UNDER DIFFERENT HOLDING TIME

Muhammet Uludağ [1]*, Mikdat Gurtaran [1], Derya Dispinar [2]

[1] Bursa Technical University, Metallurgical and Materials Engineering, Bursa-16310, Turkey
[2] Istanbul University, Metallurgical and Materials Engineering, Istanbul-34320, Turkey
**Corresponding address: e-mail: dr.uludagm@gmail.com*

Key words: A356 alloy, titanium, bifilms, holding time, casting quality

INTRODUCTION

Al and its alloys are used in different industries such as automotive, aerospace etc. In recent years, it has become important to improve the mechanical and microstructure properties of these alloys. Al alloys can be produced by using different techniques such as casting that is the most common type of production used in industries. Casting is a technique that allows for mass production and conventional according to other type of productions. Liquid metal and casting quality should be high in order to obtain good mechanical and microstructure properties from alloys which produce with casting [1]. Casting quality is affected by casting defects which occur when it is in liquid state and during solidification. Preventing the casting defects is the first and most important way to produce quality cast parts. In the application, degassing process is carried out to prevent them.

Although it was presented that hydrogen is the main reason for porosity formation in the literature [2], Dispinar and Campbell claimed that hydrogen is not main reason for porosity formation and it is only a trigger factor [3]. The main reason for porosity formation is oxide films which come from ingots or crucibles and form by folding on top of each other and dragging because of turbulence are called bifilms. Bifilms are a dangerous casting defect that creates the porosity in liquid and therefore affects the mechanical properties of the alloys negatively. On the other hand, as a result of the diffusion of hydrogen that has a very small atomic volume, casting and liquid metal quality are decreased. Hydrogen has been caused for opening of bifilms. With improving of filling and runner system for no any turbulence, quality casting can be achieved [4]. It was thought that addition of alloying elements such as titanium can also be used for the same aim [5]. Even if alloying elements are used to increase the mechanical properties of alloys with a change on microstructure [6, 7], it is thought that they also can be used to clean bifilms from melt [5]. It is estimated that Ti moves to the bottom of the crucible because of its density and it can take bifilms with the same mechanism.

In this study, effect of Ti on A356 alloys is examined under different holding times. Bifilm index and mechanical properties were investigated. However, the mechanical properties was presented only in the present work.

EXPERIMENTAL PROCEDURE

Chemical composition of A356 alloy used in this study is given in Table 1.

Table 1. Chemical composition of A356 alloy used in this study

Si	Fe	Cu	Mn	Mg	Zn	Ti	Al
7.00	0.20	0.02	0.03	0.35	0.04	0.09	Rem.

A356 ingots were cut to charge to the resistance furnace and they were melted in a SiC crucible. Ti master alloy additions were made once the melt temperature had reached 740 °C. First sample was poured after adding Ti and other samples were poured after 5, 10, 15, 20, 25, 30 minutes respectively. This process was continued for 30 min. Reduce Pressure Test (RPT) samples were collected to calculation bifilm index that can define casting quality. Tensile test mold that include six cylindrical bars was used to obtain tensile test samples. Cylindrical bars have 13 mm in diameter and 150 mm in length.

RESULTS and DISCUSSION

Data for UTS, YS and e % from the experimental study are calculated and presented in Table 2. As can be seen in the table, the highest UTS, YS and e% values were achieved for 10 min. of holding time after adding Ti. The minimum values for UTS and YS were obtained for 0 min. of holding time. For UTS results, value for 15 min. of holding time seems to close value for 10 min. of holding time. It can be understood that Ti plays an important role to clean oxide films in the liquid metal until 15 min. after addition. On the other hand, efficiency of Ti on oxide films start to decrease after 15 min. of holding time. The events in the liquid after 15 min. of holding time is complicated. It can be said that new oxide films might be occurred to form porosity that decreases mechanical properties.

Table 2. Mechanical properties obtained from experimental study

Holding time (min.)	UTS (MPa)	YS (MPa)	e %
0	246	194	1,63
5	270	218	1,63
10	312	258	1,69
15	301	253	1,50
20	247	248	1,53
25	297	249	1,53
30	291	253	1,48

Toughness calculation was made for this study and results are given in Figure 1. As can be seen on the figure, toughness values have been shown similar trend with UTS, YS and e %. The highest toughness value was resulted after 10 min. of holding time. When looked at the error bars for toughness, it can be understood that the optimum values are 10 min. and 20 min. for the maximum toughness and the most stable results respectively.

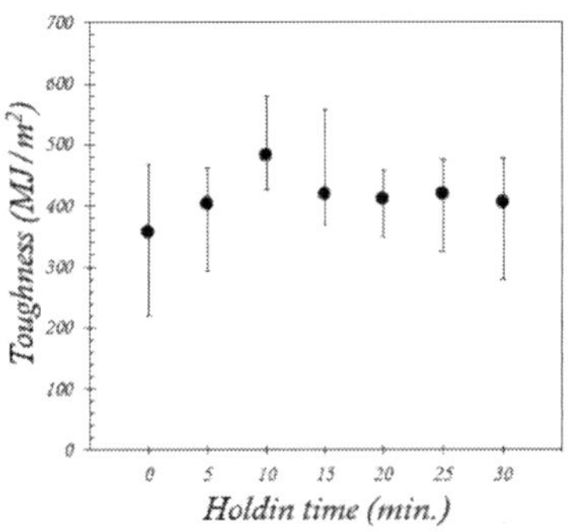

Fig. 1. Toughness values of A356 alloys under different holding times

A quality index for mechanical properties was gained by Tiryakioğlu and Campbell [8] to explain the mechanical properties in three groups. Q_T was used in this study and results are given in Figure 2. Optimum quality index is found for 10 min. of holding time and it reflects efficiency of Ti on the mechanical properties. Results of Q_T are similar to results of toughness. However, stability of the results for 30 min. of holding time is much more than the results for 20 min. of holding time. Q_T is the function of the yield strength and elongation. The difference between Q_T and toughness can be occurred because of the effect of Ti on the size, distribution, shape and location of the bifilms.

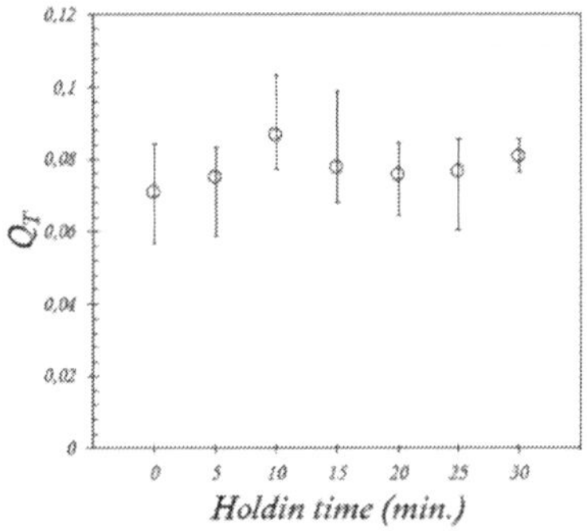

Fig. 2. QT values of A356 alloys under different holding times

CONCLUSIONS

The conclusions of this study that is about the effect of Ti on casting quality depending on holding time after addition can be summarized as follows:

1- A356 alloy melt quality is affected by Ti addition
2- The highest mechanical properties can be obtained by using Ti with 10 min. of holding time.
3- Ti plays an important role on the casting quality and mechanical properties depending on holding time.

REFERENCES

[1].Campbell, J., 2003, *Castings*. Butterworth-Heinemann.

[2].Lee, P., A. Chirazi, and D. See, 2001, *Modeling microporosity in aluminum–silicon alloys: a review*. Journal of Light Metals, **1**(1): p. 15-30.

[3].Dispinar, D. and J. Campbell, 2004, *Critical assessment of reduced pressure test. Part 1: Porosity phenomena*. International Journal of Cast Metals Research, **17**(5): p. 280-286.

[4].Campbell, J. 2004, *Castings practice–the 10 rules for casting*, Elsevier, Oxford, UK.

[5].Campbell, J., 2015, *Complete casting handbook: metal casting processes, metallurgy, techniques and design*. Butterworth-Heinemann.

[6].Bryant, M. and P. Fisher, 1993, *Grain Refining and the Aluminium Industry (Past, Present and Future)*. Aluminium Casthouse Technology.

[7].Haro-Rodriguez, S., et al., 2011, *On influence of Ti and Sr on microstructure, mechanical properties and quality index of cast eutectic Al–Si–Mg alloy*. Materials & Design, **32**(4): p. 1865-1871.

[8].Tiryakioğlu, M. and J. Campbell, 2014, *Quality index for aluminum alloy castings*. International Journal of Metalcasting, **8**(3): p. 39-42.

The 73rd World Foundry Congress
„Creative Foundry"
23 - 27 September 2018, Krakow, Poland

EFFECT OF THERMAL EXPOSURE ON HIGH TEMPERATURE MECHANICAL PROPERTIES OF SECONDARY AlSi7Cu3Mg ALLOYS

Alessandro De Mori [1]*, Alberto Fabrizi [1], Giulio Timelli [1]

[1]*University of Padova, Department of Management and Engineering, Stradella S. Nicola 3, 36100 Vicenza, Italy*
**Corresponding address: e-mail: alessandro.demori@phd.unipd.it*

Key words: Al–Si alloy; mechanical properties, thermal exposure; hardness; ageing

1. Introduction

A large number of automotive components such as cylinder heads, pistons and engine blocks are generally made by Aluminium-Silicon foundry alloys, which show a suitable combination of excellent castability, low weight, corrosion resistance and good mechanical properties.

Due to the extreme working conditions, the aforementioned components can locally reach high temperature up to 300°C. However, the degradation in the mechanical properties of Al-Si alloys occurs generally at temperatures close to 200°C [1–4].

The secondary Al process requires low energy consumption if compared to primary Al production with consequences both on greenhouse gases emissions and costs. For this reason, secondary Al alloys are increasingly being used, but the presence of impurities and small amounts of undesired elements has deleterious effects on the mechanical properties [5].

Al foundry alloys are usually heat-treated to improve mechanical properties due to precipitation hardening. Among all, the T7 heat treatment (over-aged condition) is one of the most common for engine parts. The presence of incoherent precipitates decreases strength, but guarantees greater ductility and stability of the mechanical properties during high temperature exposure [1].

The aim of this work is to evaluate the effect of thermal exposure on high temperature mechanical properties of secondary AlSi7Cu3Mg foundry alloys by means of hardness measurements and microstructural investigations.

2. Experimental procedure

The experimental AlSi7Cu3Mg cast alloy (EN AB-46300, equivalent to the US designation A320) was supplied as commercial ingots. The chemical composition, performed by optical emission spectrometry, is shown in Table 1.

Table 1. Chemical composition of the experimental AlSi7Cu3Mg alloy (Weight Percent).

Si	Fe	Cu	Mn	Mg	Ti	Cr	Al
6.7	0.45	3.21	0.29	0.32	0.1	0.032	bal.

The ingots were melted in a resistance-heated furnace at 760°C and castings were produced by using the reference die proposed in the document CEN/TR 16748:2014 [6].

The specimens were then heat treated in an air-circulating furnace. The solution treatment was performed at 485°C for 24 h and it was followed by water quench at room temperature. The artificial ageing was then conducted at 180°C for different times in a range from 1 to 10 hours.

Brinell hardness measurements were carried out using a 2.5 mm diameter indenter, a load of 62.5 kgf and 30 kgf for room- and high-temperature testing, respectively. A resistance-heated plate was used to keep the temperature constant during high temperature testing.

The previously aged specimens were held in furnace at 200°C and 300°C for one hour and then tested at high temperature. The value of 200°C was selected as actual threshold in Al-Si alloys, while 300°C is a critical value achievable in cylinder heads, as reported in [1-4].

Microstructural investigations were carried out by optical and transmission electron microscopy on specimens aged for 8 hours before and after exposure at high temperatures.

3. Results and conclusions

The hardness values of as-cast and as-quenched AlSi7Cu3Mg alloy are 90 ± 3 HB and 84 ± 3 HB, respectively.

Fig. 1 shows the hardness of the artificially aged samples as a function of the ageing time. The high temperature hardness of aged-specimens after thermal exposure at 200 and 300°C for 1 h is also reported.

Considering the results at room temperature testing, it can be observed how the ageing treatment generally increases the hardness of the alloy and the maximum value (145 ± 3 HB) is reached after 4 hours at 180°C (peak ageing condition). Then, the hardness decreases and stabilizes at about 135 HB with increasing ageing time from 4 to 10 hours.

In addition, holding the specimens for one hour at 200 and 300°C makes the effect of heat treatment disappear since hardness measurements show a constant value for all the tested ageing times.

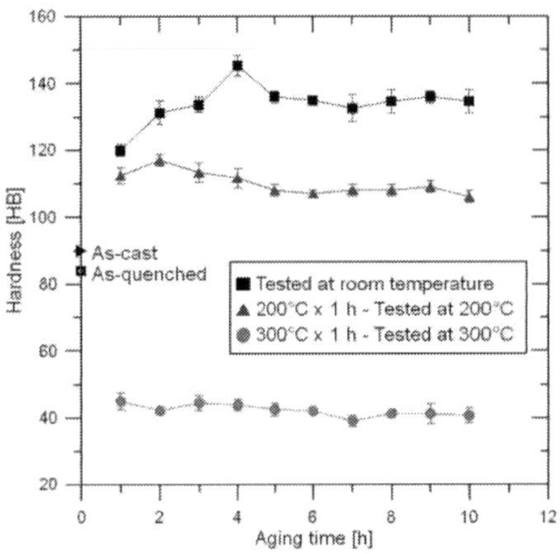

Fig. 1. Room- and high-temperature hardness of the experimental AlSi7Cu3Mg alloy.

Specimens held in the furnace for one hour at 200°C and then tested at 200°C show hardness values of about 110 HB; those held at 300°C and tested at the same temperature show significantly lower values, ~42 HB. The thermal load causes an evident decrease of the hardness and the temperature of 300°C results to be more critical than 200°C due to the coarsening phenomena of the strengthening precipitates.

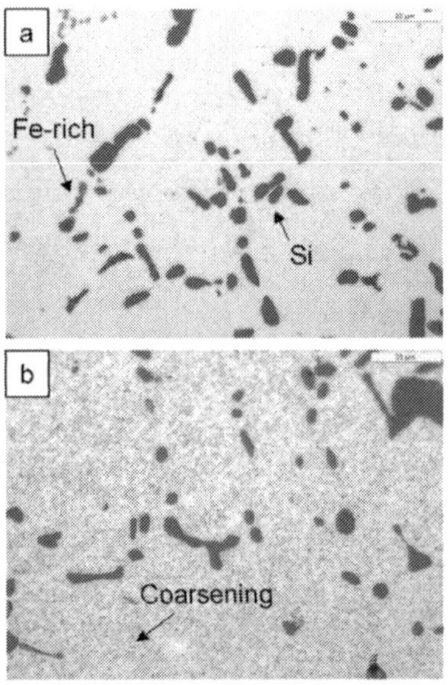

Fig. 2. Typical microstructures of the AlSi7Cu3Mg alloy (a) aged at 180 °C for 8 h and (b) after additional thermal exposure at 300°C for 1 h.

Figure 2 shows the typical microstructures of the alloy aged at 180°C for 8 hours (T7) and after a further exposure at 300°C for 1 h, respectively. The T7 ageing treatment leads to a spheroidization of eutectic Si particles and to a complete dissolution of Cu- and Mg-bearing phases, while Fe-rich phases are still present. Precipitation of finely dispersed particles throughout the α-Al matrix also occurs (Fig. 3). The same result (here not reported) appears for the T7-aged specimens exposed at 200°C for 1 h. On the contrary, Fig. 2b shows how one hour of permanence at 300°C does not further alter the morphology of eutectic Si particles, but causes a massive precipitation in the α-Al matrix, thus substantially reducing hardness.

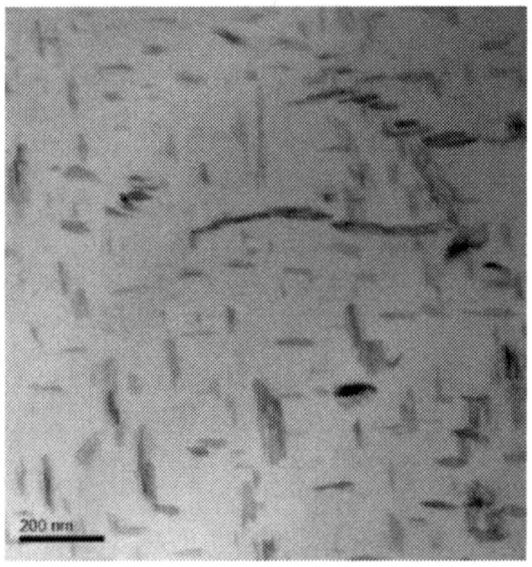

Fig. 3. TEM micrograph showing the precipitation occurred in α-Al matrix after the T7 ageing treatment.

References:

[1] P. Huter, P. Renhart & S. Oberfrank. (2016). High- and low-cycle fatigue influence of silicon, copper, strontium and iron on hypo-eutectic Al-Si-Cu and Al-Si-Mg cast alloys used in cylinder heads. International Journal of Fatigue. 82, 588–601.

[2] S.K. Shaha, F. Czerwinski & W. Kasprzak. (2016). Ageing characteristics and high-temperature tensile properties of Al-Si-Cu-Mg alloys with micro-additions of Cr, Ti, V and Zr. Materials Science & Engineering A. 652, 353–364.

[3] Y. Yang, S.-Y. Zhong & Z. Chen. (2015). Effect of Cr content and heat-treatment on the high temperature strength of eutectic Al-Si alloys. Journal of Alloys and Compounds. 647, 63–69.

[4] J. Rakhmonov, G. Timelli, F. Bonollo. (2017). Characterization of the solidification path and microstructure of secondary Al-7Si-3Cu-0.3Mg alloy with Zr, V and Ni additions. Materials Characterization. 128, 100–108.

[5] G. Timelli, G. Camicia, S. Ferraro. (2014). Effect of grain refinement and cooling rate on the microstructure and mechanical properties of secondary Al-Si-Cu alloys. Journal of Materials Engineering and Performance. 23, 611–621.

[6] CEN/TR 16748:2014. Aluminium and Aluminium alloys – Mechanical potential of Al-Si alloys for high pressure, low pressure and gravity die casting.

The 73rd World Foundry Congress
„Creative Foundry"
23 - 27 September 2018, Krakow, Poland

INCREASING THE LIFETIME OF DIE CASTING MOLDS MADE OF THE HOT WORK TOOL STEEL H13 BY MEANS OF A LOCAL SURFACE HEAT TREATMENT WITH THE ELECTRON-BEAM

Torsten Schuchardt [1*], Marlene Grohne [1], Sebastian Müller [1], Klaus Dilger [1]

[1]*Institute of Joining and Welding, TU Braunschweig, Braunschweig, Germany,*
**Corresponding e-mail: t.schuchardt@tu-braunschweig.de*

Key words: die casting dies, heat-treatment, hot work tool steel, electron beam

Heat treatment of a die casting mold represents an important step in adjusting the required mechanical properties of the material used and thus also the achievable mold service life. State of the art is an oven-based heat treatment under vacuum or inert gas atmosphere. In oven-based heat treatment, however, a conflict of objectives often exists with regard to the optimum tempering state. The requirements for a high hot yield strength and high hardness with simultaneous high toughness are contradictory and can only be achieved to a limited extent in the tempered state of the material. Hence, the primary goal is the development of a local surface heat treatment strategy using electron beam (EB) in order to increase the service life of the mold and reduce costs during production.

1. Introduction

Hot-work steels such as X40CrMoV5-1 have proven themselves for the application in die casting molds due to their well-balanced property profile. In practice, a hardness of 45 to 46 HRC is adjusted. When heat treated appropriately, this value represents a reasonable compromise between the two required material properties strength and toughness [1,2]. However, the conventional heat treatment with furnaces is associated with disadvantages [3]. On the one hand, the entire mold insert must be heat-treated, which usually requires a high amount of energy. On the other hand - due to very irregular component surfaces and different wall thicknesses - different cooling speeds result on the surface, which in turn leads to inhomogeneous mechanical material properties. Another problem are high cooling rates, which are accompanied by the tendency to hardness cracks.

2. Surface Heat-Treatment

The results of various research groups have shown that surface heat treatments can improve the mechanical properties of hot-work steels such as X40CrMoV5-1 (AISI H13). In most cases the local near-surface microstructure transformation of the material is carried out by a substitute heat source, which generates a considerably different temperature profile compared to the conventional oven-based heat treatment. The approaches can be local hardening as a solid phase process or remelting as a liquid phase process. The laser is usually the main energy source for remelting [4,5]. In addition, thermal microstructure transformation can also be induced by inductive or conductive heating, with the focus being on a solid phase process. The focus of the solid phase processes is in any case on the material behavior of the X40CrMoV5-1 during a given short-term austenitisation.

Austenitisation is based on diffusion processes in the material, which are significantly influenced by time and temperature. In short-term austenitisation, the achievable austenitisation temperature is the driving force and is increased by approx. 50 K to 100 K due to the characteristically fast heating rates during surface heat treatment [6]. Furthermore, the initial structure of the microstructure strongly influences the austenitisation conditions. Soft-annealed, coarse-grained microstructure requires significantly higher temperatures than a homogenized or quenched and tempered microstructure for complete austenitisation. This is attributed to the inhomogeneously distributed carbon content and the associated longer diffusion paths of the carbon. For steels such as X40CrMoV5-1, the special carbides for short-term austenitisation must also be taken into account. These carbides are deliberately used to ensure the high-temperature strength of the steels. However, they make austenitisation more difficult, especially in the lower temperature ranges, as grain growth is inhibited during austenitisation. A corresponding adjustment of the temperatures must be taken into account [6,7].

3. Electron-beam technique

Next to the laser beam process, electron beam technology is the second beam technology used for the thermal processing of materials. In this process, electrons with almost no mass and thus no inertia are accelerated from an electron cloud in front of the cathode to up to 2/3 of the speed of light by means of an acceleration voltage of 120 kV. Focusing and highly dynamic deflection of the electron beam is achieved by a series of electromagnetic lenses. When the electrons impact the workpiece surface, they are decelerated. During this deceleration process, the kinetic energy of the electrons is converted into heat energy, thus creating the possibility of heating or melting the metallic base or filler [8]. A temperature-dependent control system was established for the investigations, which is shown schematically in Fig. 1.

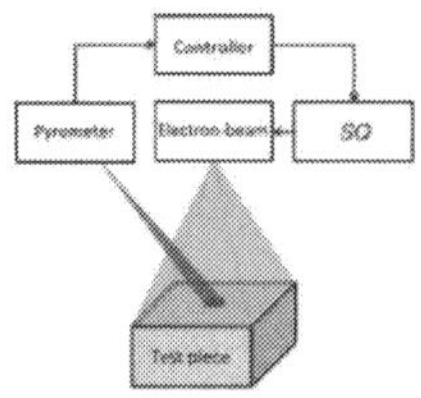

Fig. 1. Schematic block diagram of the closed-loop control

4. Researches on local heat-treatment

The aim of the investigations was to replace the conventional oven-based heat treatment of the hot-work tool steel X40CrMoV5-1 for die casting molds by a surface heat treatment by means of an electron beam process. Based on the parameters of conventional oven-based heat treatment and preliminary investigations, performance parameters for EB surface heat treatment were derived. In the first part of the study, samples of the material X40CrMoV5-1 were heated by the electron beam whose beam current (SQ) was controlled by a PID controller. The samples were prehardened to a hardness of 30 HRC. The current surface temperature was measured with a pyrometer and transmitted to the controller (cf. Fig. 1). Preliminary tests were performed to determine the emissivity (ϵ) of the given sample surface. By means of comparative temperature measurements of thermocouples and the pyrometer, the unknown emissivity for temperatures up to 1000 °C could be determined to $\epsilon = 0.2$. The parameters for the PID controllers were empirically obtained for the two steps hardening and tempering of the EB surface heat treatment.

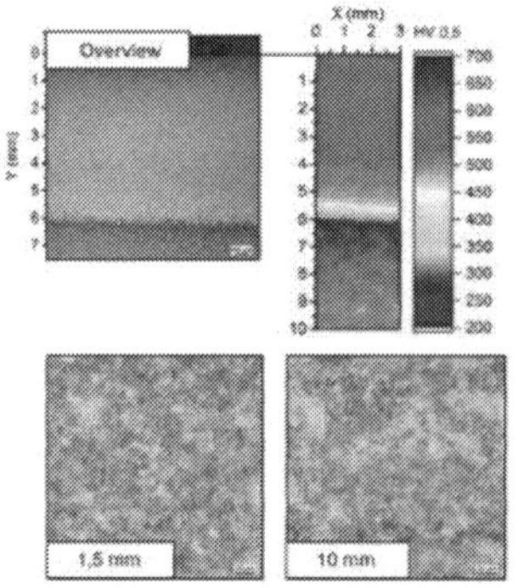

Fig. 2. Cross-section polish of the microstructure and planar hardness test of a treated test-piece

For EB hardening, the power density is 267 W/cm², which results from the high voltage SH = 80 kV, the maximum beam current SQ = 50 mA and an irradiated area of 38 mm x 38 mm. The PID controller used has reduced the beam current to the desired starting temperature, so that a lower power density was used. The cooling times from 800 °C to 500 °C after austenitisation were 100 s to partly 600 s. For further analysis, samples with a cooling time of $\Delta t_{800\text{-}500} \approx 100$ s were selected in order to achieve the highest possible comparability. For the evaluation of the heat treatments, light microscopic images of the microstructure and two-dimensional hardness measurements were performed in order to obtain a hardness curve depending on the distance of the heat-treated EB surface. EB hardening has resulted in a hardening depth of approximately 5 mm and a hardness of 55 HRC to 58 HRC, as shown in Fig. 2. At a depth of around 6 mm, tempering processes have caused an abrupt change in hardness of the prehardened material. The subsequent EB coating relaxed the martensitic structure, whereby the hardness was not significantly changed.

5. Conclusion and prospect

The experiments carried out have shown that a controlled electron beam is suitable for a specific surface heat treatment of the hot-work steel X40CrMoV5-1. For this purpose, a temperature-dependent regulation of the surface temperature has been established. With the closed loop control and a static electron beam, prehardened samples of X40CrMoV5-1 were surface-treated to a hardness of 55 HRC up to a hardening depth of 5 mm. In contrast to other technologies such as lasers or induction, the multi-beam technology enables a more effective heat input and also the opportunity to apply a surface treatment to steep edges of a highly inhomogeneous mold surface. The next step of the investigations will be the application of the results obtained to real prehardened die casting molds.

References

[1] Zwick G. (2009) Materialbezogene Einflüsse auf die Standzeit einer Druckgießform. GIESSEREI 96(2): 58–60.

[2] Deterding D. (1989) Gebräuchliche Warmarbeitsstähle zum Einsatz für Druckgießformen mit besonderem Hinweis auf den Warmarbeitsstahl Nr. 1.2367. GIESSEREI 76(24): 848–849.

[3] Zieger B. (2009) Latest state of the art at the heat treatment of die casting dies made of hot works steels. Proceedings of 8th International Tooling Conference:1025–1036, Aachen, Germany

[4] Telasang G, Dutta Majumdar J, Padmanabham G, et al. (2014) Structure–property correlation in laser surface treated AISI H13 tool steel for improved mechanical properties. Materials Science and Engineering: A 599: page 255–267.

[5] Jia Z-x, Liu Y-w, Li J-q et al. (2015) Crack growth behavior at thermal fatigue of H13 tool steel processed by laser surface melting. International Journal of Fatigue 78:61–71.

[6] Stiele H, Krause Christian (2013) Kurzzeitaustenitisieren - Werkstoffkundliche Grundlagen zum Randschicht-erwärmen: Short time austenitization - Material basics of surfacer heating. Elektrowärme international 71(4):47–53

[7] Müller K. (1999) Werkstoffkundliche Qualifizierung des Randschichthärtens mit Laserstrahlung. Zugl.: Bayreuth, Univ., Diss., 1999. Randschichthärten mit Laserstrahlung, vol 2. Utz Wiss, München

[8] Schultz H. (2000): Elektronenstrahlschweißen, Fachbuchreihe Schweißtechnik, Band 93.

The 73rd World Foundry Congress
„Creative Foundry"
23 - 27 September 2018, Krakow, Poland

EFFECT OF QUENCHING MEDIUMS ON MECHANICAL PROPERTIES OF Al-Cu ALLOY

Muhammet Uludağ [1]*, Mikdat Gurtaran [1], Derya Dispinar [2]

[1] *Bursa Technical University, Metallurgical and Materials Engineering, Bursa-16310, Turkey*
[2] *Istanbul University, Metallurgical and Materials Engineering, Istanbul-34320, Turkey*
**Corresponding address: e-mail: dr.uludagm@gmail.com*

Key words: Quenching mediums, Al-Cu alloys, mechanical properties, cooling rate, precipitation hardening, aging

INTRODUCTION

Use of Al cast alloys are quite common in crucial applications such as automotive and aerospace industries. Al alloys have advantageous properties such as low density, recyclability, good workability and low cost. Hence, these alloys are used in automotive industries in order to produce various parts such as engine block and cylinder head [1]. Al-Cu alloys are one of the irreplaceable materials because of its superior properties compared to other alloys. One of the reasons of using of these alloys in different industries commonly is that precipitation hardening (T6) is applicable [2-4]. Besides precipitation hardening process improves mechanical properties of these alloys, it increases machinability of these alloys. Also tensile and yield strength can be increased by using of T6 heat treatment [5]. T6 heat treatment is affected by chemical composition of an alloys, thermal requirements and types of casting methods that are used. Precipitation hardening process has been formed in 3 main steps which are solution treated, quenching and aging [6].The concentration of vacancy is increased by quenching. The existing secondary phases is solutionized with the solution treated process at a temperature below the solvus curve. T6 heat treatment is applicated to alloys showing partial solid solubility. Solid solubility increases with increasing temperature or decreases with decreasing temperature. This system is obtained in metals that have dissolved secondary phases in each case in the liquid state and partially in the solid state. At room temperature or slightly higher temperatures, increasing of hardness depending on the aging time is called as "hardening" and the heat treatment that is realized to hardening is called as "aging" in metal science [7]. By the aging process, secondary phases are precipitated in vacancies which are raised by quenching process [8]. Quenching process can be made by using different mediums such as water at room temperature, boiling water and oil mediums [9]. Heat transfer coefficient difference between material and medium is provided with different quenching mediums.

In this study, effect of three different quenching mediums on the mechanical properties of the alloy have been studied.

EXPERIMENTAL STUDY

Al-Cu alloy used in the study was provided from Eti Aluminyum Company in Turkey and the chemical composition of the alloy is given in Table 1. Solutionizing process was made at 537 °C for 30 minutes by using 3 different quenching mediums (Water Room Temperature (WRT), Boiling Water (BW) and Oil (O)). The artificial aging process was carried out at 195 °C for 12 hours.

Table 1. Composition of the Al-Cu alloy used in study

Si	Fe	Cu	Mn	Mg	Zn	Ti	Ni	Pb	Al
0,30	0,30	4,5	0,10	0,05	0,10	0,25	0,10	0,05	Rem.

Quenching mediums can be ranked as water at room temperature, boiling water and oil from fast to slow in terms of cooling rate. Cooling rate is important because it effects the amount of precipitation of the secondary phases. Heat treatment-free and heat-treated samples were prepared for tensile test by machined in CNC machine. All samples were applied tensile test and the relationship between effects of quenching mediums on mechanical properties was investigated.

RESULTS and DISCUSSIONS

It was seen that the mechanical values are the lowest for non-treatment alloy (Fig. 1, 2 and 3). The fact that the mechanical properties of the material was improved by the quenching process has also been proved by this study.

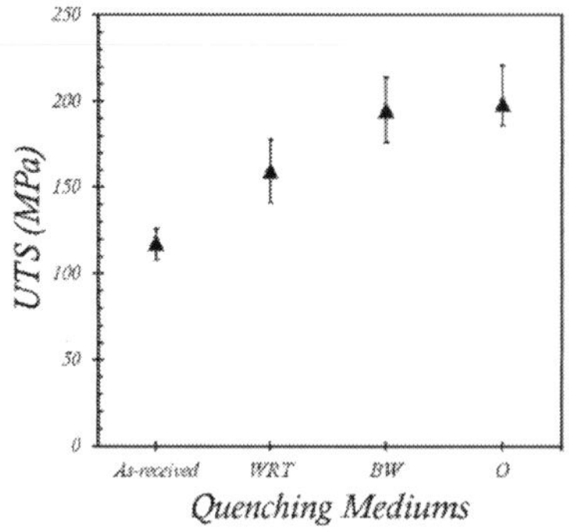

Fig. 1. UTS values for different quenching mediums

The highest UTS values were achieved in boiling water and oil mediums that occur the low cooling rate. When higher cooling rate was achieved by the quenching process which realized at room temperature, UTS values are much lower (Figure 1).

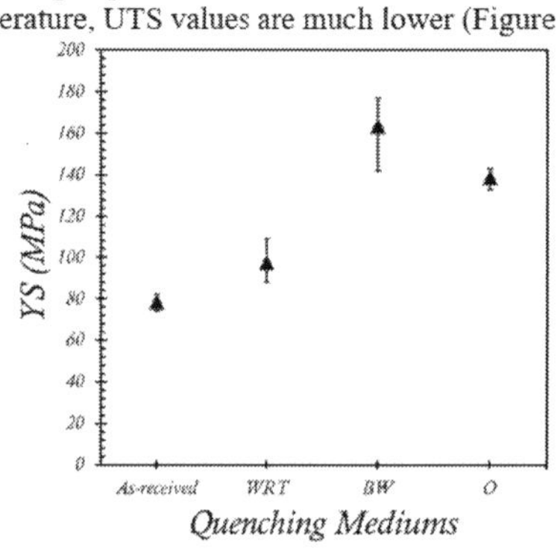

Fig. 2. YS values for different quenching mediums

It was obtained that trend of yield strength values are almost the same as UTS's under different quenching mediums. While the highest YS value was seen in boiling water medium, the lowest value was seen in room temperature medium (Figure 2).

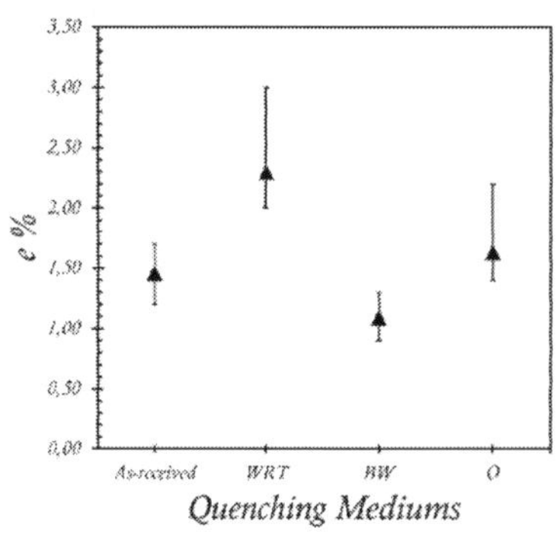

Fig. 3. Elongation % values for different quenching mediums

As can be seen in Fig. 3 clearly that the highest elongations were achieved by quenching process at room temperature. Even if boiling water medium has better results for UTS and YS, it has the worst result for e %.

CONCLUCIONS

1- Cooling rate differences occur with the difference between quenching medium and material temperature, and due to heat transitions. These factors have an effect on the mechanical properties of the material.
2- Change in the mechanical properties can be patterned thanks to the reasons of cooling rate differences.
3- The effect of different quenching mediums on all mechanical properties is not the same.

REFERENCES

[1] Liu G-l, Si N-c, Sun S-c, Wu Q-f. (2014).Effects of grain refining and modification on mechanical properties and microstructures of Al–7.5 Si–4Cu cast alloy. Transactions of Nonferrous Metals Society of China.24:946-53.

[2] Boyer HE. (1984).Practical heat treating.

[3] Committee AIH. (1991). ASM handbook: Heat treating: Asm Intl.

[4] Doğan M. (1989).Alüminyum Alaşımlarının Isıl İşlemi.

[5] Aydın B, Özçatalbaş Y. (2002).AA2014 Alaşımının İşlenebilirliğine Yaşlanma Süresinin Etkisi. 11 Uluslararası Metalurji ve Malzeme Kongresi.05-9.

[6] Callister Jr WD, Rethwisch DG. (2012). Fundamentals of materials science and engineering: an integrated approach: John Wiley & Sons.

[7] Savaşkan T. (1999). Malzeme Bilgisi ve Muayenesi: Derya Kitabevi.

[8] GÜVEN ŞY, DELİKANLI YE. (2012).AA 2024 Alüminyum Alaşımında Çökelme Sertleşmesinin Mekanik Özelliklere Etkisi. SDU Teknik Bilimler Dergisi.2.

[9] Sharma A, Das S. (2009).Study of age hardening behavior of Al–4.5 wt% Cu/zircon sand composite in different quenching media–A comparative study. Materials & Design.30:3900-3.

The 73rd World Foundry Congress
„Creative Foundry"
23 - 27 September 2018, Krakow, Poland

THE EXAMINATION OF THE MODIFICATION EFFECT OF STRONTIUM IN AlSi8Cu3 FOUNDRY ALLOY

MONIKA Tokár[1]*, GYÖRGY Fegyverneki [2], TAMÁS Mende [3], VALÉRIA Mertinger [4]

[1,2]University of Miskolc, Institute of Foundry Engineering, 3515 Miskolc-Egyetemváros, Hungary
[3,4]University of Miskolc, Institute of Physical Metallurgy, Metalforming and Nanotechnology, 3515 Miskolc-Egyetemváros, Hungary
*Corresponding address: e-mail: monika.tokar@uni-miskolc.hu

Key words: aluminium alloys, silicon, modification, strontium, thermal analysis

1. Abstract

Experiments under operating conditions were carried out using AlSi8Cu3 alloy to examine the beneficial modification effects of excess strontium concentration. The aim of the experiments was to determine the additional AlSr10 quantity to achieve the beneficial Sr content for the recommended modification. The quantity of added strontium concentration and the method to regulate the strontium concentration was changed. The level of the modification of test bars casted during thermal analysis and final castings was determined with cooling curves and standard images.

2. Introduction

Al-Si based alloys are the most commonly used alloys in aluminium light metal foundries, especially in the automobile industry. The aim of the modification is to refine the silicon of the eutectic structure with the addition of a master alloy. Sodium, strontium and antimony are the most common elements for the modification of silicon in the eutectic structure. In case of the modified eutectic structure, the silicon in the structure becomes fine and spheroidal, increasing tensile strength, elongation and formability [1, 2].

3. Experimental part

AlSi8Cu3 (DIN 226.10) alloy was examined during the experiments. The results of the chemical analysis can be seen in Table 1.

Table 1. Chemical composition of the AlSi8Cu3 ingots

Elements	**Si**	**Cu**	**Mg**	**Ti**	**Sr**
wt.%	9.06	2.45	0.31	0.11	0.0011
Elements	**Fe**	**Mn**	**Pb**	**Sn**	**Zn**
wt.%	0.49	0.45	0.03	0.009	0.54

AlSr10 (pure aluminium wire with 10 wt. % Sr) master alloy was used to change the strontium concentration in the melt. The experiments were carried out under operating conditions following the technological steps seen in Fig. 1. Test bars were casted after the degassing treatment.

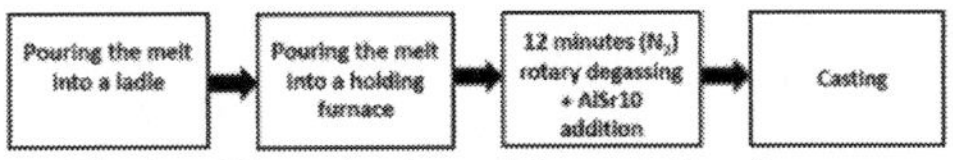

Fig. 1. The technology of the operational process

The strontium concentration of the base melt was changed and strontium master alloy was added during the rotary degassing treatment (Table 2.).

Table 2. The values of the strontium concentration

Sr concentration of the base melt	Sr concentration with AlSr10 master alloy addition during the rotary degassing treatment
103 ppm	100 ppm
219 ppm	40 ppm
242 ppm (regulated by the manufacturer)	-

The cooling curves were determined with an MK type data collector. The level of modification is usually calculated from the ΔT supercooling value of the cooling curves based on the following relation [3]:

$$\Delta T_{E,G}^{Al-Si} = T_{E,G,non\ modified}^{Al-Si} - T_{E,G,modified}^{Al-Si} \qquad (1)$$

where:

$T^{Al-Si}_{E,G,\ non-modified}$ – solidification temperature of the non-modified eutectic structure

$T^{Al-Si}_{E,G,\ modified}$ – solidification temperature of the modified eutectic structure

$\Delta T^{Al-Si}_{E,G}$ – degree of supercooling

The maximum value of the solidification temperature of the eutectic structure was determined in case of both non-modified ($T_{E,G}^{Al-Si}$ $_{non-modified}$) and modified ($T_{E,G}^{Al-Si}$ $_{modified}$) AlSi8Cu3 alloys. The degree of supercooling ($\Delta T_{E,G}^{Al-Si}$) was calculated from two temperatures based on equation (1). The values of supercooling are illustrated in Fig. 3. Based on literature, the structure of eutectic silicon is considered non-modified when ΔT is below 9 °C and modified if the value is higher than that [3].

Based on the table on the qualification of the silicon modification of Al-Si alloys prepared by the American Foundry Society (AFS), the level of modification can be easily determined by a simple comparison [1].

After preparing microstructures from the test bars, 15 photos were taken of the non-etched microsections with 500x magnification using a Zeiss optical light microscope. These were compared to the images from the rating table of the AFS. The level of modification was determined based on the examined images.

4. Results

The strontium concentrations after the used technological steps are dependent on the pre-treatment of the melts, which is shown in Figure 2. It can be observed that when the starting strontium concentration was lower (103 ppm) and further strontium addition was necessary to achieve the standard Sr quantity, the added strontium either contributes to compound formation or has no effect due to melting loss. Thus, it has no effect on modification. This fact is confirmed by the highly intense decrease in strontium concentration in case of strontium addition between the 1st and 2nd tests after degassing, compared to the melt with higher strontium concentration.

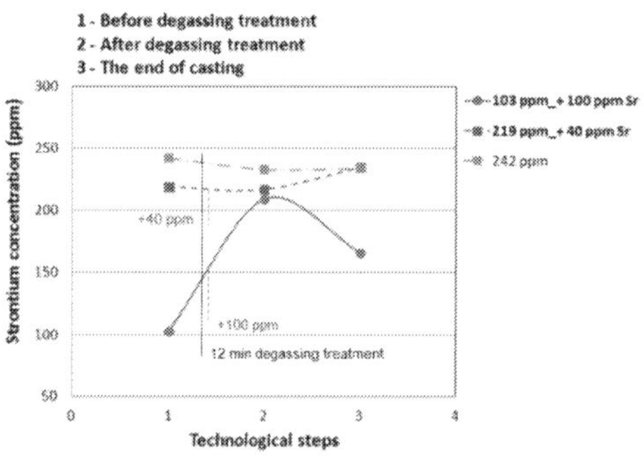

Fig. 2. Changes in strontium concentration in relation to the technological steps

As it can be observed in Figure 3, the initial strontium concentration is indicated in the first technological step. If the initial strontium concentration was lower (103 ppm), the ΔT value decreased at the 2nd technological step. This corresponds to the assumption that strontium addition at this technological step contributes to compound formation and has no modifying effect.

In case of the highest initial strontium concentration (242 ppm), when no further strontium was added, the behaviour of the curve is reversed, as ΔT increases at the 2nd technological step. In case of the other curves, the ΔT value is decreased at the 2nd step compared to the 1st step, even though the strontium concentration of the melt is significantly higher. Therefore, the ΔT value decreased (given that the efficiency of the ΔT method is accepted, the level of modification also decreased), even though the concentration of the modifying element (Sr) was significantly increased. To validate the results of the ΔT method, structural analysis had to be carried out, the results can be seen in Figure 3.

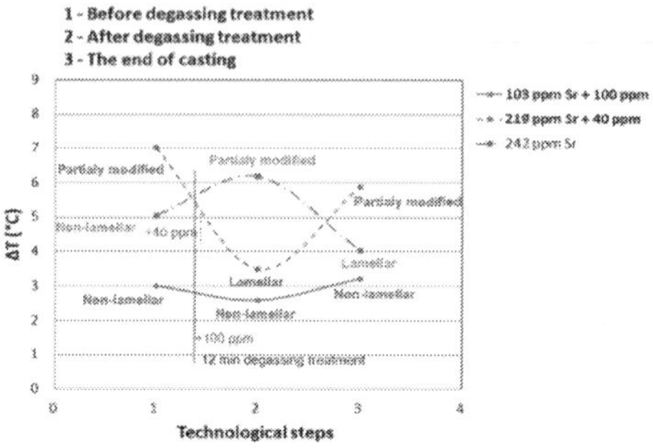

Fig. 3. The ΔT supercooling values in case of various strontium concentrations at the technological steps, indicating the levels of modification in case of microstructure

Figure 4. shows the levels of modification in case of microstructure after degassing treatment.

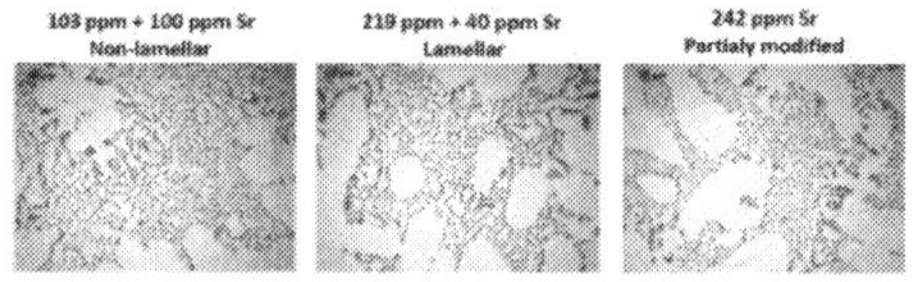

Fig. 4. The structure of eutectic silicon after degassing treatment

5. Submission

The experiments carried out under operating conditions with various initial strontium concentrations revealed that the pre-treatment of the melt has a significant effect on the level of modification. It can be stated that not the exact strontium concentration is relevant, as not the melt with the most strontium content had the highest level of modification. Some of the strontium in the system contributes to the modification but some of it is not beneficial. A decreasing tendency can be observed in the ΔT in case of extra strontium addition. The results showed that the partly modified structure can be identified in case of 5 °C of higher ΔT. Furthermore, here is correlation between the results of the ΔT method and the structural analysis. Thus, the ΔT method is suitable for the determination of the level of modification. The additional 40 ppm strontium is required to compensate the strontium loss of the treatments and not to increase the strontium concentration.

References

[1] John E. Gruzleski, Bernard M.Closset (1990): The treatment of liquid Aluminum-Silicon Alloys, Des Plaines, Illinois USA, American Foundrymen's Society, ISBN 0-87433-121-8

[2] A.Pacz, US Patent No. 1387900, (1921)

[3] M. Djurdjevic, H. Jiang, J. Sokolowski (2001): On-line prediction of aluminum-silicon eutectic modification level using thermal analysis. Materials Characterization 46, pp. 31

The 73rd World Foundry Congress
„Creative Foundry"
23 - 27 September 2018, Krakow, Poland

MANGANESE IN COMPACTED GRAPHITE IRON

Grzegorz Gumienny*, Barbara Kurowska, Paweł Just

Lodz University of Technology, Department of Materials Engineering and Production Systems, Stefanowskiego 1/15 Street, 90-924 Łódź, Poland
**Corresponding address: e-mail:grzegorz.gumienny@p.lodz.pl*

Key words: Theory of Crystallization, Compacted Graphite Iron, Manganese, DTA Method

The paper presents the effect of manganese on the crystallization process, microstructure and selected properties of the compacted graphite iron (CGI). Until the middle of the 20th century CGI was considered as cast iron with unsuccessful nodular graphite. However, this material is more and more willingly used by constructors [1]. Its mechanical properties (tensile strength 300 – 575 N/mm^2, yield stress 210 – 400 N/mm^2, elongation 0 – 2%, and hardness 140 – 260 HBW), result from the shape of graphite – intermediate between flake and nodular [2].

1. Introduction

High mechanical properties, plastic and good thermal conductivity cause that this material is becoming more and more popular among producers of high quality castings. In comparison to cast iron with flake graphite CGI is characterized by higher strength properties, greater plasticity, thermal shock resistance, corrosion resistance, low section sensitivity. Compared to ductile cast iron it is characterized by higher thermal shock resistance, higher thermal conductance, better damping capacity as well as better castability. Undoubted advantage of CGI is relatively low production cost in comparison to ductile iron resulting from less expenditure on master alloy [3, 4]. Due to the large contact surface of graphite with the matrix, CGI tends to form ferrite both directly and indirectly [3].

Manganese is one of the basic elements in cast iron. In the liquid state it dissolves in it without restrictions. Above a concentration of about 1.2% it is considered in the cast iron as an alloying element. Its maximum concentration in cast iron alloys for non-magnetic castings and resistant to gases and chemicals is about 20% [5]. Manganese is carbide-forming element, at concentrations greater than 0.7% it is possible to crystallization carbides on the boundaries of the eutectic grains [6]. This element reduces the temperature range of eutectic transformation according to a stable and metastable system. It causes the shift of the eutectic transformation point to the right, and the eutectoid to the left and increases the durability of austenite. It is a pearlite forming element too [5, 7]. Manganese dissolves in cementite to form isomorphite (Fe, Mn_3C). There are discrepancies in the effect of manganese on a cast iron graphitization. Some authors reports that in the concentration of 0.8 - 1.2% there is no effect, other publications indicate that it counteracts graphitization in the entire concentration range. It promotes the formation of interdendritic graphite separations. According to A. Roul manganese makes degeneracy in nodular graphite [8]. Manganese to a concentration of approx. 1.2%, increases hardness, tensile strength and decrease in elongation. It also has an adverse effect on the cast iron properties by increasing its shrinkage and brittleness.

From the above it follows that there are some information on the effect of manganese on the microstructure and properties of cast iron in the literature [7, 8]. However, there are few references to the action of Mn in CGI. The shortcomings in this field prompted the Authors to complete the data in this research area.

2. Test methodology

The chemical composition of tested cast iron is presented in Table 1.

Table 1 The chemical composition of tested cast iron

Chemical composition, wt.%					
C	Si	Mn	Mg	P	S
3.28-4.02	1.83-2.18	0.04-1.39	0.014-0.025	max. 0.07	max. 0.02

The compacted graphite was obtained using Inmold technology. Schematic layout of elements in the mould is presented in Figure 1.

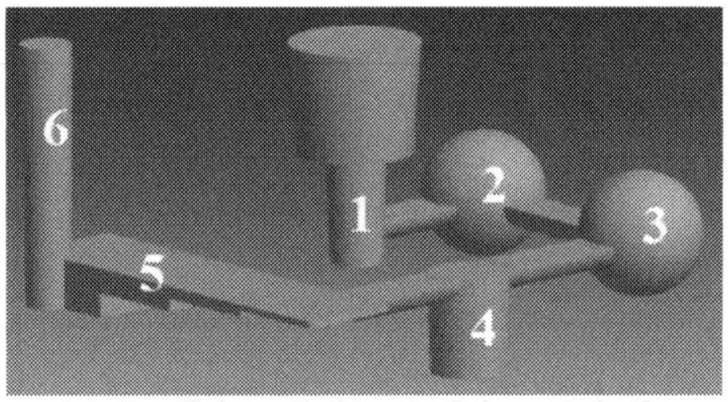

Fig. 1. Schematic layout of elements in the mould: 1 – downsprue, 2 – reaction chamber, 3 – mixing chamber, 4 – control chamber, 5 – stepped test casting, 6 – flow – off

In a spherical reaction chamber (2) a magnesium master alloy LAMET 5504 containing about 5.5% Mg and 0.4% La was placed. For better mixing of the master alloy with liquid cast iron, a spherical mixing chamber (3) is placed behind the reaction chamber. There was also a control chamber behind the mixing chamber. In its thermal center, an S-type thermocouple was placed. The voltage signal from the thermocouple through the compensating cable went to the U/f frequency converter. Then the signal went to the computer, where curves of the derivative and thermal analysis (DTA) were recorded. This made it possible to assess the effect of manganese on the temperature of phase transformations in CGI.

3. Conclusions

There was no significant effect of manganese on the temperature of the eutectic transformation. However, its addition has a significant effect on the temperature of the eutectoid transformation. This element decreases the temperature of eutectoid transformation by about 30°C per 1% in the tested range of the chemical composition.

The effect of manganese on the microstructure of CGI matrix as well as its hardness and microhardness of pearlite and ferrite were investigated.

Figure 2 (a, b) shows the microstructure of CGI containing about 0.3% (a) and 1.4% Mn (b).

a)

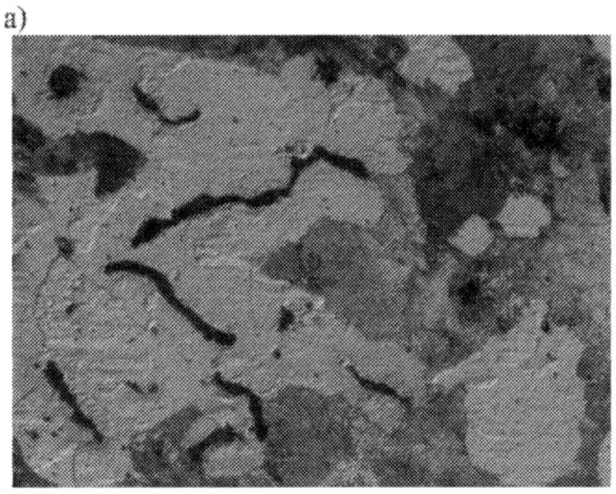

b)

Fig. 2. Microstructure of CGI containing about 0.3% (a) and 1.4% Mn (b)

With the increase in the Mn concentration, the pearlite surface area increased in the entire range of the tested casting wall thickness. In castings with a wall thickness of 3 and 6 mm, carbide separations appeared at a concentration of about 0.6% Mn, in a casting of 12 mm - at a concentration of about 1% Mn, while for the thickness of 24 mm, carbide precipitations did not occur in the concentrations tested. At a concentration of Mn 1.39% in castings with a wall thickness of 6, 12, 24 mm there was a small amount of ferrite. Manganese increases the hardness of CGI by approx. 61 HB per 1%, as well as perlite microhardness approx. 34 μHV0.1 per 1% Mn. There was no significant effect of manganese on the microhardness of ferrite.

References

[1] Dawson S. (2009). Compacted graphite iron – A material solution for modern diesel engine cylinder blocks and heads. China Foundry. Vol. 6. No. 3. 241 – 246.

[2] Atilim University, CLAAS GUSS GmbH (28.03.2018). Vermicular graphite cast iron. Access 30.03.2018. http://www.atilim.edu.tr.

[3] Pietrowski S. (2000). Compendium of knowledge about compacted graphite iron. Solidifications of Metals and Alloys. No 44. 279 – 292.

[4] Pietrowski S. (1998). Mechanism of crystallization of compacted graphite in cast iron. Solidifications of Metals and Alloys. No 37. 97 – 104.

[5] Kalata Cz., Kobyliński S. & Rudpol F. (1958). Cast Iron. PWN.

[6] Nayar A. (1997). The metals databook. Tata McGraw-Hill.

[7] Rao L., Tao W., Wang S. & Geng M. (2014). Influence of the composition ratio of manganese and copper on the mechanical properties and the machining performance of ductile iron. Indian Journal of Engineering & Materials Sciences. Vol. 21. 573 – 579.

[8] Roula A. & Kosnikov G.A. (2008). Manganese distribution and effect on graphite shape in advanced cast irons. Materials Letters. No. 62. 3796-3799.

The 73rd World Foundry Congress
„Creative Foundry"
23 - 27 September 2018, Krakow, Poland

INFLUENCE OF EVALUATED TEMPERATURE OF INTERCRITICAL TREATMENT ON MECHANICAL PROPERTIES OF AUSTEMPERED DUCTILE IRON

Piotr Nawrocki[1]*, Kamil Wasiluk[2], Dawid Myszka[3]

[1,3]Warsaw University of Technology/Faculty of Production Engineering, Narbutta 85, 02-524 Warsaw, Poland,
[2] Warsaw University of Technology/ Faculty of Materials Science and Engineering, Wołoska 14, 02-507 Warsaw, Poland
**Corresponding address: e-mail: p.nawrocki@wip.pw.edu.pl*

Key words: Austempered ductile iron (ADI), intercritical temperature, impact resistance, dilatometric tests, bainitic transformation

1. Introduction

ADI cast iron is characterized by a combination of high strength properties along with good wear resistance and good ductility.

The paper attempts to evaluate the influence of step heat treatment on the microstructure and properties of ductile cast iron with an increased content of Mo, Ni and Cu elements. When the DI was partially austenitized within the intercritical eutectoid regions interval, where austenite + ferrite + graphite coexist, the microstrucure in this case would be consisted of ausferrite and proeutectoid ferritic islands. As a result of the low austenitizing temperature, the austenite has relatively low carbon amount, which raises the Ms temperature during quenching, resulting in increased martensite formation in alloyed and unalloyed intercritically austempered ductile iron. The ferrite proeutectoid will then occur in a smaller amount and with a different morphology for cast iron containing alloying elements, compared to unalloyed. However, other processes take place when in the intercritical region ausferritic structure is austenitized.

The literature describes the application of intercritical ranges [1,2,3,4] however, the nature of the heat treatment process following after the first isothermal treatment is not fully understood.

2. Experimental procedure

The research material was ductile cast iron with a ferrite-pearlite matrix, which chemical composition is presented in Table 1.

Table 1. Chemical composition of ductile iron [wt.%]

Chemical composition [mass pct]						
C	Si	Mn	Cu	Mo	Ni	Mg
3,55	2,5	0,35	0,5	0,35	< 2	0,055

Y-shaped casts were made at the Foundry Institute in Krakow. The image of microstructure of as cast iron is presented in Figure 1.

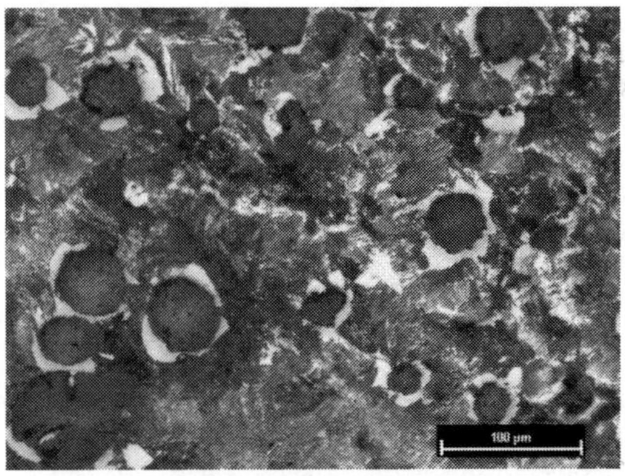

Fig. 1. As-cast microstructure of Mo-Ni-Cu ductile iron; (etched with 5% nital)

The obtained castings were cut into dilatometry and Charpy test samples, which were subject to a thermal process according to the scheme shown in Figure 2.

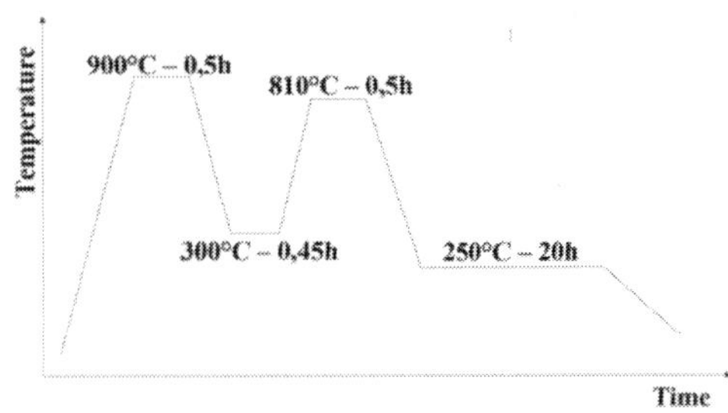

Fig. 2. Schemes of conducted heat treatment process of Mo-Ni-Cu ductile iron

For heat treated samples, the following tests were performed: dilatometry, microstructure observation using light microscopy and scanning electron microscopy, hardness and Charpy impact test at room temperature. Hardness measurements were also made for dilatometer samples.

3. Results and discussion

The aim of the authors was to refinement of microstructure through heat treatment in the lowest range above the Ac_3 temperature. That is why a process consisting of several thermal treatments was designed. First, the isothermal quenching process was carried out to obtain an incomplete bainitic transformation and to obtaining the right proportion of ferrite and austenite. After obtaining 50% of bainitic transformation, the material was heated to the intercritical temperature determined on the basis of dilatometer plots (Fig.3). In the next step, tested material was quenched to an isothermal temperature of 250°C and withstood for to complete the 100% bainite transformation and to stabilized the austenite.

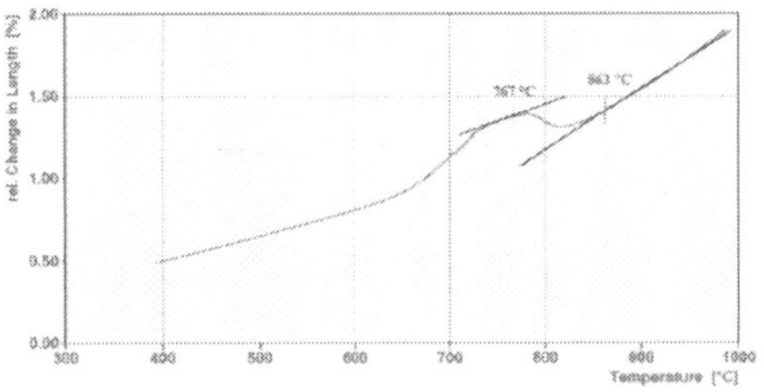

Fig. 3. Dilatometric diagram for the tested Mo-Ni-Cu ductile iron

Tables 2 present the obtained results from hardness and impact measurements, in the form of mean values from the measurement along with the standard deviation. A comparison to the ISO 148-1 standard indicates that the obtained values significantly exceed the minimum and indicate very good impact resistance for a given, very high hardness. The reason is a very ultra-fine mixture of bainitic ferrite and austenite, the kinetics of which will be described in the article (Fig.4).

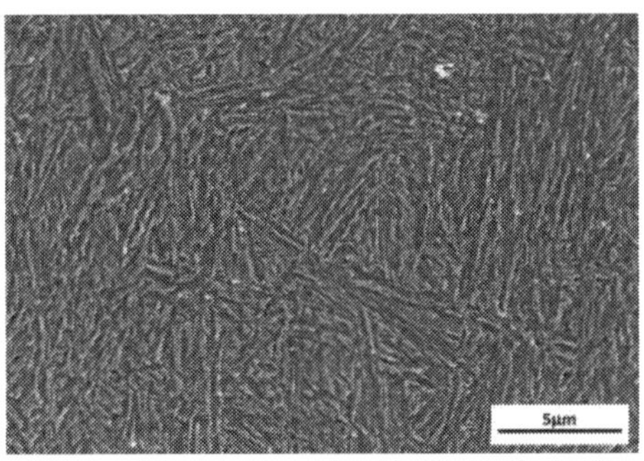

Fig. 4. Microstructure of matrix Mo-Ni-Cu ductile iron after heat treatment; SEM

Table 1. Average hardness values for heat treatment of Mo-Ni-Cu ductile iron

	HV1	HV0.2	K [J]
Regular sample	485±48	500±48	45±9,5
Dilatometric sample	552±54	622±23	-

4. Conclusion

The conducted research allowed to achieve the assumed aim of obtaining a fine grain structure of the ductile iron ausferritic matrix. The proposed gradual heat process for alloy ductile iron enables high impact resistance in combination with high hardness, which can give a possibility of using in special constructional solutions.

Acknowledgments

This work was supported by the National Centre for Research and Development NCBiR project no PBS3/B5/45/2015.

References

[1] Kobayashi, T. & Yamada, S. (1996). Effect of holding time in the (α+γ) temperature range on toughness of specially austempered ductile iron. Metallurgical and Materials Transactions A. 27(7). 1961-1971.

[2] Putatunda, S.K., Panneerselvam, S. Alshwigi, M. (2015). Development of nanostructured austempered ductile cast iron (ADI). 28th ASM heat treating society conference. 20-22 October 2015 (71-75). USA: ASM International.

[3] Jianghuai Y., & Putatunda S.K. (2004). Influence of a novel two-step austempering process on the strain-hardening behaviour of austempered ductile cast iron (ADI). Materials Science and Engineering A. 382. 256-279.

[4] Mahmoud F. H. (2010). Effect of second step austempering temperature and time on the impact properties of two-step austempered SG-cast iron. Journal of Al Azhar University Engineering Sector. 5(14). 59-65.

The 73rd World Foundry Congress
„Creative Foundry"
23 - 27 September 2018, Krakow, Poland

MICROSTRUCTURAL AND MECHANICAL CHARACTERIZATION OF HIGH-ALUMINIUM ZINC ALLOY MODIFIED WITH 2.0% TiB_2

Matheus Roberto Bellé[1]*, Willian Martins Pasini[1], Lucas Caraffini[1], Túlio Sérgio do Nascimento[2], Leonardo Pereira[1], Régis Fabiano do Amaral[1], Vinicius Karlinski de Barcellos[1].

[1]*Federal University of Rio Grande do Sul, School of Engineering, Foundry Laboratory, Bento Gonçalves Ave. 9500, Postal code: 15.021, Porto Alegre, Brazil.*

[2]*Federal Institute of Science, Technoloy and Education of Rio Grande do Sul, Avelino Antônio de Souza Street 1730 Caxias do Sul,*

**Corresponding address: e-mail: matheus.belle@ufrgs.br*

Key words: Zinc-based alloy, TiB_2 particles, dendritic pattern refinement, metal matrix composites, mechanical properties.

1. Introduction

High-Aluminum Zinc Alloys have been used as an alternative in the general industry due to its low melting point, thereby demanding less energy and cost. These alloys have great castability and fluidity which allows it to be casted in many ways (continuous, semi-continuous, centrifugal, investment, squeeze and semi-solid casting). Considering the zinc-based alloys group, ZA-27 alloy exhibits the most suitable specifications, such as excellent bearing, wear and corrosion resistance and good tribological/mechanical properties. However, there are major disabilities: low operating/application temperature and solidifies with coarse dendritic structure in sand casting [1].

Metal Matrix Composites (MMC) with second phase particles present an improvement on the mechanical properties and wear response as compared to zinc-based alloys. TiB_2 is an attractive ceramic reinforcing particle for ZA-27 alloy matrix composites due to its superior wear resistance, high hardness, chemical stability and similar density to matrix material (avoids settling process) [2,3].

In this paper, TiB_2 particles were used as second-phase reinforcement in ZA-27 alloy to evaluate its behavior regarding the microstructure refinement and composites formation in the material to improve the mechanical properties.

2. Experimental

Commercial aluminium was primary melted in a crucible in a resistive furnace and afterwards copper wire and commercial Zamak 5 ingots were charged to achieve the final chemical composition. ZA-27 MMC was produced by addition of TiB_2 powder in ZA27 liquid phase and by mechanical stirring process. Conventional ZA-27 alloy and ZA-27 MMC with 2 wt% of TiB_2 were poured at 750°C in a sand mold. The tensile specimens were casted directly in a mold with a pattern based on ASTM B108-15 geometry (Fig. 1). The microstructures of the ZA-27 and ZA-27 MMC were characterized using metallographic techniques and the mechanical properties were measured with tensile and impact tests.

Fig 1. Sand mold for tensile specimens.

3. Results and Discussion

The results of the tensile and impact tests as function of TiB_2 concentration are presented (Fig. 3). The addition of 2 wt% reinforcement particles increased the ultimate tensile strength (UTS) and the absorved energy of the material.

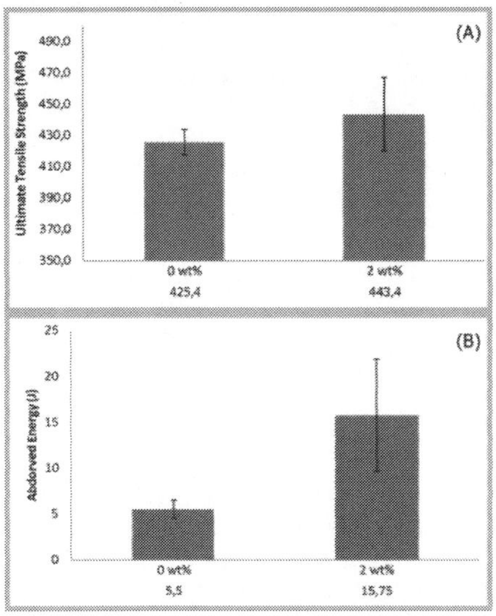

Fig 2. (A) Ultimate Tensile Strength results and (B) Impact Tests Results

These improvements are a result of the reinforcement particles role as barriers to dislocation motion and as heterogeneus nucleation sites during the solidification, resulting in a more refined microstructure (Fig. 3).

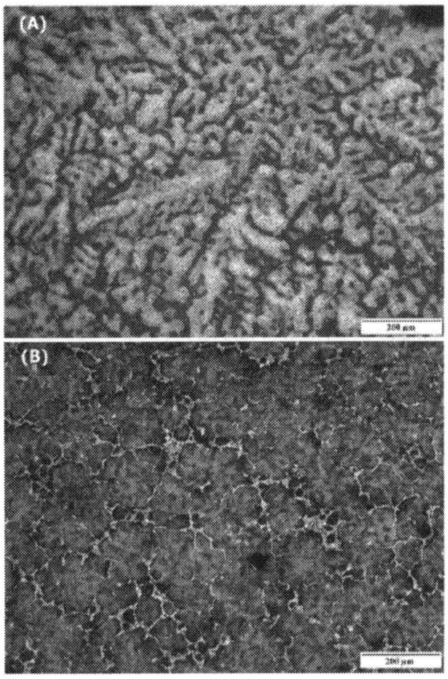

Fig 3. Effects of TiB2 particles on the microstructure of ZA27 alloy: (A) without TiB_2 and (B) with TiB_2 addition. (CrO_3:Na_2SO_4:H_2O etching).

As presented in Fig. 4(A), the as-cast structures of ZA-27 alloy consist of primary α-Al phase dendritic structure and α+η eutectoid in the interdendritic regions along with η-Zn phase. The addition of TiB_2 particles changed the microstructure from dendritic pattern to a petal-like pattern, as shown in Fig. 4(B) [1, 3, 4].

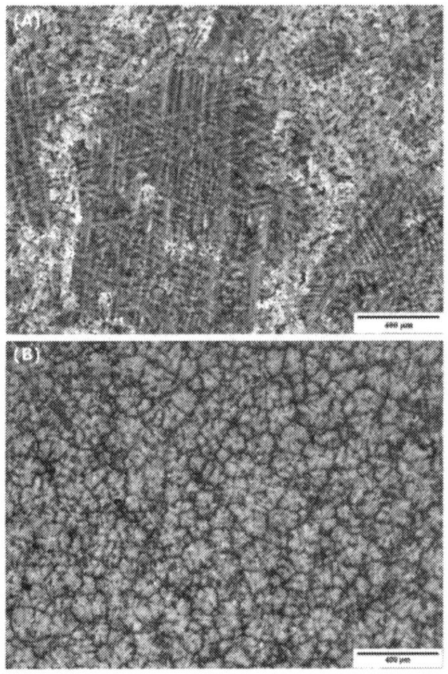

Fig 4 Effects of TiB2 particles on the microstructure of ZA27 alloy: (A) without TiB2 and (B) with TiB2 addition. (HNO_3:H_2O etching)

4. Conclusion

The TiB_2 particles promoted morphological modification in the interdendritic region and in the primary α-Al phase, changing the coarse dendritic pattern to a petal-like pattern. This effect caused the increase on the mechanical properties of the alloy.

Acknowledgments

The authors acknowledge financial support provided by CNPq (The Brazilian Research Council) and FAPERGS (The Scientific Research Foundation of the State of Rio Grande do Sul).

References

[1] Krajewski, W., Zak, P., Orava, J., et al. (2012). Structural Stability of the High-Aluminium Zinc Alloys Modified with Ti Addition. Archives of Foundry Engineering, 12(1), pp. 61-66.

[2] Chen F., Wang T., Chen Z., et al, (2015) Microstructure, mechanical properties and wear behavior of Zn−Al−Cu−TiB2 in situ composites, Trans. Nonferrous Met. Soc. China 25,pp 103−111.

[3] Li Z.G., (2014). Fabrication of in-situ TiB_2 particulates reinforced zinc alloy matrix composite, Materials Letters 121 pp 1–4

[4] Pollard W.A., Pickwick K.M., Jubb J.T et al. Packwood (1974) The grain refinement of zinc-aluminum alloys by titanium, Canadian Metallurgical Quarterly, 13:4, 535-543

The 73rd World Foundry Congress
„Creative Foundry"
23 - 27 September 2018, Krakow, Poland

EFFECT OF Al, Cr AND Mo ADDITION ON MECHANICAL PROPERTIES OF Nb-Si BASED ALLOYS

Tsuyoshi Uemura[1]*, Kazunori Asano [2]

[1,2]*Kindai University/ Faculty of Mechanical Engineering, 3-4-1 Kowakae, Higashiosaka, Osaka, Japan,*
**e-mail: tys254wqw@gmail.com*

Key words: Nb-Si based alloy, microstructure, mechanical properties, compression strength, heat treatment

1. Abstract

Nb-24Ti-18Si-Al-Cr-Mo alloys were prepared by arc-melting, and the effects of the concentration of Al, Cr and Mo on the microstructure and mechanical properties of the alloys have been clarified. As the concentration of Mo increased, the proof strength of the alloy increased while the fracture strain of the alloy decreased. This would be due to the dissolution Mo into the Nb solid solution. Although the proof strength of the alloy increased due to the low concentration of Al and Cr, this effect was reduced due to the addition of 10 at. % of these elements. The fracture strain of the alloy significantly decreased due to the formation of the intermetallic compounds consisting of Cr or Al.

2. Introduction

Ni based super alloys have been used for aircraft engine turbine blades. However new heat resistance materials having higher melting points have been required for improving the efficiency of engines. Nb-Si based alloy, consisting of αNb_5Si_3 and Nb solid solution (Nb_{ss}), which are formed by the eutectoid transformation ($Nb_3Si \rightarrow Nb_{ss} + \alpha Nb_5Si_3$), is expected higher melting point and better strength at elevated temperature than Ni based super alloys. However, the alloy is poor in toughness at ambient temperature and oxidation resistance at elevated temperature. It is reported that the addition of elements to the alloy is effective for improving the properties[1, 2]. Based on these reports, we prepared Nb-24Ti-18Si-5Al-5Cr-5Mo (at.%) alloy and clarified that the microstructure and mechanical properties of the alloy [2]. These results indicate that the optimum concentration of these elements can improve the mechanical properties of the alloy. In this study, we prepared the alloy containing the various concentration of Al, Cr and Mo, and the effects of these elements addition on the microstructure and mechanical properties of the alloy have been clarified.

3. Experimental Procedure

The alloys that having the chemical composition as shown Table 1 were prepared by arc-melting. The concentration of Al, Cr and Mo in Nb-24Ti-18Si-5Al-5Cr-5Mo alloy (base alloy) was varied. As well as the as-solidified alloys, the heat-treated alloys to promote the eutectoid transformation mentioned above (HT) were also prepared. Microstructure of the alloys was observed using the BEI, and quantitative analysis of each phase were conducted using the FE-EPMA. The XRD was conducted to identify the constituent phases in the alloys. Test pieces were (3×3×6 mm) cut form the ingot, and the compression test was carried out at ambient temperature. 0.2 % proof strength was measured by the offset method using the compression stress-strain curves.

Table 1. Chemical composition of the alloys.

Alloys	Chemical composition (at.%)
Base alloy	Nb-24Ti-18Si-5Cr-5Al-5Mo
2A	Nb-24Ti-18Si-5Cr-2Al-5Mo
10A	Nb-24Ti-18Si-5Cr-10Al-5Mo
0C	Nb-24Ti-18Si-10Al-5Mo
10C	Nb-24Ti-18Si-10Cr-5Al-5Mo
15C	Nb-24Ti-18Si-15Cr-5Al-5Mo
0M	Nb-24Ti-18Si-5Cr-5Al
2M	Nb-24Ti-18Si-5Cr-5Al-2Mo
10M	Nb-24Ti-18Si-5Cr-5Al-10Mo

4. Results and Discussion

Microstructure

Fig. 1 shows the microstructure of the alloys. In the Fig. 1, the constituent phases in the alloys, which identified by the XRD patterns and the quantitative analysis, are also described. The $\beta(Nb,Ti)_5Si_3$ in the alloy 10A (① in Fig.1(b)) was larger than that in the base alloy (① in Fig.1(a)). This would be due to the fact that the formation of the $\beta(Nb,Ti)_5Si_3$ was promoted by the increase in the addition of Al. When the addition amount of Cr was greater than 10at. %, the $(Cr,Si)_2(Nb,Ti)$ was formed in the alloy (⑥ in Fig.1(c)). On the other hand, addition of Mo didn't strongly affect the microstructure of the alloy.

After the heat treatment, the $\beta(Nb,Ti)_5Si_3$ and $(Nb,Ti)_3Si$ disappeared in all alloys except for the alloy 10A_HT, and the formation of the $\alpha(Nb,Ti)_5Si_3$ was observed the fine granular $(Nb,Ti)_{ss}$ was formed in the $\alpha(Nb,Ti)_5Si_3$. On the other hand, in the alloy 10A_HT, the $\alpha(Nb,Ti)_5Si_3$ was not observed while the $\beta(Nb,Ti)_5Si_3$ remained, and the $(Nb,Ti)_3Al$ was observed. The

($(Cr,Si)_2(Nb,Ti)$ in the alloy 10C_HT was more than the as-solidified alloy 10C. After the heat treatment, amount of the Ti-rich $(Nb,Ti)_5Si_3$ increased as the addition of Mo increased.

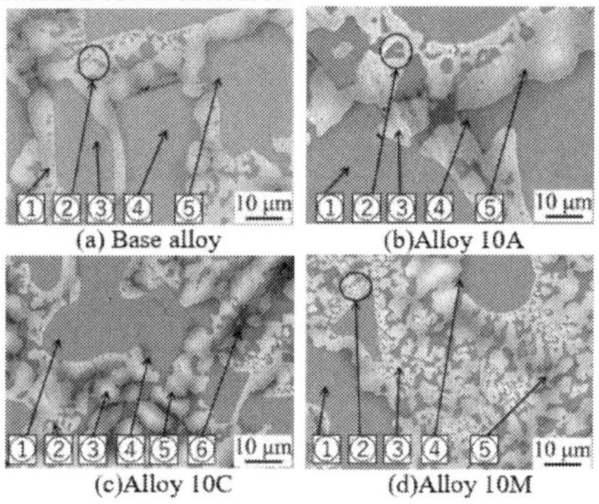

Fig. 1. BEI of as-solidified alloys. *(①β$(Nb,Ti)_5Si_3$ ②$(Nb,Ti)_3Si$ ③$(Nb,Ti)_{ss}$ ④Ti-rich $(Nb,Ti)_5Si_3$ ⑤Ti-rich $(Nb,Ti)_{ss}$ ⑥$(Cr,Si)_2(Nb,Ti)$)*

Compression Properties

Table 2 shows the 0.2 % proof strength and fracture strain of the as-solidified alloys. The proof strength and fracture strain of the alloy increased due to the addition of 5 at. %Al, while the fracture strain decreased due to the addition of 10 at. %Al. The proof strength of the alloy slightly increased due to addition of 5 at. %Cr. When 15 at. % Cr was added to the alloy, the proof stress was not clearly detected from the stress-strain curve and the fracture strain further decreased. The increase in the concentration of Mo, the proof strength of the alloy increased, while decreased the fracture strain of the alloy. The decrease in the fracture strain of the alloy due to the increase in the addition of 10 at. %Al would be due to the fact that the β$(Nb,Ti)_5Si_3$ grew larger as the increase in the concentration of Al. The addition of Cr promotes the solid solution hardening of $(Nb,Ti)_{ss}$ by dissolving Cr into $(Nb,Ti)_{ss}$. However, when the concentration of Cr was greater than 10 at. %, the $(Cr,Si)_2(Nb,Ti)$ was formed in the alloy, and it leads to the decrease in the fracture strain of the alloy. On the other hand, the increase in the proof strength and the decrease in the fracture strain of the alloy by the addition of Mo would be due to the solid solution hardening of $(Nb,Ti)_{ss}$ by dissolving Mo into $(Nb,Ti)_{ss}$.

The addition of 10 at. %Al and heat treatment significantly decreased the proof strength and fracture strain of the alloy. This would be due to the residual β$(Nb,Ti)_5Si_3$ after the heat treatment or the formation of the brittle $(Nb,Ti)_3Al$ in the alloy during the heat treatment. The addition of 10 at. %Cr and heat treatment decreased the fracture strain of the alloy. This would be due to the increase in the $(Cr,Si)_2(Nb,Ti)$ in the alloy during the heat treatment. For the heat-treated alloy, the fracture strain of the alloy decreased as the concentration of Mo increased. This would be due to the fact that the Ti-rich $(Nb,Ti)_5Si_3$ was more as the concentration of Mo increased. These results indicate that there is an optimum concentration of Al, Cr and Mo for the Nb-24Ti-18Si-Al-Cr-Mo system.

Table 2. Mechanical properties of as-solidified alloys.

Alloys	Proof strength (MPa)	Fracture strain (%)
Base Alloy	1950	1.4
2A	1875	1.0
10A	2000	0.42
0C	1900	1.6
10C	1700	0.84
15C	-	0.082
0M	1800	2.4
2M	1900	2.1
10M	2250	0.72

5. Conclusions

1. Although the fracture strain of the alloy increased due to the addition of 5at. %Al, it decreased due to the addition of 10 at. %Al. When 10 at. % Al was added to the alloy and exposed to the heat treatment, the fracture strain significantly decreased due to the formation of the (Nb,Ti)3Al.
2. The proof strength increased due to the addition of 5 at. %Cr. This would be due to the solid solution hardening of (Nb,Ti)ss. When the concentration of Cr increased, the (Cr,Si)2(Nb,Ti) was formed in the alloy, then the proof strength and fracture strain significantly decreased. When 10at. % Cr was added to the alloy and exposed to the heat treatment, the (Cr,Si)2(Nb,Ti) was more, then the fracture strain further decreased.
3. The proof strength of the alloy increased as the increase in the concentration of Mo, while the fracture strain of the alloy decreased. This would be due to the solid solution hardening of (Nb,Ti)ss. For the heat-treated alloy, the fracture strain of the alloy decreased as the in the concentration of Mo increased. This would be due to the increase in the Ti-rich (Nb,Ti)5Si3.

References

[1] J. Geng. (2006). Development of Nb_{ss}/Nb_5Si_3 based in-situ composites. Paper of Metallurgy Research Group School of Engineering University of Surrey.

[2] Z. Sun, X. Guo, Y. He, J. Guo, Y. Yang & Y. Austin Chang. (2010). Investigation on the as-cast microstructure of Nb–Nb silicide based multicomponent alloys. Intermetallics. 18. 992-997.

[3] T. Uemura & K. Asano. (2017). Effects of Cr, Mo and Ti Addition on Mechanical Properties of Nb-based Alloys. The 14th Asian Foundry Congress, Nov.7-10 2017, Korea Foundry Society.

The 73rd World Foundry Congress
„Creative Foundry"
23 - 27 September 2018, Krakow, Poland

QUALITY INDEX - A WAY OF ILLUSTRATING DIVERSIFIED PROPERTIES OF A HEAT-TREATED AL-ZN-MG-CU ALUMINIUM ALLOYS DEPENDING ON HOMOGENIZATION PARAMETERS

Zofia Kwak[1]*, Edward Czekaj[2], Aldona Garbacz-Klempka[1]

[1] *AGH University of Science and Technology, Faculty of Foundry Engineering, Reymonta 23 Str., 30-059 Kraków, Poland*
[2] *Foundry Research Institute, Zakopiańska 73 Str., 30-418 Kraków, Poland,*

*Corresponding address: e-mail: kwakzosia@gmail.com

Key words: Innovative Foundry Technologies and Materials, Al-Zn-Mg-Cu alloy, Quality Index, Heat Treatment, Homogenization

Introduction

Rising client expectations in relation to the quality of products require developing new technologies that improve the quality of casting. Aluminium, due to large possibilities of controlling its mechanical properties, holds the key position in the industry. The advantage of Al-Zn-Mg-Cu alloys are undoubtedly good strength, light-weight and resistance to corrosion. In additions, they are characterized by perfect energy absorption and good plasticity. Since these qualities are needed in all engineering areas, mentioned alloys can find many applications in all of them, [1, 2].

Presented study was conducted for 7xxx series aluminium alloy according to PN-EN 573-3:2010 – Polish version. Different homogenization variants and subsequent tests were carried out for samples taken from non-homogenized ingots, obtained by Semi-continuous casting method. The samples were subjected to various conditions of homogenization, controlled by the temperature and time of the process, including two-stage homogenization. The aim of this publication is to present how the macro and microstructure are shaped, strength parameters of the material, correlation between basic properties of 7xxx series aluminium alloys, i.e. relative elongation E, yield strength YS and ultimate tensile strength UTS depending on the homogenization parameters. On the basis of the obtained mechanical properties results, a quality index for the Al-Zn-Mg-Cu alloy was created.

Generally it is known, that operating durability of construction elements depends both on the strength and plastic properties of their material. Therefore, for several years now, in Polish and foreign specialist literature, there has been the concept of quality index (QI), which combines these two important qualities of construction material. That concept was first introduced in 1980 by M. Drouzy, S. Jacob and M. Richard , originally it was written as Q and defined in the following way (1) [3]:

$$QI = Q = UTS + d \cdot \log(A) \qquad (1)$$

where: UTS – ultimate tensile strength;

A – elongation at the moment the sample fractures;

d – experimentally chosen coefficient d, depending on the alloy type.

Quality indicators known from literature as a means to estimate quality are also intended to help engineers-founders look for higher and higher quality. Gradually, guidelines for their use and interpretation are also provided. Heat-treatment that is used to shape mechanical properties of alloys, i.e. tensile strength, elongation or fatigue resistance, largely affects the original quality of the alloy. Therefore, in order to solve problems related to the quality of aluminium castings, and more precisely low plasticity, castings are subjected to cooling or heat treatment. We know that the cooling, in our case, during semi-continuous casting, taking place in crystallizers, causes that crystallization takes place faster. The consequence of this is the smaller spacing of the arms between the secondary dendrites (SDAS). We know from the foundry literature that the increase in elongation is dictated by shrinking SDAS. [4]

Methodology and Research

The chemical profiles of the tested samples were made using a spark spectrometer ARL 3460 OES. The Brinell hardness test was carried out on a ZWICK Roell ZHU250 hardness tester according to the PN-EN ISO 6506-1. The study of mechanical properties was carried out on the ZWICK Roell Z300 tensile machine in compliance with PN-EN ISO 6892-1.

The tested aluminium alloy was obtained on the SECO-ALMEX brand line. The line consisted of: induction melting furnace „SIEMENS", a resistance furnace with concrete lining SECO- WAWRICK and Almex casting line. Metal refining during casting was carried out with the LIQUID ALUMINIUM REFINING SYSTEM device, the modification was carried out in the gutter - modifier directly fed to the liquid metal. The temperature of liquid metal during pouring reached 670-682°C.

Samples were taken from ingots according to the scheme below (Fig. 1).

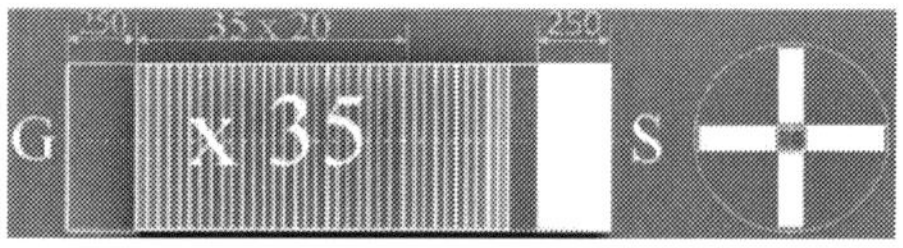

Fig. 1. Sampling scheme from the ingot (markings in the drawing: G – top of the ingot, S – bottom of the ignot)

Samples for tensile strength tests were prepared according to PN-EN 10002-1 + AC1.

Mechanical properties (*UTS, YS* and *E(A), HB*) of the experimental alloy were, determined during the tensile strength test and Brinell hardness test, at the room temperature. They were tested at the stages presented in Table 1.

Table 1. Homogenization parameters for aluminium alloys base on Al-Zn-Mg-Cu

Version of Homogenization	Temperature [°C]	Time [h]	Cooling
1	As cast		
2	420 °C	8 h	with furnance
3	440 °C	8 h	with furnance
4	440 °C	24 h	with furnance
5	400 °C	8 h	outdoors
6	400 °C	24 h	outdoors
7	1) 400 °C 2) 460 °C	¼ h 8 h	outdoors
8	1) 460 °C 2) 480 °C	¼ h 10 h	outdoors
9	1) 440 °C 2) 470 °C	¾ h 5,5 h	outdoors
10	1) 450 °C 2) 470 °C	¾ h 5.5 h	outdoors
11	1) 460 °C 2) 480 °C	1 h 5 h	outdoors
12	1) 450 °C 2) 485 °C	¼ h 5 h	outdoors

Based on the mechanical properties (UTS, YS and E (A), HB) at the room temperature, using the (1) equation, the QI index values were calculated. The obtained data made it possible to prepare interesting summary (Fig. 2), ranking the data from the highest to the lowest value.

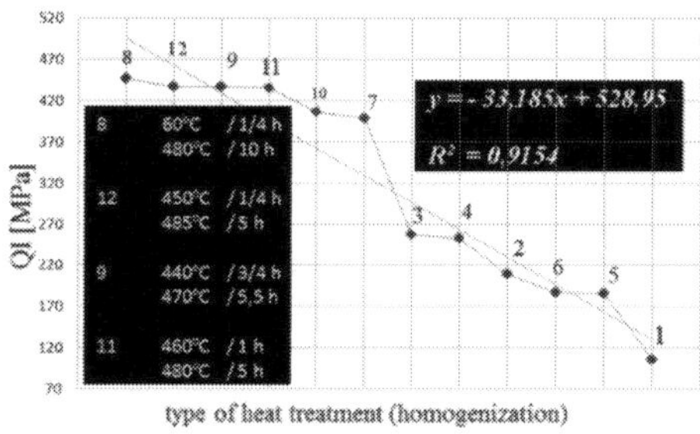

Fig. 2. Ranking of the heat treatment for Al-Zn-Mg-Cu with use of Quality Index

Conclusions

The highest values of the quality index were observed for the 8th stage of homogenization - two-stage homogenization at 460 and 480 degrees. The conducted studies show that the QI can be a simple, relatively inexpensive and effective way to comprehensively evaluate the quality of casting aluminium alloys.

Proper homogenization can bring very high quality of the aluminium alloy if it is needed, or find sufficient values of those properties.

References

[1] Z. Kwak, A. Garbacz-Klempka, M. Perek-Nowak & Ł. Szymański (2016) Correlation between mechanical properties of selected 7xxx series aluminium alloys obtained by semi-continuous casting. Metallurgy and Foundry Engineering MaFE. 42 (1), pp. 69–78.

[2] . Kwak, A. Garbacz-Klempka & M. Perek-Nowak (2016) Microstructure of selected 7xxx series aluminium alloys obtained by semi-continuous casting. Transactions of the Foundry Research Institute. 56(3), pp 221–232.

[3] E. Czekaj, J. Zych, Z. Kwak & A. Garbacz-Klempka (2016) Quality index of the AlSi7Mg0.3 aluminium casting alloy depending on the heat treatment parameters. Archives of Foundry Engineering. 16(3), pp. 25–28.

[4] M. Tiryakioglu & J. Campbell (2014) Quality index for aluminium alloy casting. International Journal of Metalcasting. 8(3), pp. 39-42.

The 73rd World Foundry Congress
„Creative Foundry"
23 - 27 September 2018, Krakow, Poland

NON-DESTRUCTIVE TESTING METHODS IN THE ANALYSIS OF METAL ARTEFACTS

Aneta Sosnowska[1]*, Aldona Garbacz-Klempka [2]

[1]AGH University of Science and Technology,
Faculty of Management, Gramatyka 10, Str, 30-067 Kraków, Poland
Faculty of Foundry Engineering,, Reymonta 23 Str., 30-059 Kraków, Poland
[2] AGH University of Science and Technology, Faculty of Foundry Engineering,
Historical Layers Research Centre, Reymonta 23 Str., 30-059 Kraków, Poland,
**Corresponding address: e-mail: asosnowska1702@gmail.com*

Key words: Archaeometallurgy, Lost-wax casting, Tin bronzes, Computer Modelling, Bronze Age

Abstract

Foundry is a technique that was started about 6000 years ago, and without this technique it is very difficult to imagine today's industry. Branches of industry in which castings are used include mining, car industry or construction. Due to the fact that from the castings produced more and more better properties are required, and at the same time the absence of any defects, hence the research methods for analyzing castings and detecting possible defects existing in them must be developed. Non-destructive methods allow to control the quality of castings without disturbing their structure, giving information about the properties of materials and allow to detect defects inside and on the surface of the tested element. The non-destructive testing methods (NDT) are:

- ultrasonic methods,
- whirling currents,
- radiographic methods,
- magnetic methods,
- visual methods,
- penetration methods,
- tightness testing.

In this work special attention was focused on modern research techniques, ie radiography and computed tomography, which were used to discover the history of foundry [1,2]. The subject of the research are archaeological objects from the Archaeological Museum in Poznań, including finds originating from Komorowo and Biskupin (Great Poland) and from the Copper Museum in Legnica from Grzybiany (Lower Silesia) [3-6]. During the tests, the following test equipment was used:

- MU2000 X-ray device from YXLON International GmbH,
- X-ray fluorescence spectrometer with energy dispersion (ED-XRF) - Spectro Midex model,
- NIKON SMZ 745T stereomicroscope with a NIKON Digital Sight DS-U2 12V camera.

The use of macroscopic observations (Fig.1), X-ray defectoscopy (Fig.2) and X-ray fluorescence spectrometry (Fig.3-4) has shown that it is possible to combine modern solutions with old casting techniques to obtain quality information about historic objects. Among archaeological finds, a very large group are foundry forms (fragments of forms in other words) and metal decorations that have been subjected to observations. The main assumption of the work was to conduct research to examine the structure and surface of the artefacts, reveal the shape of the mold cavity taking into account the type of castings possible in this form, as well as to examine the contents of found fragments of molds and foundry products and to know their chemical composition.

Fig. 1. Fragment of the casting mold from Komorowo (Great Poland), Archaeological Museum in Poznań

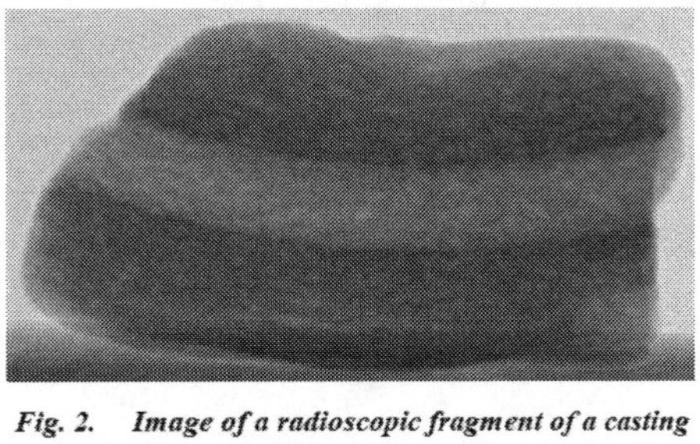

Fig. 2. Image of a radioscopic fragment of a casting mold, originating from Biskupin (Great Poland), Archaeological Museum in Poznań

Fig. 3. The mapping area. Fragm. of a casting mold from Komorowo (Great Poland).

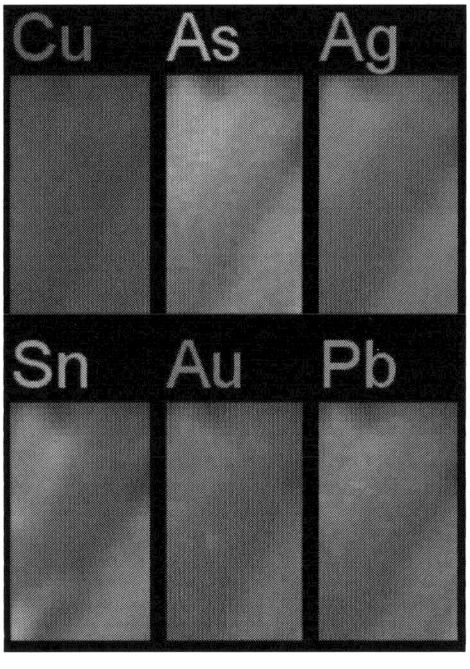

Fig. 4. Element Map for Cu, As, Ag, Sn, Au and Pb using XRF. Fragm. of a casting mold from Komorowo (Great Poland).

Summary

The obtained results allowed to determine the casting technology used in foundry workshops, including the method of smelting models. The results developed on the basis of the performed tests constitute a material that gives qualitative information about the structure, composition, material and shape of historic objects. They can therefore be used as a source of information for the purpose of product reconstruction. In turn, the reconstruction of foundry objects may contribute to increasing knowledge about metallurgical products from the Bronze Age and the beginnings of the Iron Age.

Acknowledgments (optional)

The research was conducted as part of the National Science Center, Poland project number 2017/26/E/HS3/00656 Biography of Bronze. Archeometallurgical study on the foundry technology of the Lusatian culture community in Great Poland.

References

[1] Sosnowska A., (2018). Nowoczesne metody badawcze w analizie zabytków metalowych-Modern research methods in the analysis of metal monuments , praca dyplomowa inżynierska pod kierunkiem Aldony Garbacz-Klempki, AGH University of Science and Technology, Faculty of Foundry Engineering, Kraków.

[2] Ciliberto E., Spoto G. 2000. Modern analytical methods in art and archaeology, Toronto.

[3] Piaskowski, J. (1957). Technika odlewnictwa w grodzie kultury łużyckiej w Biskupinie. Tarnów: Stowarzyszenie Techniczne Odlewników Polskich.

[4] Rzadkosz, S., Zych, J., Garbacz-Klempka, A., Kranc, M., Kozana, J., Piękoś, M., Kolczyk, J., Jamrozowicz, Ł., Stolarczyk, T. 2015. Copper Alloys in Investment Casting Technology. METALURGIJA. 54 (1), 293-296.

[5] Garbacz-Klempka, A., Kowalski, Ł., Kozana, J., Gackowski, J., Perek-Nowak, M., Szczepańska, G. & Piękoś, M. (2016). Archaeometallurgical investigations of the Early Iron Age casting workshop at Kamieniec. A preliminary study. Archives of Foundry Engineering 16 (3), 29–34.

[6] Garbacz-Klempka, A., Kozana, J., Piekoś, M., Kwak, Z., Długosz, P. & Stolarczyk. T. (2015). Ceramic moulds for precision casting in the bronze age and computer reconstruction of the casts. Archives of Foundry Engineering 15 (spec. iss.1), 21-26.

The 73rd World Foundry Congress
„Creative Foundry"
23 - 27 September 2018, Krakow, Poland

EFFECT OF PREDEFORMATION AND SEMISOLID ISOTHERMAL TREATMENT TIME ON THE GLOBULAR GRAINS IN MRI230 ALLOY DURING SIMA PROCESS

Caroline Almeida Santos Fraga[1*], Sérgio Luiz Telles Bartex, Lucas Caraffini, Vinicius Karlinski de Barcellos

[1]Federal University of Rio Grande do Sul, Engineering School, Foundry Laboratory, Bento Gonçalves Av., 9500, ZIP Code 91501-970, Porto Alegre, RS, Brazil

***Corresponding address: e-mail:** *caroline.almeida@ufrgs.br*

Key words: Globularization, SIMA process, semisolid, average grain size, shape factor

1. Introduction

The increasing tendency for near net shape production of industrial parts, taking advantage of costs reduction and quality improvement, is turning semisolid processing into viable technologies.[1] The strain-induced melt activation (SIMA) process is the simplest method between semisolid state metal processing [1,2]. In this work, the microstructure development in the semisolid state of MRI230 magnesium alloy prepared by SIMA was analyzed aiming to compare the morphology of the formed globules through the average grain size (D) and shape factor (SF).

2. Materials and Methods

MRI230 magnesium alloy was used in this study with the measured composition (%wt) of 6.45%Al, 2.25%Ca, 0.84%Sn, 0.27%Mn and 0.25%Sr, balance Mg. The liquidus temperature of the alloy is 603°C and solidus temperature is 522°C. The solid fraction calculated, corresponding to semisolid temperature (580°C) was 0.51. The alloy ingot was machined into six samples with 125mm x 12mm x 55mm. Semisolid slurry was prepared by SIMA. During this process, as-cast samples were firstly preheated at 250°C for 30 min and rolled up to different levels of predeformation, 4% e 7% respectively. Semisolid isothermal treatment (SSIT) was held at 580°C for both predeformations. The SSIT time was varied in 20 min, 40 min and 60 min with the aim to compare the morphology of the formed globules. After, the samples were quickly quenched into water to keep microstructural morphology of semisolid slurry. The temperature was controlled using type K thermocouples. Samples for microstructure characterization were prepared by standard metallography techniques. The microstructure of the alloy was observed using a optical microscope. The average grain size and shape factor of the solid phase were measured using a digital image analysis system. Average grain size (D) and shape factor (SF) of solid grains were calculated in each case by applying Eqs (1) and (2).

$$D = \frac{\sum_1^N \sqrt{\frac{4A}{\pi}}}{N} \quad (1)$$

$$SF = \frac{1}{(\sum_1^N P^2/4\pi A)/N} \quad (2)$$

In these equations, D, SF, A, N and P are average grain size, shape factor, area, the number and perimeter of solid particles, respectively. SF is a value close to 1, corresponding to a better globularity of solid particles.

3. Results and Discussion

Fig. 1 and Fig. 2 show microstructure of 4% and 7% predeformation with magnification 50x for figures (a), (b), (c) and 100x for (d), (e) and (f) respectively, for different SSIT time. Fig.1 shows microstructure evolution of MRI230 semisolid slurry prepared by 580°C 4% deformation for (a) 20 min (c) 40 min (d) 60 min.

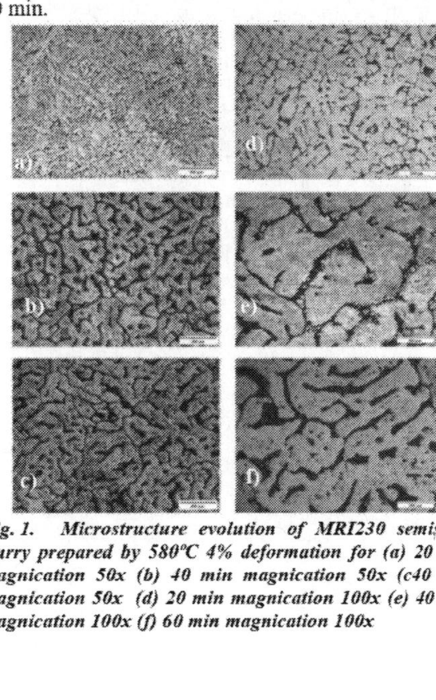

Fig. 1. Microstructure evolution of MRI230 semisolid slurry prepared by 580°C 4% deformation for (a) 20 min magnication 50x (b) 40 min magnication 50x (c40 min magnication 50x (d) 20 min magnication 100x (e) 40 min magnication 100x (f) 60 min magnication 100x

As result, the plastic deformation was insufficient to obtain microstructure with grains refined by dynamic recrystallization. A number of coarse dendrites still existed in its microstructure. As the holding time increases, the grain size also increases. This is a consequence of liquid diffusion in the solid phase.

Fig. 2 shows the microstructure with 7% predeformation for (a) 20 min, (b) 40 min, (c) 60 min. In this case, coarse dendrite disappeared and globular grains were produced, indicating the successful preparation of semisolid slurry. As the SSIT time increases, the globules size slightly increases and the microstructure becomes more homogeneous.

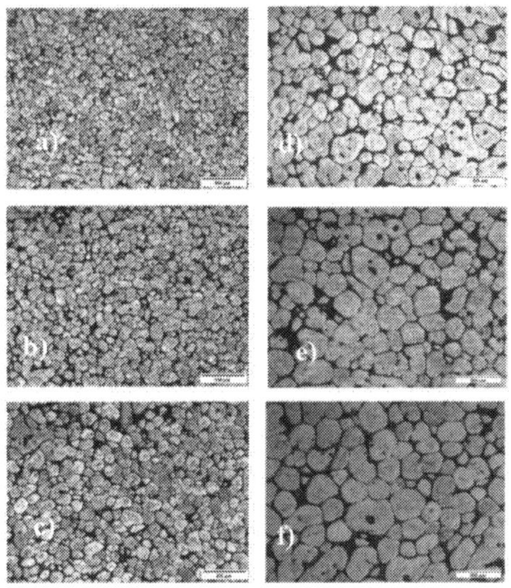

Fig. 2. Microstructure evolution of MRI230 semisolid slurry prepared by 580°C 7% deformation for (a) 20 min magnication 50x (b) 40 min magnication 50x (c) 40 min magnication 50x (d) 20 min magnication 100x (e) 40 min magnication 100x (f) 60 min magnication 100x

Fig. 3 shows the effect of predeformation and holding time on average grain size (D) and shape factor (SF) of solid grains. When the deformation is 4% and the alloy was held for 20 min, 40 min e 60 min, the average grain size of semisolid slurry was 204 µm, 425 µm and 582 µm, respectively. The primary solid phase coarsened quickly with the increase of holding time. Furthermore, the grain size of alloy semisolid slurry prepared by SSIT was very coarse due to dendrite coarsening mechanism of as-cast material. The shape factor presents a slightly increase, 0.43 to 0.50, and remains practically the same, indicating low globularization. For 7% predeformation, Fig. 3 shows the holding time increase, the average grain size has a slightly increase, 92 µm for 20 min up to 112 µm for 60 min. Shape factor did not show considerable variation (close to 0.85) for all SSIT times and indicating high globularization.

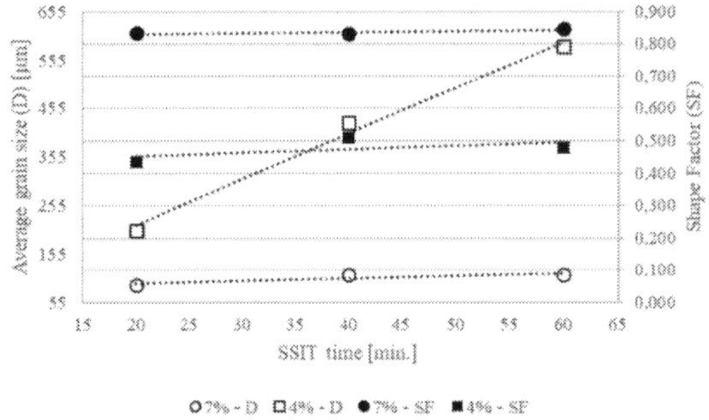

Fig. 3. Effects of predeformation and SSTI time on average grain size and shape factor

4. Conclusion

The effect of predeformation and holding time in the semisolid temperature has been investigated in MRI230 alloy through SIMA process. The following results can be concluded:

i. 4% deformation was not efficient to promote grain globularization, showing the worst shape factor, 0.43 for 20 min holding time and the biggest average grain size, 582µm for 60 min SSIT time.

ii. 7% deformation was able to promote homogeneous globular microstructure with the best shape factor 0.85 and average grain size 112 µm for 60 min SSIT time.

5. References

[1] Jiang Jufu,Wang Ying, Qu Jianjunb, Du Zhiminga, Sun Yid, Luo Shoujinga. (2010). Microstructure evolution of AM60 magnesium alloy semisolid slurry prepared by new SIMA. Journal of Alloys and Compounds 497 62–67

[2] B. Shayegh Boroujeny, M.R. Ghashghaei, E. Akbari. (2017) Effects of SIMA (Strain Induced Melt Activation) on microstructure and electrochemical behavior of Al-Zn-In sacrificial anodes. Journal of Alloys and Compounds 731 354e363

[3] Yan Xua, Lianxi Hua, Jianbo Jiaa,b, Bo Xuc. (2016) Microstructure evolution of a SIMA processed AZ91D magnesium alloy based on repetitive upsetting-extrusion (RUE) process. Materials Characterization 118 309–323

The 73rd World Foundry Congress
„Creative Foundry"
23 - 27 September 2018, Krakow, Poland

WETTABILTY, REACTIVITY AND INTERFACES IN Gd/Y_2O_3 SYSTEM

Patrycja Turalska[1*], Natalia Sobczak[1,2], Grzegorz Bruzda[1], Ivan Kaban[3], Norbert Mattern[3], Jürgen Eckert[4,5]

[1]*Foundry Research Institute, 73 Zakopianska Str., 30-418 Krakow, Poland*
[2]*Institute of Precision Mechanics, 3 Duchnicka Str., 01-796 Warsaw, Poland*
[3]*IFW Dresden, Institute for Complex Materials, Helmholtzstraße 20, 01069 Dresden, Germany*
[4]*Erich Schmidt Institute of Materials Science, Austrian Academy of Sciences, Jahnstraße 12, 8700 Leoben, Austria*
[5]*Department Materials Physics, Montanuniversität Leoben, Jahnstraße 12, 8700 Leoben, Austria*

**Corresponding author: patrycja.turalska@iod.krakow.pl*

Key words: Gd, yttrium oxide, wettability, reactivity, interfaces

The sessile drop method combined with classical contact heating procedure was applied for investigation of high temperature interaction between liquid Gd and dense polycrystalline yttrium oxide under an inert gas (Ar) atmosphere at 1362°C. The contact angles were determined from the images of drop/substrate couples recorded with a CCD camera.

1. Introduction

Development of metallic materials, especially alloys containing highly reactive elements such as Gd, is inextricably linked to the search for new refractory materials for melting and casting processes. Also high-temperature testing of Gd and its alloys using container-assisted methods (e.g. calorimetric, dilatometric and thermogravimetric tests) is difficult because of high melting temperature, high oxygen affinity and chemical aggressiveness of Gd [1,2]. In both cases, the characteristic feature is the relatively long time of contact between the liquid metal and the container material. Literature data and thermodynamic analysis show, that the best and the most stable material for applications in liquid-assisted processes using Gd is Y_2O_3 ceramics.

The paper is focused on the investigation of high temperature interaction between liquid Gd and yttrium oxide. It contains an analysis of the structure and composition of phases formed at the drop/substrate interface.

2. Experimental

Pure Gd (99.99%) and dense polycrystalline Y_2O_3 substrate made from Y_2O_3 powder by HP-HT synthesis were used in the high temperature tests. Directly before testing, the substrate (R_a~120 nm) was preheated in air at 1700°C for 2 h.

The wettability test was performed using the experimental setup described in details in [3,4]. Wetting behavior and reactivity between pure Gd and Y_2O_3 substrate were investigated by means of the sessile drop method combined with classical contact heating procedure [3,5]. The high temperature test was conducted under an inert gas (Ar, 99.9992%) at T_{exp}=1362°C, i.e. about 50°C above the melting point of Gd. During the test the contact angle values were recorded using a Microtron 1310 high resolution digital camera and calculated based on the drop image analysis using the ASTRA software [6,7].

Post-test microstructure examinations of the sample surface and cross-section were carried out using an optical microscope (OM) and a scanning electron microscope equipped with energy dispersive X-ray spectroscopy analyzer (SEM+EDS).

3. Results

Table 1 contains the contact angle values recorded during the wettability test of the Gd/Y_2O_3 system. Immediately after its melting, pure Gd wets Y_2O_3 substrate forming the contact angle of θ_0^*~52°. After 2 min heating of the system, the contact angle decreases to a value of θ_0=33°. A further heating to 1362°C does not affect the Gd/Y_2O_3 contact angle, showing the final value of θ_f^*= 33°.

A deep crater in the substrate under the drop (Fig. 1a) indicates, that the value of θ_f^* determined for the Gd/Y_2O_3 represents an apparent contact angle. The contact angle measured on the cross-section of the sample was equal to θ_f = 70°. This value could be erroneous due to the fact, that the contact angle was measured on a solidified sample after the wettability test. However, this does not change the fact that the contact angle of liquid gadolinium on Y_2O_3 is more than twice the value of the apparent contact angle.

The structural investigations performed on the cross-sectioned Gd/Y_2O_3 couple showed: 1) the presence of the precipitates of new phase in the initially pure Gd drop (Fig. 1a); 2) the formation of a deep crater under the drop (Fig. 1a) and 3) the occurrence of a liquid metal penetration zone into the substrate (LMP) due to good wetting between Y_2O_3 and liquid Gd (Fig. 1b).

Table 1. The contact angles values in Gd/Y_2O_3 system

T_{exp}, °C	Contact angle θ, °	
1362	θ_0^*	52
	θ_0	33
	$\theta_{f'}$	33
	θ_f	70

θ_0^* – contact angle immediately after melting
θ_0 – contact angle in t=0s at T_{exp}
$\theta_{f'}$ – final apparent contact angle calculated from the drop image at the test temperature
θ_f – final value measured on solidified and cross-sectioned sample

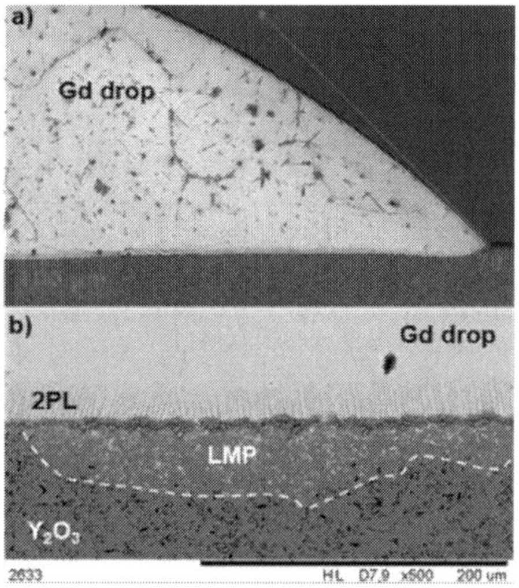

Fig. 1. Structure of solidified Gd/Y_2O_3 couple: a) OM image; b) SEM image

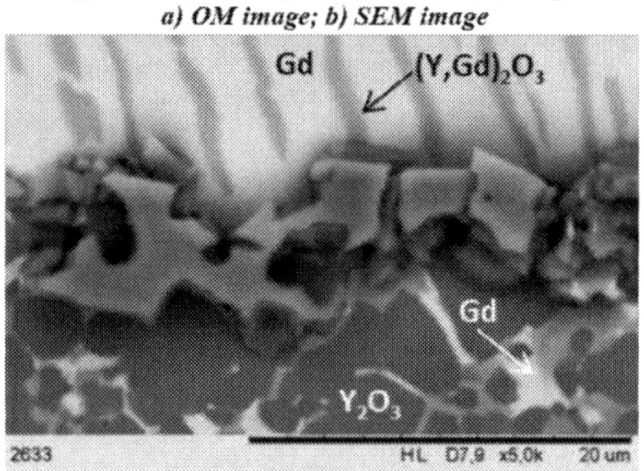

Fig. 2. SEM analysis of the Gd/Y_2O_3 interface

The EDS analysis showed the dissolution of the substrate in liquid Gd and confirmed the presence of new phases formed in Gd/Y_2O_3 system (Fig.2). At the drop/substrate interface oxide layer containing both Gd and Y was identified. From the analysis of Gd_2O_3-Y_2O_3 phase diagram it was suggested that it was formed by *dissolution-reprecipitation* mechanism. The needle-shaped precipitations, formed from the drop side, constituted the solid solution of $(Y,Gd)_2O_3$ surrounded by Gd.

Furthermore, the presence of O and Y in the initially pure Gd evidenced that the mechanism of interaction in Gd/Y_2O_3 system is dominated by a strong dissolution of Y_2O_3 in the liquid metal resulting in the formation of a deep crater in the substrate.

4. Summary

The results of the wettability test of the Gd/Y_2O_3 system performed under an inert atmosphere at a temperature of 1362°C evidence that liquid gadolinium shows good wetting of yttrium oxide substrate. The dissolutive wetting mechanism is responsible for a good wetting between liquid Gd and Y_2O_3 substrate. This is related to the specific situation, where the unlimited solubility occurs both between pure Gd-Y elements and their oxides.

Acknowledgments

This study was done in the frame of collaboration program between the German Academic Exchange Service DAAD and the Ministry of Science and Higher Education of Poland. Financial support from the National Science Centre of Poland (program HARMONIA, project No. 193624) and the German Academic Exchange Service DAAD (Project No. DAAD-56269397) is acknowledged. The authors acknowledge Dr. Rafał Nowak and M.Sc. Janina Radzikowska for technical assistance in structural characterization.

References

[1] I. Kaban, R. Nowak, O. Shuleshova, B. Korpala, G. Bruzda, A. Siewiorek, J.H. Han, N. Sobczak, N. Mattern & J. Eckert. (2012). Sessile drop study of Gd–Ti monotectic alloys on ceramic substrates: phase transformations, wetting, and reactivity. Journal of Materials Science. 47, 8381–8386.

[2] N. Sobczak, R. Nowak, A. Siewiorek, B. Korpala, G. Bruzda, I. Kaban, O. Shuleshova, J.H. Han, N. Mattern & J. Eckert. (2014). High temperature study of monotectic transformation, wetting and reactivity of liquid Gd-Ti alloys. Proceedings of the 71st World Foundry Congress, 19th-21st 2014 May. Bilbao, Spain.

[3] N. Sobczak, J. Sobczak, R. Asthana & R. Purgert. (2010). The mystery of molten metal. China Foundry. 7/4, 425-437.

[4] N. Sobczak, R. Nowak, W. Radziwill, J. Budzioch & A. Glenz. (2008). Experimental complex for investigations of high temperature capillarity phenomena. Materials Science Engineering. A495, 43-49.

[5] N. Sobczak, M. Singh & R. Asthana. (2005). High-temperature wettability measurements in metal/ceramic systems – Some methodological issues. Current Opinion in Solid State Materials Science. 9(4), 241–253.

[6] L. Liggieri & A. Passerone. (1989). An automatic technique for measuring the surface tension of liquid metals. High Temperature Technologies. 7(2), 82–86.

[7] ASTRA Reference Book. (2007). IENRI. Report.

The 73rd World Foundry Congress
„Creative Foundry"
23 - 27 September 2018, Krakow, Poland

WETTING BEHAVIOUR AND REACTIVITY OF MOLTEN HIGH CHROMIUM CAST IRON WITH ZrO_2-Al_2O_3 CERAMIC

Mihail Kolev [1]*, L. Drenchev [1], J.J. Sobczak [2], P. Turalska [2], A. Kudyba [2], G. Bruzda [2], N. Sobczak [2,3]

[1] Institute of Metal Science, Equipment and Technologies with Hydroaerodynamics Centre, Sofia 1574 67 Shipchenski prohod Str., Bulgaria
[2] Foundry Research Institute, 73 Zakopianska Str., 30-418 Krakow, Poland
[3] Institute of Precision Mechanics, 3 Duchnicka Str., 01-796 Warsaw, Poland
*Corresponding address: mihail1kolev@gmail.com

Keywords: MMC, high chromium cast iron, Al_2O_3-ZrO_2, sessile drop method, wetting kinetics, reactivity

1. Introduction

The composite materials are the basis of technological progress in contemporary society. The metal matrix composites (MMCs) mostly used as functional and structural materials in engineering and equipment design. Understanding high-temperature wetting behavior in liquid metal/solid ceramic systems is critical for improving industrial liquid phase-assisted processes and the quality of the final product [1,2].

Therefore, the investigation of the wettability phenomena at elevated temperatures constitutes one of the most important scientific aspects of high-temperature liquid-assisted materials processing stimulated by the demands of modern metallurgy and foundry industry.

Recently, it has been reported that utility properties of iron-based alloys can be improved by introduction of a ceramic phase (e.g. [3-5]).

This study is focused on high temperature interaction between molten cast iron and selected ceramic (Al_2O_3-ZrO_2) in order to clarify their wetting behavior and reactivity that affect the structure and properties of MMCs produced from cast iron and reinforced with Al_2O_3-ZrO_2 particles by casting processes (Fig. 1).

Fig. 1. Cross-section of high chromium cast iron/$(Al_2O_3+ZrO_2)_p$ composite

2. Experimental

The materials used in this study were:

1) high chromium cast iron containing (wt%): 27 Cr, 10 Mg, 2.95 C, Fe – bal.;

2) Al_2O_3-ZrO_2 substrate containing (wt%): 50 Al_2O_3, 34-40 ZrO_2, 13 SiO_2, and ~1%NaO_2.

The substrate was cut from the Al_2O_3-ZrO_2 ceramic produced by the Institute of Metal Science, Equipment and Technologies with Hydro-aerodynamics Centre. Wettability study was conducted by a sessile drop method combined with classical contact heating procedure using the experimental complex described in [6]. Testing conditions were as follows: flowing gas atmosphere (Ar, 850–900 hPa), 1450°C/15 min. During wettability test, the drop/substrate images were recorded with a high-speed high-resolution CCD camera. Solidified and cross-sectioned sessile drop couple was investigated by SEM+EDS analysis.

3. Results

The results of wettability test (Figs. 2,3) show that under testing conditions of this study, the selected cast iron alloy does not wet the Al_2O_3-ZrO_2 substrate (the contact angle Θ>90°). Immediately after melting cast iron sample, Θ~140° while the first contact angle formed at the test temperature (1450°C, time t=0s) is Θ_0~122° and in 15 min, it decreases up to a final value of Θ_f~118°, showing better wetting properties compared to conventional cast iron [7]. After the solidification, the drop was found detached from the substrate with together with small fragments of the substrate (Figs. 3a,c,h). Most probably, it took place during cooling due to the mismatch of CTEs of the alloy and the ceramic.

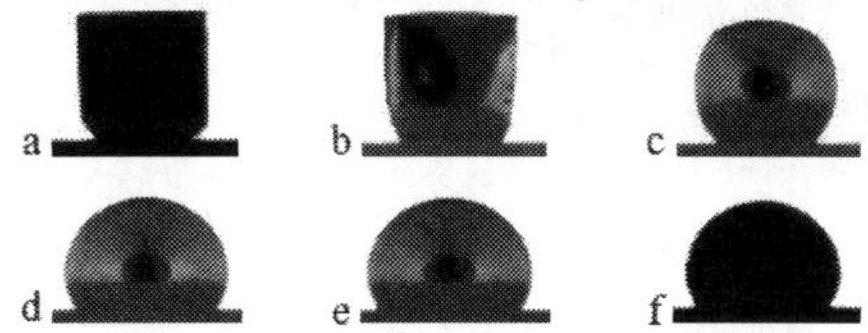

Fig. 2. Images of cast iron/Al_2O_3-ZrO_2 couple recorded during wettability test:
a) the beginning of heating, 21°C;
b) the beginning of melting of cast iron sample, 1282°C;
c) the formation of the drop, 1292°C;
d) the start of the test, 1450°C;
e) the end of the test, 1450°C (t = 15 min);
f) the end of the wettability test, 1122°C.

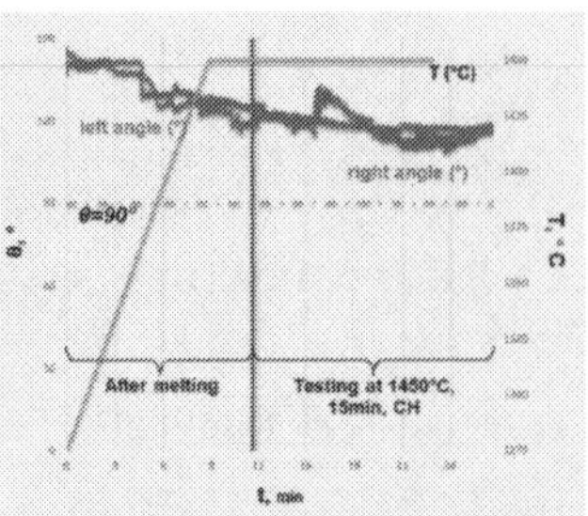

Fig. 3. Wettability kinetics of high chromium cast iron/(Al_2O_3-ZrO_2)$_P$ composite

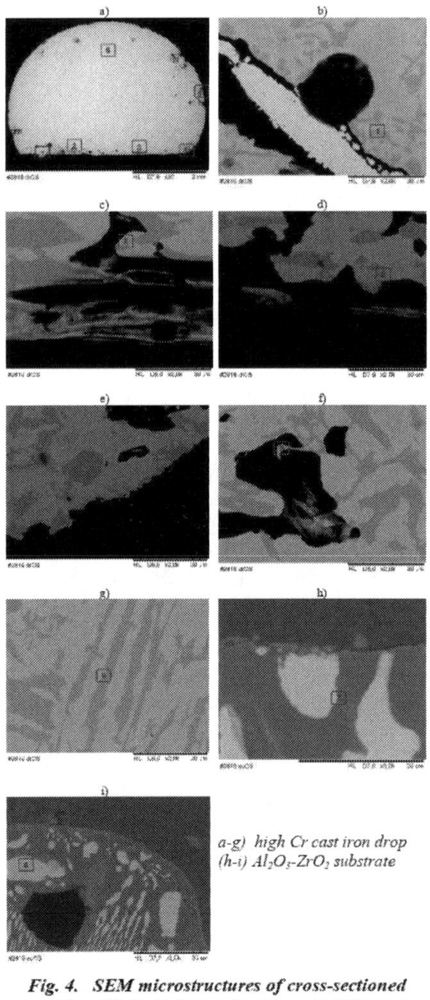

Fig. 4. SEM microstructures of cross-sectioned cast iron/Al_2O_3-ZrO_2 couple after wettability test (labelled points of EDS measurements)

Table 1 and Figure 4 summarize the most representative results of SEM structural characterization accompanied with local EDS analysis of chemical composition (at.%) made in different areas of the cross-sectioned sessile drop sample (in marked points of Fig. 3).

Table 1. Results of SEM+EDS analysis (at.%)

Point	Zr	C	O	Fe	Cr	Al	Si	Na	Mn
1	20.0	48.7	27.0	2.0	1.3	0.5	0.5	-	-
2	1.8	40.6	2.9	16.9	36.2	-	0.1	0.5	0.9
3	1.8	45.9	3.6	13.9	32.9	0.1	0.3	0.7	0.7
4	1.2	71.4	16.0	6.2	4.4	0.0	0.6	0.2	-
5	1.9	42.4	-	17.2	36.3	0.2	0.5	0.5	1.0
6	2.0	31.6	1.7	21.9	40.7	0.1	0.4	0.6	1.0
7	7.1	24.0	22.3	-	0.3	7.5	7.1	0.7	2.0
8	17.7	50.6	27.9	-	0.01	3.4	0.2	0.1	0.1

4. Summary

Under testing conditions of this study, high chromium cast iron alloy shows better wetting properties and higher reactivity with the Al_2O_3-ZrO_2 ceramic, compared to conventional gray cast iron. Further improvement of physico-chemical compatibility between cast iron matrix and ceramic reinforcement for successful production of MMC by casting techniques is recommended.

Acknowledgments

The authors are grateful for financial support of National Science Centre of Poland within OPUS program (UMO-2015/17/B/ST8/03391). This work was partially supported by the European Regional Development Fund within the OP "Science and Education for Smart Growth 2014 - 2020", Project CoE "National center for/of mechatronics and clean technologies" (No BG05M2OP001-1.001-0008-C01).

MK wishes to thank for research fellowship from European Virtual Institute on Knowledge-based Multifunctional Materials AISBL (KMM-VIN).

References

[1] Sobczak N., Singh M., Asthana R. (2005) High-temperature wettability measurements in metal/ceramic systems – Some methodological issues, Current Opinion in Solid State Materials Science 9 (4–5): 241–253.

[2] Sobczak N., Sobczak J., Asthana R., Purgert R. (2010) The mystery of molten metal, China Foundry 7 (4): 425–437.

[3] Hathaway R.M., Rohatgi P., Sobczak N., Sobczak J. (1998) Ferrous composites: a review. In: N. Eustathopoulos, N. Sobczak (Eds.). Proceedings of 2nd Int. Conf. High Temperature Capillarity, 1997, Cracow, Poland, Foundry Research Institute, pp. 267–276.

[4] Dulska N. Studnicki A., Szajnar J. 2017. Reinforcing cast iron with composite insert, *Archives of Metallurgy and Metals* 62 (1): 355–357.

[5] Kan W.H., C. Albino, D. Dias-da-Costa, et al. (2018) Microstructure characterisation and mechanical properties of a functionally-graded NbC/high chromium white cast iron composite, Materials Characterization 136: 196–205.

[6] Sobczak N., Nowak R., Radziwill W., Budzioch J., Glenz A. 2008. Experimental complex for investigations of high temperature capillarity phenomena. Materials Science Engineering A 495 (1–2): 43–49.

[7] Kolev M., Drenchev L., Sobczak J., Turalska P., Kudyba A., Bruzda G., Sobczak N. (2018) High temperature interaction between molten conventional gray cast iron and Al_2O_3-ZrO2 ceramic – this volume

The 73rd World Foundry Congress
„Creative Foundry"
23 - 27 September 2018, Krakow, Poland

THE EFFECT OF BORON CONTENT ON WETTING KINETICS IN Si-B alloy/h-BN SYSTEM

Wojciech Polkowski [1,*], Natalia Sobczak [1,2*], Grzegorz Bruzda[1], Rafał Nowak[1], Donatella Giuranno[1,3], Artur Kudyba[1], Adelajda Polkowska[1], Olga Shuleshova[4], Ivan Kaban[4]

[1] *Foundry Research Institute, Zakopiańska 73 Str., 30-418 Krakow, Poland*
[2] *Institute of Precision Mechanics, Duchnicka 3 Str., 01-796 Warsaw, Poland*
[3] *CNR - Institute of Condensed Matter Chemistry and Technologies for Energy, Via E. De Marini, 6 16149 Genova, Italy*
[4] *IFW Dresden, Institute for Complex Materials, Helmholtzstraße 20, 01-069 Dresden, Germany*

**Corresponding e-mail: grzegorz.bruzda@iod.krakow.pl*

Key words: AMADEUS Project, Si-B alloys, hexagonal boron nitride, wettability, interfaces

1. Introduction

Due to extraordinarily high melting points and latent heat values of silicon and boron, Si-B alloys have been assumed as excellent candidates to be used as phase changes materials (PCMs) for applications in latent heat thermal energy storage (LHTES) and conversion systems working at ultra-high temperatures (up to 2000°C). The main idea of every LHTES based systems is that the latent heat (i.e. that is absorbed during melting and released upon consecutive solidification) can be stored and converted into other energy form e.g. an electricity. The basic principle of AMADEUS Project [1], founded by European Commission within Horizon 2020 Programme, assumes that the utilization of latent heat of Si-B alloys will significantly overcome energy density limitations of *state-of-art* molten salt based LHTES systems. Nevertheless, in order to successfully accomplished the Project goals, some scientific and technological challenges need to be *a priori* faced and solved. Beside of a development of new ultra-high temperature devices for a heat-electricity conversion, a proper ceramic materials able to withstand a long-term heating, holding and cooling in contact with molten Si-B alloys, need to be selected. Thus, an implementation of these materials to real devices must be preceded by careful examinations of their high temperature interaction with refractories which can be used for building PCMs containers.

Firstly, we have assumed that an achievement of long life-times and high reliability of the PCM container requires its inertness toward contacting Si-B alloys under predicted operating conditions. In other words, a proper selection of refractories should be based on two main criteria: (i) an absence of wetting and (ii) a negligible reactivity in contact with Si-B melts.

However, due to pioneering nature of the research, there is a prominent lack of both theoretical and experimental data reported on interaction between molten Si-B alloys and ceramics at any temperature. On the other hand, high temperature capillarity phenomena (i.e. wetting, spreading and infiltration) have been so far widely examined for Si/ceramic systems in terms of a fabrication of photovoltaic grade silicon [2,3]. Although, all reported works deal with temperatures below 1500°C, it is concluded that at temperature near melting point of Si (T_m=1414°C) most of commonly applied in practice ceramics (e.g. oxides, carbides, borides or nitrides) are well wetted by molten Si [2,3]. This behavior originates from a high chemical affinity of Si to oxygen, carbon and nitrogen leading to an easy formation of reaction products at contacting interfaces. In fact, the only one ceramic showing non-wetting behavior with molten Si is hexagonal boron nitride (h-BN). Reported values of the equilibrium contact angle θ measured for Si/h-BN couples at T<1500°C are within the range of 95-145° [e.g. 4,5] (i.e. they correspond to the lack of wetting[1]). In our previous work [6], by using a sessile drop experiments we documented that the non-wetting behavior is maintained up to 1650°C, while at higher temperatures a non-wetting-to-wetting transition takes place. We have also evidenced that the involved reactivity mechanism under static argon atmosphere is mostly based on h-BN dissolution in molten Si (the dissolution rate depends on the temperature) followed by a reprecipitation of h-BN platelets during cooling. Additionally, the interaction was accompanied by a formation of gaseous product (most probably N_2), what was reflected by a lack of completely equilibrated conditions.

In this work, we used the sessile drop method experiments to examine the effect of boron content

[1] The contact angle θ is a metric describing wettability in a considered liquid/solid system. The θ<90° corresponds to a wetting behavior, while θ>90° to a lack of wetting.

on the high temperature interaction (wetting and spreading) during contact heating of Si-B alloys with hexagonal boron nitride (h-BN) substrates at temperatures up to 1750°C.

2. Materials and methods

The nominal compositions of Si-B alloys (wt%) were selected in accordance to Si-B binary phase diagram [7] as: hypoeutectic (Si-1B), eutectic (Si-3.2B) and hypereutectic (Si-5.7B). Additionally, ultra-highly pure Si (7N) was used as the reference material. The Si-B alloys were fabricated by the electric arc melting technique using properly weight mixtures of pure elements. Commercially available h-BN sinters were used as the substrates. The sessile drop experiments were performed with a dedicated experimental complex [8] working under argon atmosphere, in the temperature range between 1450°C and 1750°C (Fig. 1). More details on experimental procedure are shown elsewhere [6]. During the tests, the images of Si-B/h-BN couples were recorded at 100 fps by using a high speed CDD camera. After the tests, the acquired images were post-processed to compile movies and calculate wetting kinetics.

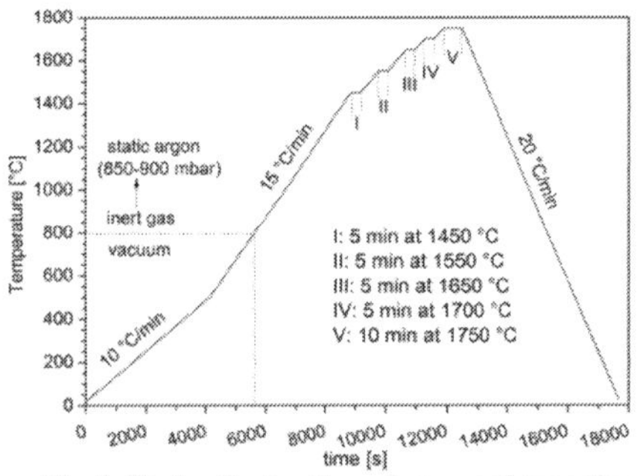

Fig. 1. The heating/cooling scheme used upon the wetting test of Si-B/h-BN couples (more detailed description is shown in [6]).

3. Results and conclusions

The wetting kinetics curves (showing a change of θ vs. time) calculated for pure silicon and Si-B alloys subjected to contact heating with h-BN substrates, are shown in Fig. 2. By comparing the results obtained for Si-B alloys to the behavior of pure silicon on the h-BN substrate (described in details in [6]), it might be concluded that the addition of boron decreases the wettability in the system. For all Si-B alloys, the contact angle values were very high (above 130°) in the whole examined temperature range. The highest value of 150° was received for the eutectic composition of Si-B alloy (Si-3.2B), what beside its lowest melting point and theoretically highest available latent heat [9] allows concluding that this material should be an optimal choice for application as PCM in the ultra-high temperature LHTES system.

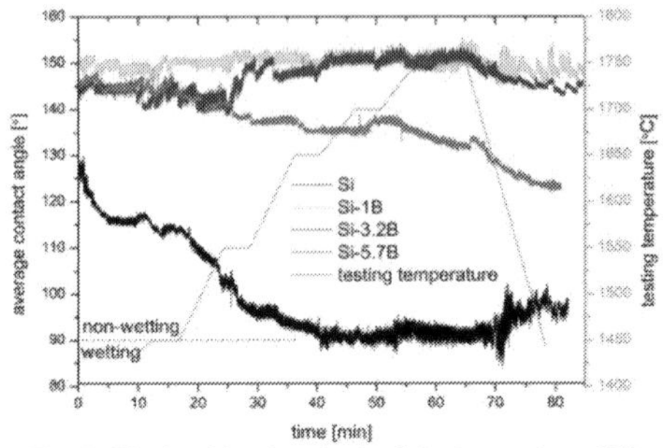

Fig. 2. Wetting kinetics obtained during testing of Si-B/h-BN couples.

Acknowledgments

The project AMADEUS has received funds from the European Union's Horizon2020 research and innovation program, FET-OPEN action, under grant agreement 737054. The sole responsibility for the content of this publication lies with the authors. It does not necessarily reflect the opinion of the European Union. Neither the REA nor the European Commission are responsible for any use that may be made of the information contained therein.

References

[1] Datas, A. Ramos, A. Marti, C. del Canizo, A. Luque (2016) Ultra high temperature latent heat energy storage and thermophotovoltaic energy conversion. Energy 107, 542-549

[2] B. Drevet, N. Eustathopoulos (2012) Wetting of ceramics by molten silicon and silicon alloys: a review. Journal of Materials Science 47, 8247–8260

[3] Z. Yuan, W.I. Huang, K. Mukai (2004) Wettability and reactivity of molten silicon with various substrates. Applied Physics 78, 617–622

[4] J.A. Champion, B.J. Keene, S. Allen (1973) Wetting of refractory materials by molten metallides, Journal of Materials Science 8, 423-426

[5] B. Drevet, R. Voytovych, R. Israel, N. Eustathopoulos (2009) Wetting and adhesion of Si on Si_3N_4 and BN substrates, Journal of European Ceramic Society 29, 2363–2367

[6] W. Polkowski, N. Sobczak, R. Nowak, et al. (2017). Wetting behavior and reactivity of molten silicon with h-BN substrate at ultrahigh temperatures up to 1750°C. Journal of Materials Engineering and Performance. https://doi.org/10.1007/s11665-017-3114-8

[7] R.W. Olesinski, G.J. Abbaschian (1984) The B-Si (Boron-Silicon) system. Bulletin of Alloys Phase Diagrams 5, 479-484

[8] N. Sobczak, R. Nowak, W. Radziwill, J. Budzioch, A. Glenz (2008) Experimental complex for investigations of high temperature capillarity phenomena. Materials Science and Engineering: A 495, 43–49

[9] A. Datas, A.B. Cristobal, C. del Cañizo et al. (2017) AMADEUS: Next Generation Materials and Solid State Devices for Ultra High Temperature Energy Storage and Conversion. SolarPaces 2017 Conference, Santiago, Chile, 2017. Article submitted to AIP Conference Proceedings

TECHNOLOGY

The 73rd World Foundry Congress
„Creative Foundry"
23 - 27 September 2018, Krakow, Poland

TECHNICAL CAPABILITIES FOR LOW-TEMPERATURE WASTE HEAT RECOVERY IN FOUNDRIES

Tanja Ludwig [1]*, Gotthard Wolf [2]

[12]*Technical University Freiberg, Foundry Department, Bernhard-von-Cotta-Str. 4, 09599 Freiberg, Germany*
**Corresponding address: e-mail: tanja.ludwig@gi.tu-freiberg.de*

Key words: waste heat recovery, thermal store, portable heat transfer, heat capacity, hot slag

1. Motivation

Energy-intensive branches like the foundry industry generate a large amount of waste heat. The utilization of this kind of energy depends on different impact factors, such as temporal availability or temperature level. The latter is of particular importance for the present research project, which is supported by the European Union (European Social Fund) and the Saxonian Government since the low-temperature waste heat is a major research feature. In practice, the re-use of waste heat is already widespread in the higher temperature area (>500°C), while little attention is paid to the utilization of waste heat sources in the low-temperature area (80-250°C). Nevertheless, low-temperature waste heat contains a potential, which should be defined with the help of the linking by engineer-scientific, and economic-scientific methods closer (**Fig. 1**).

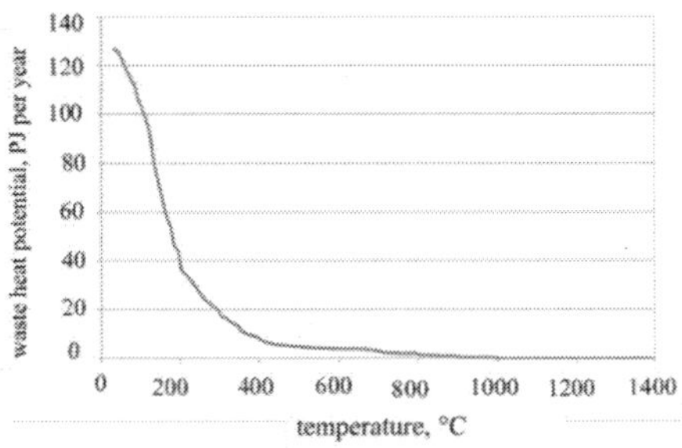

Fig. 1. Waste heat potential in selected industries in Germany in 2008 [1]

Fig. 1 shows a bottom-up analysis of the calculated waste heat amount from exhaust streams of analyzed industrial plants. At each temperature the cumulative waste heat amount that appears above this temperature, is shown. The reference temperature amounts to 35°C, at which the waste heat potential comes to 127 PJ/a. The biggest potential of waste heat is up to a temperature of 200°C [1]. For example, a city of approximately 800.000 habitants like Krakau, consumes about 6 PJ per year, assuming that 1 person consumes about 2000 kWh per year [2].

Waste heat recovery in foundries is very well implemented in most of the companies. Heat sources are well known and already used for typical parts as heating up the water for domestic use or supporting the internal heat system. However, it could be reused more if there were suitable heat consumer or if the thermal potential could be transported flexible to a place where the heat is actual used. Therefore there is much more waste heat than possibilities to use the whole amount, so the waste heat is discharged into the environment. Based on this difficulty the concept to build a portable and economical heat transfer is created in the present project.

2. Technology

Most of the waste heat technologies need pipelines and medium circulations inter alia for generation of electricity or for recuperative blast preheating. The key point of the present project is the construction of a portable heat system, which transfer useful heat from an everyday existing heat source to a suitable consumer, in a very easy way.

The principle is comparable with the common heater at home, where soapstone is often used as the storage material. Some chosen stone types and the specific heat capacity are listed in **Table 1.**

Table 1. Various specific heat capacities from different stone types [3, 4]

stone type	**specific heat capacity,** $\frac{J}{kgK}$
fireclay	840 (100°C)
granite	890 (20°C)
limestone	740 (20°C)
basalt	1200 (20°C)

stones

Stones as mentioned in **Table 1** are very suitable for thermal stores because of their high heat capacity. The construction of the thermal store, the selection of the right types and the loading time are currently being developed at the Foundry Department in Freiberg.

3. Calculation

Transferred heat energy is indicated by the following formula:

$$Q = c_p \cdot m \cdot \Delta T \quad (1)$$

$$\Delta T = T_{waste\ heat} - T_{environment} \quad (2)$$

In the above-mentioned equation

Q, kJ is the heat energy,

cp, $\frac{\text{kJ}}{\text{kg K}}$ is the specific heat capacity,

m, kg is the mass and

T, K is the temperature [1].

This formula calculates the transferred heat amount from the waste heat source to the consumer. The specific heat capacity defines the energy capacity of a solid or a fluid per mass. It depends on the material as well as temperature. In general, it is the ratio of the reversible supplied heat to the caused temperature increase [5].

In the present project, the transferred heat energy will be calculated once from the hot slag container to the lower stones in the thermal store and second from the stones to the environment. Therefore, the formula has to be expanded by additional parameters, which characterise the thermal radiation and the free convection. Heat transfer through radiation is given by the equation:

$$\dot{Q} = \varepsilon \cdot A \cdot \sigma \cdot (T_1^4 - T_2^4) \quad (3)$$

Whereby

$\dot{Q}$, W is the transferred heat flux through solid-state radiation between two surfaces,

ε is the emissivity,

A, m^2 is the area and

σ, $\frac{\text{W}}{\text{m}^2\text{K}^4}$ is the temperature of each solid-state [6].

After loading by thermal radiation, the thermodynamically process of free convection dominates in the heater and is given by the following equation:

$$\dot{Q} = \alpha \cdot A \cdot (T_{solid} - T_{environment}) \quad (4)$$

This equation contains beside the area and the delta temperature the heat transfer coefficient α, $\frac{\text{W}}{\text{m}^2\text{K}}$ [6]. It is an empiric description of the heat transfer over interfaces in a fluid phase or from this phase out [5]. At the end, the useful heat depends on:

$$Q = f(T, t, \alpha, \varepsilon, \sigma, c_p, m). \quad (5)$$

References

[1] Brückner, Sarah. (2016). Industrielle Abwärme in Deutschland – Bestimmung von gesichertem Aufkommen und technischer bzw. wirtschaftlicher Nutzbarkeit. Falkutät für Maschinenwesen der Technischen Universität München

[2] Statista (2014). Städte mit dem höchsten Pro-Kopf-Stromverbrauch in Deutschland im Jahr 2014. 15.12.2017, https://de.statista.com/statistik/daten/studie/217664/umfrage/staedte-mit-dem-hoechsten-stromverbrauch-in-deutschland/

[3] Riehl, I., Dög, A., Wulf, R., Mahler, A., Barth, G. (2011). Wärme- und Stoffübertragung. Prof. Dr.-Ing. habil. U. Groß. Institut für Wärmetechnik und Thermodynamik. Technische Universität Bergakademie Freiberg

[4] Al-Addous, Mohammad. (2006). Berechnen der Größe von Wasserspeichern zum saisonalen Speichern von Wärme bei der ausschließlichen Wärmeversorgung von Häusern mit Solarkollektoren. Fakultät für Maschinenbau, Verfahrens- und Energietechnik der Technischen Universität Bergakademie Freiberg

[5] Herwig, H. (2000). Wärmeübertragung A-Z, Systematische und ausführliche Erläuterungen wichtiger Größen und Konzepte.

[6] Walter, Prof. Dr.-Ing. habil. G. (2004). Arbeitsblätter Wärmetechnische Berechnungen. Professur für Gas- und Wärmetechnische Anlagen. Technische Universität Bergakademie Freiberg.

The 73rd World Foundry Congress
„Creative Foundry"
23 - 27 September 2018, Krakow, Poland

DESIGN OPTIONS TO INCREASE THE SERVICE LIFE OF CAST ACCESSORIES FOR THERMAL AND CHEMICAL TREATMENT FURNACES

Artur Bajwoluk [1*], **Paweł Gutowski** [1],

[1] *West Pomeranian University of Technology. Szczecin, Faculty of Mechanical Engineering and Mechatronics, Al. Piastów 17, Szczecin 70-310, Poland,*
**Corresponding address: e-mail: artur.bajwoluk@zut.edu.pl*

Key words: cast grate, heat treatment equipment, thermal stresses

1. Introduction

High temperature changing rapidly in cycles, strong carburizing atmosphere and heavy mechanical load contribute to the creation of an environment favoring the destruction of the accessories of carburizing furnaces, such as grates, baskets, etc. As a result of these extra harsh conditions, the mechanical properties of castings suffer degradation caused by changes in the structure of the heated and rapidly cooled material, additionally enhanced by the build up of a carburized layer as a result of carbon diffusion progressing towards the core of the casting. The consequence of these changes is the loss of the initially satisfactory elastic and plastic properties. The cast material becomes brittle with reduced strength and impact resistance. Rapid temperature changes are the additional source of large micro- and macroscopic stresses, leading to the formation and accumulation of deformations combined with the onset and development of cracks that eliminate castings from further use. [1-4]

Both deformation and development of cracks can be counteracted in many ways, e.g. by applying protective coatings or by selecting optimum design solutions [2-5]. In this article, the authors present quantitative results of the numerical analysis using FEM to illustrate the effect of changes in the design of cast grates on the value and distribution of stresses generated during cooling.

2. Research object

The examined grates are openwork, thin-walled castings usually made of creep-resistant high-nickel alloys (over 20% Ni), which ensures a stable austenitic structure within the whole range of operating temperatures.

Figure 1a shows a sample grate before use, while Figure 1b shows typical crack in the grate corner that will eliminate it from further use. The occurrence of such cracks is the effect of a variable field of temperature in which the grates are to operated. During heating or cooling the batch of heat-treated products, high temperature gradients are formed across and along the grate walls, in the area of the thermal nodes and in the transition area between the inner part of the grate resting directly under the heat-treated batch and the more remote outer part.

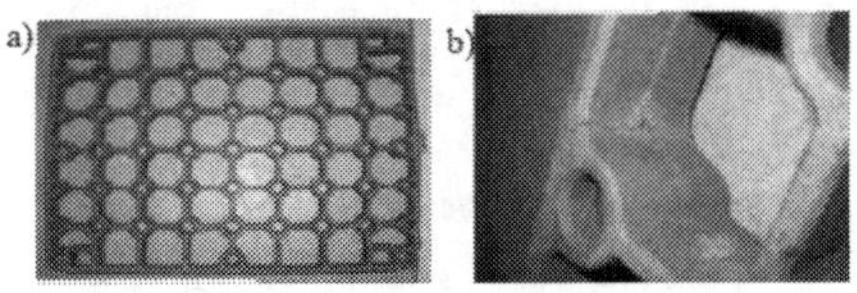

Fig. 1. Cast grate a) before operation, b) crack in the corner formed during operation

3. Wall connections and thermal nods

To reduce to minimum the level of thermal stresses, an important issue in the design of a grate is proper connecting of its ribs. Different geometries of the connections contribute to differences in the heat flow and transfer. The connections of the grate ribs are exposed to slower temperature changes than the connected walls with the resulting higher temperature gradient and, consequently, higher level of thermal stress. Fig. 2 presents typical designs of wall connections used in grates, i.e.: X, T and Y, and their modified counterparts, i.e. T_m and Y_m with technological recesses, the task of which is to reduce thermal nodes by optimizing the cross-sections of these connections.

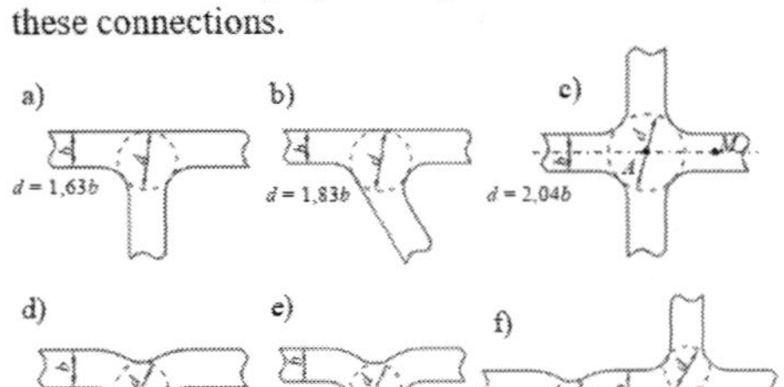

Fig. 2. Connections of grate walls: a) T, b) Y, c) X; and their modified versions: d) T_m, e) Y_m, f) double T_m

The effect on temperature gradient of the new design solution of the grate ribs connection is illustrated by respective graphs in Fig. 3, showing changes in the temperature difference during cooling between the connection center (point *A* Fig. 2c) and the wall center at a distance of 3*b* from *A* (point *M*

Fig. 2c). The graphs were plotted for the wall thickness b = 8 mm. It was assumed that the tested elements were cooled in oil from the temperature of T_o = 900°C to the temperature of T_1 = 40°C.

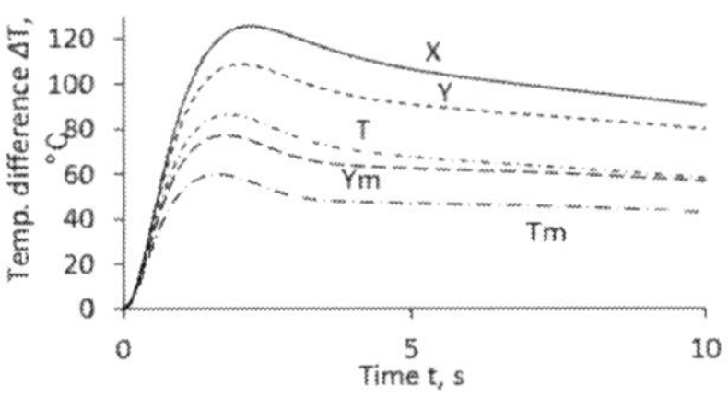

Fig. 3. Changes in the temperature differences between points A and M for various types of connections

It is easy to notice that owing to the use of technological recesses, the maximum temperature differences arising in the connections of the grate ribs during cooling have been significantly reduced.

4. Changes in flexibility of outer ribs

The most common location of cracks eliminating grates from further use is the area of the corners. Hence the need for numerical analysis to reduce to minimum the occurrence of thermal stresses in these areas. The paper presents the results of numerical calculations of minimization of these stresses by changing the type of connections and by reducing the stiffness of the external grate contour. Since grates have symmetric configurations, calculations have been limited to ¼ of their entire area. A quarter of the basic grate with X type connections is shown in Fig. 4a, whereas Fig. 4b shows its counterpart after first modification, where X connections have been replaced by T_m type connections. Fig. 4c and 4d illustrates further modifications introduced to the grates in which relievers (U-shape ribs) are used to reduce the stiffness of the contour.

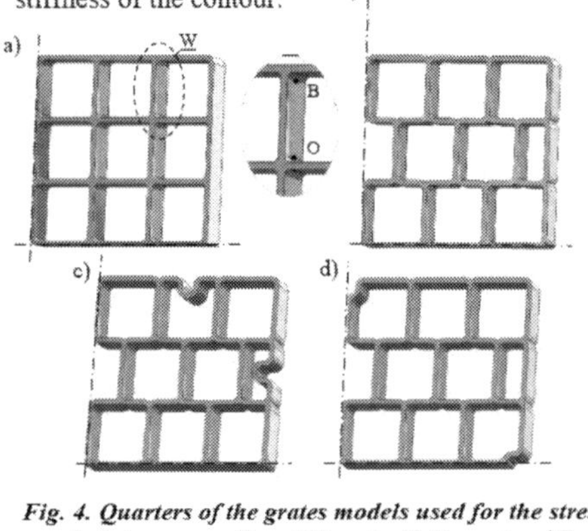

Fig. 4. Quarters of the grates models used for the stress analysis in corner ribs: a) X-grate, b) T_m-grate, c) T_{m1}-grate, d) T_{m2}-grate

To evaluate the impact of the proposed design solutions, Fig. 5 shows the results of stress distribution analysis along the *OB* ribs in the grate before and after the introduction of the changes. It was assumed in the calculations that the grates are symmetrically loaded with the charge of a cuboidal shape located parallel to the grate edge and reaching half of the *OB* ribs. During cooling, the charge resting on the grate is acting like a heat accumulator, slowing down the grate cooling process. From the graphs shown in Figure 5, it clearly follows that in the case of stresses originating from this source, the mere change in the type of internal connections obtained by shifting some ribs can not have any more important effect on the obtained results, while the introduction of relievers significantly reduces the level of stresses arising in the corner ribs during cooling.

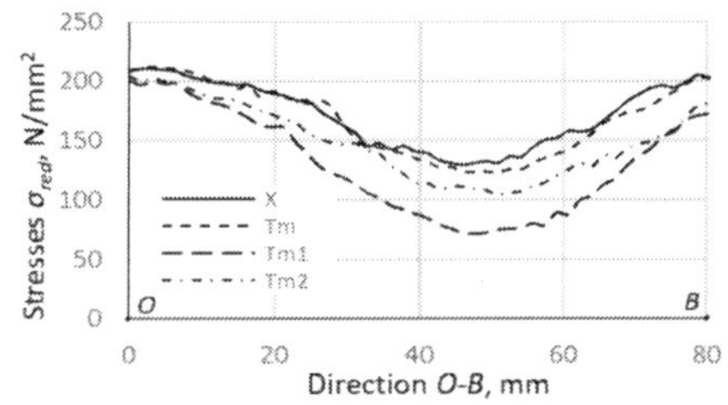

Fig. 5. Distribution of reduced stresses in OB rib of the tested grate design

5. Summary

The obtained results clearly indicate the positive impact of used design solutions introduced to the cast grate. The first modification, i.e. re-designing of the connection, markedly reduces thermal nodes formed during cooling, while relievers introduced in the outer contour of the grate significantly reduce stresses due to the thermal effect of charge and structure rigidity.

These results also justify the need for further optimization of calculations using modern computational techniques (e.g. FEM) to obtain quantitative data which, depending on the operating conditions, can assist the selection of optimal design solution to maximize the service life of the used equipment.

References

[1] Davis, J.R. (Ed.). (1997). Industrial Applications of Heat-Resistant Materials. In Heat Resistant Materials (67-85). ASM International.

[2] Lai, G.Y. (2007). High-Temperature Corrosion and Materials Applications. ASM International.

[3] Piekarski B. (2012). Creep-resistant castings used in heat treatment furnaces. Szczecin: Wydawnictwo Uczelniane Zachodniopomorskiego Uniwersytetu Technologicznego, (in Polish).

[4] Bajwoluk A., Gutowski P. (2015). The effect of pallet component geometry on temperature gradient during cooling. Archive of Foundry Engineering. 15(1), 5-8. DOI: 10.1515/afe-2015-0001.

[5] Piekarski B. (2006). Anticarburising coatings applied on parts of carburising furnaces. Archives of Foundry. 6(18), 473-478 (in Polish).

The 73rd World Foundry Congress
„Creative Foundry"
23 - 27 September 2018, Krakow, Poland

SUPPRESSING THE BASKET DEFORMATION PROCESS DURING HEAT TREATMENT

Artur Bajwoluk[1]*, Andrzej Drotlew[1], Paweł Gutowski[1], Bogdan Piekarski[1]

1West Pomeranian University of Technology. Szczecin, Faculty of Mechanical Engineering and Mechatronics, Al. Piastów 17, Szczecin 70-310, Poland,
**Corresponding address: e-mail: artur.bajwoluk@zut.edu.pl*

Key words: creep-resistant castings for heat treatment plants, baskets, thermal stresses

1. Introduction

Baskets forming part of furnace equipment are used for the heat treatment of small-size heavy components, the bulk treatment of which in facilities of other types would be practically impossible. This equipment is also used for the heat treatment of low-demand parts, when the way they are laid in the basket does not affect final outcome of the conducted process, e.g. during stress relief annealing. The use of baskets raises furnace efficiency and improves loading and unloading of the batch [1-3].

As a result of operation under the demanding conditions of a heat treatment process, the walls of the baskets undergo deformation with the imminent risk of crack formation. This article describes various methods applied in order to minimize in practice the deformation of baskets.

2. Research object

The tested basket (Fig. 1) consists of several cast components, including base and side walls. The interlocking elements are assembled using fixed connections (welds) for the side walls and the base, and separable connections for the side walls.

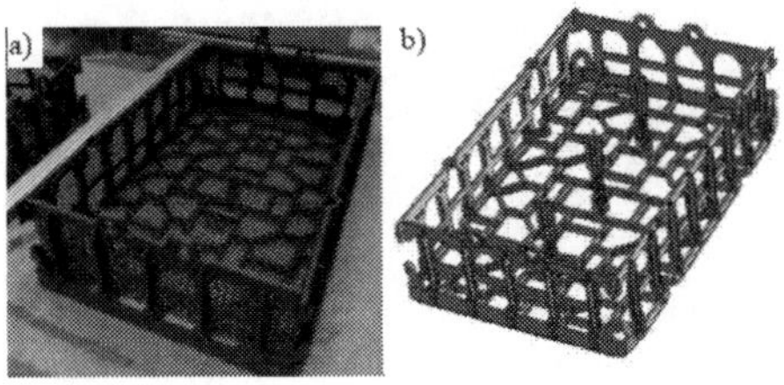

Fig. 1. Basket of 870 × 555 × 200 mm dimensions assembled from cast components. The cast material is GX40NiCrSiNb38-19: a) real object, b) 3D model

Due to its openwork construction, the basket has a relatively small weight. Baskets are designed to be stacked on top of each other with the help of transport handles located in the middle of the shorter sides. A grid and posts can be placed inside the basket (Fig. 1) to prevent deformation of its bottom when one basket is placed on top of the other and loaded with a heavy charge [3].

In the tested basket, after about one month of operation, a visible deformation of its side walls appeared (Fig. 2) which caused that the basket had to be withdrawn from operation. The walls were distorted losing their original shape, and there was a clear inclination of their centre towards the inside of the basket.

Fig. 2. Deformation of the basket side walls

As claimed by the user, so large deformation of the side walls occurred too quickly. Therefore the basket producer was requested to introduce such changes to the original design of the basket that would reduce its deformation and, at the same time, make its service life longer than 6 months.

In order to increase the durability of the basket by minimizing the deformation arrising during its operation, numerical strength analysis of the design was performed. The analysis was based on a FEM method.

3. Deformation analysis

The first step in the calculations was to conduct a heat flow analysis using a three-dimensional model of the basket structure subjected to a cooling cycle. The obtained temperature distribution in the basket walls enabled to perform the analysis of the stress and strain distribution in the basket.

Fig. 3 shows a top view of the basket model that reveals the deformation of its side walls as a result of cooling. It is obvious that the calculated deformation corresponds to the deformation that appears in the real basket (Fig. 2).

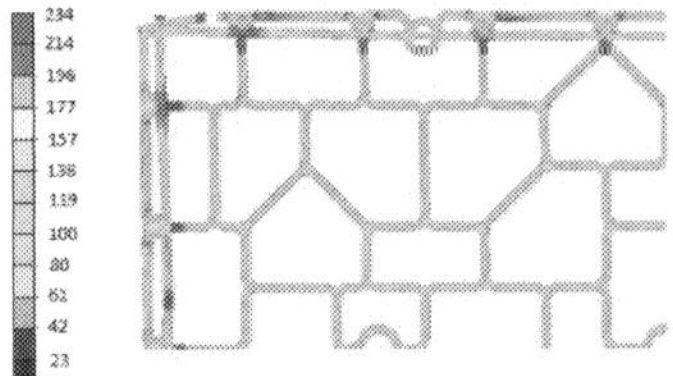

Fig. 3. The results of analysis of stress and deformation generated in the basket during cooling

As in the real object, the center of the side walls is inclined towards the inside of the basket. The compatibility of model deformation with the real deformation confirms that the adopted model and its loading pattern are correct. Hence the conclusion follows that this model is fully applicable in further studies of the behavior of the re-designed basket structure in terms of the deformation of its side walls.

On base of the obtained results it may be assumed that the deformation of the side walls is caused by too high rigidity of their connection with the basket base. Due to the presence of thermal nodes (Fig. 4), these areas are particularly susceptible to stress and strain concentration.

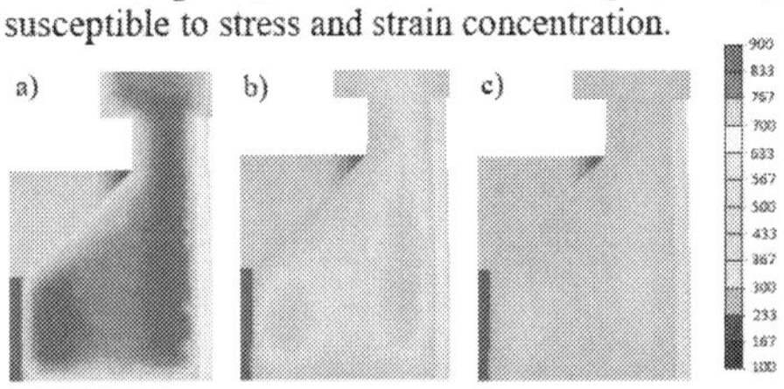

Fig. 4. Change of temperature distribution in connection formed between the side wall pillar and basket base after: a) 1 s, b) 3 s, c) 5 s of cooling time

Moreover, in these areas, concentrate deformations due to the presence of ribs of different thicknesses and non-uniform cooling of the basket base under the effect of heat accumulated in the charge. The impact of these factors makes pillars in the initial stages of the cooling process at the places of their connection with the base bend to the outside of the basket first (Fig. 5a) and towards the inside of the basket next (Fig. 5b). During subsequent heat treatment cycles, the deformations tend to accumulate, and after a certain time, the overall deformation is too large and eliminates the basket from further use.

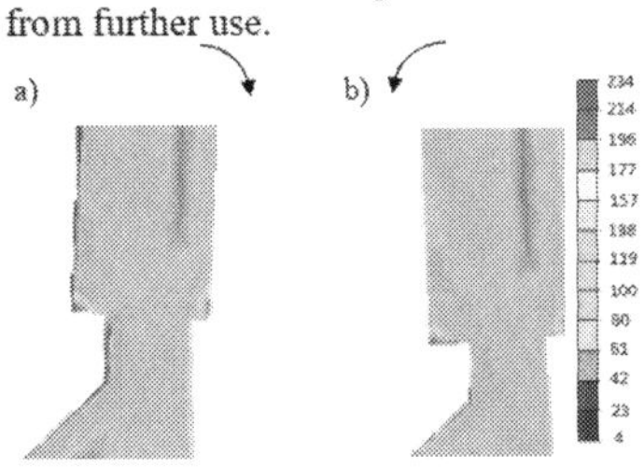

Fig. 5. Deformations of pillars of the basket side walls during cooling: a) stage I, b) stage II

4. Modification of the design

Based on the results of the numerical analysis of the stress and strain distribution in the model of basket undergoing the process of cooling, its design was modified by making undercuts in vertical pillars of longer side walls (Fig. 6).

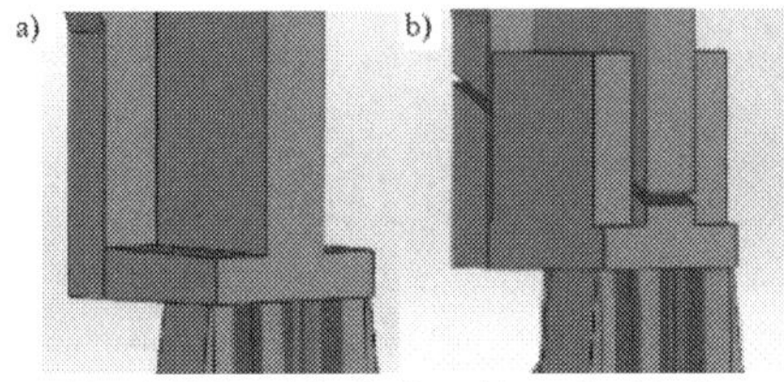

Fig. 6. Pillar of the basket side wall: a) before redesigning, b) after redesigning

This modification decreased the rigidity of connection between the side walls and base. Conducted calculations showed that such change results in significant reduction of the deformation of the side walls observed in the original construction. The modified baskets were put into operation. During the exploitation it has been noted, that the deformations of their side walls were much lower according to basic design.

5. Summary

The paper presents a study of the real problem that concern too large and too quickly occurring deformations in the side walls of the basket operating in a furnace for heat treatment.

Based on the results of the conducted simulation analysis, it has been found that the source of the deformation is too rigid connection of the side walls of the basket with its base. To cope with this problem, undercuts were made in the side wall pillars, and then the geometrical stability of the basket was checked under production conditions. Asked for his opinion, the user has confirmed that the redesign significantly reduced the deformation rate of the side walls during operation and improved the reliability and durability of the basket, which means that it has fulfilled its task correctly.

References

[1] Davis, J.R. (Ed.). (1997). Industrial Applications of Heat-Resistant Materials. In Heat Resistant Materials (67-85). ASM International.

[2] Dossett J.L., Boyer H.E. (2006). Practical Heat Treating. ASM International, 2nd ed.

[3] Drotlew A., Piekarski B., Słowik J. (2017). The design of cast technological equipment for heat treatment furnaces. Archive of Foundry Engineering. 17(3), 31-36. DOI: 10.1515/afe-2017-0086.

The 73rd World Foundry Congress
„Creative Foundry"
23 - 27 September 2018, Krakow, Poland

A NOVEL MEASUREMENT METHOD TO STUDY THE THEMRAL ASPECTS OF MOULDING MIXTURE DECOMPOSITION

József Tamás Svidró [1]*, Attila Diószegi [1],

[1]Jönköping University School of Engineering, Department of Materials and Manufacturing, Gjuterigatan 5 Box 1026, SE-55111 Jönköping, Sweden
**Corresponding e-mail: jozsef.svidro@ju.se*

Key words: cold-box, greensand, moulding mixture, thermal analysis

A wide variety of moulding mixtures are extensively used in the process of sand casting of metal components. The sector is continuously developing in production volume, moreover the expectations from the customers are increasing on monthly basis, also from the quality point of view. Even though mould and core manufacturing are well-organized processes in most foundries, technological problems still appear which may lead to casting defects. These trends are forcing metal casters to come up with fitting strategies to solve their daily production challenges, while their suppliers are expected to keep up the continuous development of their existing foundry products and to find innovative solutions. Research on moulding materials and their properties must also take the step forward accordingly to generate the necessary new knowledge.
This paper summarizes the latest results of a novel measurement method suitable for studying the degradation characteristics of different moulding materials from a new perspective. The fundamentals of the method are based on thermal analysis, focusing on the heat absorption behaviour of sand mixtures regardless of binder type/amount or manufacturing process.

1. Introduction

Thermal analysis in the form of recording and interpretation of temperature variation in time of a cooled or heated material is widely used today in foundry technology [1]. The broad areas of application mainly cover works based on cooling curve analysis of both non-ferrous and ferrous alloys [2, 3]. Recently, studies applying traditional methods such as TG-DSC-DTG were conducted to examine materials applicable as bonding agents in foundry purpose mould and coremaking technologies [4]. Other sand related works deal with the measurement of thermophysical properties in greensand moulds applying real foundry conditions [5, 6].
The new measurement method is based on Fourier thermal analysis (FTA), used in an inverse way to interpret heat distribution versus time curves recorded in sand samples. The aim was to obtain new understanding of the thermal aspects of degradation processes in various moulding mixtures. Thus, heat absorbtion behavior by the mixture decomposition were measured/calculated.

2. Experimental setup and materials

The foundation of the experimental setup is the application of spherical sand samples of 40, 50 and 60mm in diameter. Two N-type thermocouples were placed in the specimens; in the geometrical centre and in a lateral measuring point near the sample wall (Figure 1). Quartz glass pipes of 1 mm wall thickness and various dimensions were used for thermocouple protection. The size variations of the specimens and their immersion into liquid metal (as a heat source) during the test ensured real foundry conditions.

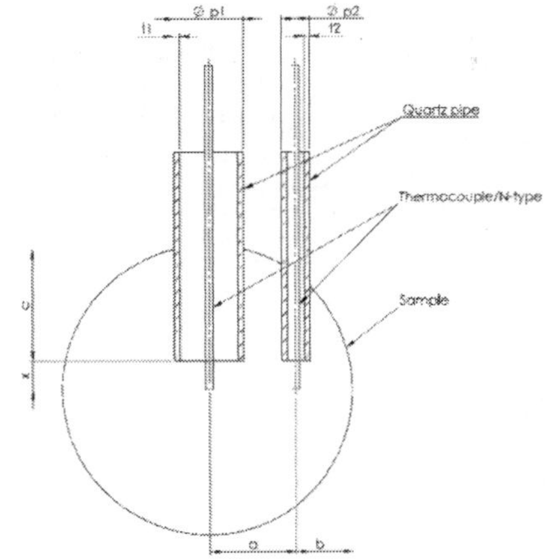

Fig. 1. Sample geometry

The moulding mixtures investigated were greensand, epoxy/SO_2 and polyurethane cold-box (PUCB) chemically bonded systems. They were prepared with the identical type of foundry purpose quartz sand, according to the compositions described in Table 1. Composition variables were set to be similar to the typical industrial application of each mixture. Specimens were not coated before the measurement. The epoxy/SO_2 and the PUCB mixtures were tested without drying, however, the greensand samples were dried for 2 hours at 105°C in order to increase their integrity for the thermocouple placement procedure. The samples were then immersed into liquid cast iron of 1300±10°C and the temperature distribution in both measuring points was recorded.

Table 1. Production parameters of the studied mixtures

Mixture	Composition/ wt% (by the weight of sand)
Greensand	8% bentonite, 3.3% water, 3% coal dust
Epoxy/SO_2	1% epoxy resin + 30%* hydroperoxide *by the weight of the resin (cured by SO_2)
PUCB	0.9% phenolic resin + 0.9% isocyanate (cured by amine vapor)

3. Results and discussion

Based on the initial measurements of temperature distribution inside the samples, Fourier thermal analysis was run and the total absorbed heat values (L) were first calculated for the 50mm specimens, shown on Figure 2. These values correspond to the amount of heat taken away from the melt by the different decomposition processes of the bonding agents and the possible phase transitions in the sand. Greensand absorbed significantly higher amount of heat compared to the lower values of epoxy/SO_2 and PUCB mixtures. The heat absorption behaviour was further investigated by determining its temperature dependency by plotting the rate of heat absorption versus the temperature inside the samples. Figure 2 shows the various heat absorption/degradation profiles of the studied mixtures, including key features such as the strong initial heat absorption of the greensand at lower sand temperatures, or the multi-staged decomposition of the PUCB sample exhibiting multiple maximum peaks up to 450°C.

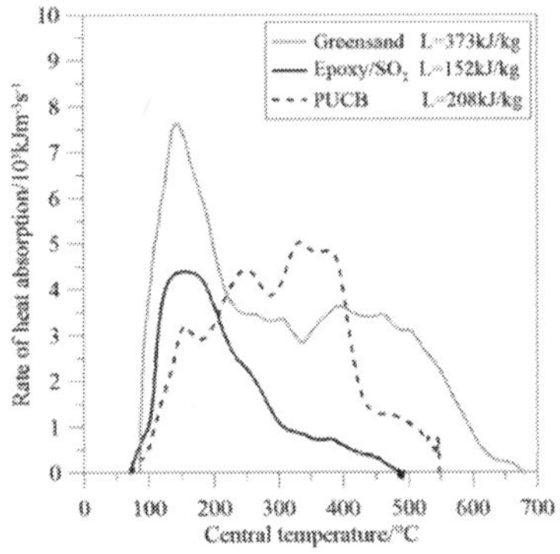

Fig. 2. Heat absorption/degradation profiles of the studied mixtures versus temperatures in the centers of the samples, 50mm diameter [7-9]

4. Conclusion

In this work, conventional moulding mixtures were studied by a new method based on Fourier thermal analysis. Results showed that greensand, epoxy/SO_2 and PUCB mixtures absorbed various amounts of heat from the liquid metal and have diverse degradation profiles. Heat absorption behaviour can be directly interpreted as the cooling ability of the mixtures. Maximums of heat absorption at specific sand temperatures is expected to eventuate in improved cooling ability of the cores, which is expected to shorten the solidification time in the casting, affecting its final microstructural morphology and may result in large variations in mechanical properties. Future work is therefore planned to further study and prove the relationship between the heat absorption capacity of sand mixtures and the mechanical properties of castings.

Acknowledgments

The present work was financed by the Swedish Knowledge Foundation. Cooperating parties in the project are Jönköping University, Scania CV AB and Volvo Powertrain Production Gjuteriet AB.

References

[1] D. M. Stefanescu. (2015). Thermal Analysis - Theory and applications in metal casting. International Journal of Metalcasting. 9 (1), 7-22.

[2] M. Krupinski, B. Krupinska, Z. Rdzawski, K. Labisz, T. Tanski. (2015). Additives and thermal treatment influence on nonferrous alloys. Journal of Thermal Analysis and Calorimetry. 120 (3), 1573-1583.

[3] A. Diószegi, V.L. Diaconu, V. Fourlakidis (2016). Prediction of volume fraction of primary austenite at solidification of lamellar graphite cast iron using thermal analyses. Journal of Thermal Analysis and Calorimetry. 124 (1), 215-225.

[4] B. Grabowska, P. Malinowski, M. Szucki, L. Byczynski (2016). Thermal analysis in foundry technology. Journal of Thermal Analysis and Calorimetry. 126 (1), 245-250.

[5] P.K. Krajewski, G. Piwowarski (2015). Range of thermal conductivity changes of wet green foundry sand during casting solidification. Archives of Metallurgy and Materials. 60 (3B), 2391-2395.

[6] P.K. Krajewski, G. Piwowarski, P.L. Zak, W.K. Krajewski (2014). Experiment and numerical modelling the time of plate-shape casting solidification vs. thermal conductivity of mould material. Archives of Metallurgy and Materials. 59 (4), 1405-1408.

[7] K. Lidumnieks, J.T. Svidró, A. Diószegi (2017). The effect of various production parameters on the heat absorbing capacity of greensand. International Journal of Cast Metals Research. 1-7.

[8] J.T. Svidró, A. Diószegi, J. Svidró, T. Ferenczi (2017). Thermophysical aspects of reclaimed sand addition to the epoxy-SO_2 coremaking system studied by Fourier thermal analysis. Journal of Thermal Analysis and Calorimetry. 130 (3), 1779-1789.

[9] J.T. Svidró, A. Diószegi, J. Svidró, T. Ferenczi (2017). The effect of different binder levels on the heat absorption capacity of moulding mixtures made by the phenolic urethane cold-box process. Journal of Thermal Analysis and Calorimetry. 130 (3), 1769-1777.

The 73rd World Foundry Congress
„Creative Foundry"
23 - 27 September 2018, Krakow, Poland

DEVELOPMENT OF CASTING GENOME FOR AUTOMOTIVE PISTON CASTING: A STUDY

Rupshree Ozah[1]*, Muralidhar Manapuram[2]

[1,2]North Eastern Regional Institute of Science and Technology,Itanagar,Arunachal Pradesh-791109,India,
**Corresponding author (PhD scholar: e-mail: rupshreeozah91@gmail.com*

Key words: Piston casting, casting chromosome, casting codons, casting genes, casting genome

1. Introduction

The total of casting facilities in the world is about 104,129,257 metric ton in 50th census of World Casting Production includes 37 nations from four continents [1]. India's foundry industry is the second biggest globally in terms of production capacity (10,770,000 metric ton) after China (45,600,000 metric ton). The various types of castings which are produced are ferrous, non-ferrous, aluminum alloy, gray cast iron, ductile iron, and steel etc. for application in automobiles, railways, pumps compressors & valves, diesel engines, cement/ electrical/ textile machinery, aero & sanitary pipes & fittings etc & castings for special applications. However, grey iron castings are the major share approx 40 % of total castings produced.

An automobile is a self-propelled vehicle driven by an internal combustion engine and is used for transportation of passenger and goods on ground. It is a combination of many parts and piston plays a major role as a part of an engine. Piston is the key part of the engine as it works under high temperature, high pressure, corrosive and wearing conditions while running with high speed. It is fitted to each combustion cylinder as a face to receive gas pressure and transmit the thrust to the connecting rod of an IC engine.

Casting is the first step in making any product as it is the oldest manufacturing process. Castings are engineered metal components used for mass production formed by designing a mould and introducing materials in a molten state into mould cavity then allow them to solidify. The reciprocating motion of piston in order to satisfy all these severe conditions, piston materials must have high strength, high toughness and light weight. Combustion engine will generate severe stress on the piston crown, sidewall, and the piston's top rings.

2. Casting genome

A genome is an organism's complete set of DNA, including all of its genes. Each genome contains all of the information needed to build and maintain that organism. In humans, a copy of the entire genome has more than 3 billion DNA base pairs which contained in all cells that have a nucleus. After investigating by many scientists and engineers' human genome project, genetic application has been expanded to various engineering areas of manufacturing in general and foundry in particular. The development of one billion codons with one-lakh different genes with twenty three pairs of chromosomes has results into the capability to produce future children with high physical and intellectual abilities.

Production of defect free casting is a primary need of foundry industry. Defects of casting can vary time to time, shift to shift and so on. In order to produce sound-free casting, an attempt has been made to develop a casting genome map with different chromosomes, genes and codons for the manufacture of good quality of castings.

To improve toughness, wear resistance of metal matrix composites relative to aluminum alloys can significantly improve the strength and stiffness of the composites on one hand, but it drastically decreases the toughness and ductility on the other hand [2]. In this paper, a casting genome map has been developed considering four chromosomes, twelve genes and several codons for production of quality casting [Table 1].

3. Present work

Casting genome is classified into casting chromosomes, casting genes, casting codons and casting arrays for specific foundry applications [3]. Casting genome map is presented in figure consist of general structure of global foundry industry. Castings arrays have been developed from the selected casting codons [Table 2].

Al-Si Automotive piston casting codon Array = {A1 A11 A12 A13 A14 A15 A16 A17 A18 A19 A110 A111 A112 A113 A114 A115 B1 B2 B3 C1 C2 C3 C4 C5 D1 D2 D3 D4 D5 D6 E1 E2 E3 F1 F2 F3 F4 G1 G2 G3 G4 G5 G6 G7 H1 H11 H12 I1 I2 J1 J2 K1 K2 L1 L2}.

4. Conclusion

The main use of casting genome is to improve the productivity and yield of casting by reducing number of defective castings in foundry. It is expected that this work of generating a casting genome map, identification of casting condons and casting array development will help for preservation of tactic knowledge and may lead to produce casting of required quality with best productive path.

References

[1] (2016) Global Casting Production Stagnant, 50th census of world casting production, internation jpurnal of modern casting, 25-29.

[2] Hashem F. El-Labban, M. Abdelaziz and Essam R.I. Mahmoud. (2016). Preparation and characterization of squeeze cast-Al–Si piston alloy reinforced by Ni and nano-Al2O3 particles, Journal of King Saud University – Engineering Sciences, (28), 230–239.

[3] Muralidhar Manapuram. (2007). Studies on status of global foundry industry and development of a framework for a holonic casting manufacturing, PhD thesis, IIT kharagpur.

Table 1: Casting genome map foundry industry

Casting chromosomes	Casting genes			Casting codons	Casting arrays
Material chromosomes W	Alloy Composition A	Purity genes B	Properties genes, C	$A_1, A_2, A_3, \dots A_n$ $B_1, B_2, B_3, \dots B_n$ $C_1, C_2, C_3, \dots C_n$	$A_1, B_1, C_1, D_1, E_1, F_1, G_1, H_1, I_1, J_1, K_1, L_1$
Application chromosomes X	Dimensions Genes D	Internal Soundness genes E	Surface condition genes, F	$D_1, D_2, D_3, \dots D_n$ $E_1, E_2, E_3, \dots E_n$ $F_1, F_2, F_3, \dots F_n$	$A_2, B_2, C_2, D_2, E_2, F_2, G_2, H_2, I_2, J_2, K_2, L_2$
Process chromosomes Y	Moulding type & preparation Genes, G	Core type & preparation genes H	Metal filling genes, I	$G_1, G_2, G_3, \dots G_n$ $H_1, H_2, H_3, \dots H_n$ $I_1, I_2, I_3, \dots I_n$	$A_3, B_3, C_3, D_3, E_3, F_3, G_3, H_3, I_3, J_3, K_3, L_3$
Geometry chromosomes Z	Weight & size of the casting Genes J	Thickness (minimum, maximum, average) of casting genes, K	Holes and Cavities genes, L	$J_1, J_2, J_3, \dots J_n$ $K_1, K_2, K_3, \dots K_n$ $L_1, L_2, L_3, \dots L_n$	$A_4, B_4, C_4, D_4, E_4, F_4, G_4, H_4, I_4, J_4, K_4, L_4$

Table 2: Casting genome of a piston (micro level)

Casting Chromosome	Casting Genes	Casting Codons
Material chromosomes, W	Alloy composition, A	A1 = Aluminium alloy, A11 = Si 11.56%, A12 = Fe 0.53%, A13 = Cu 1.03%, A14 = Mg 0.9%, A15 = Mn 0.18%, A16 = Ni 1.55%, A17 = Ti 0.06%, A18 = Cr 0.04%, A19 = Pb 0.04%, A110 = Al 84.11%, A111 = Density of aluminium silicon alloy at ambient temperature in kg/m^3:2700, A112 = Thermal conductivity of aluminium at ambient temperature in J/mKs:125, A113 = Specific heat of aluminium alloy at ambient temperature in J/mKs:910, A114 = Coefficient of linear expansion of aluminium in µm/mK:22.5, A115 = Shrinkage allowance of aluminium : 1.3%
	Purity genes, B	B1 = Composition: Si ± 0.05, B2 = Pouring temperature 720 ± 20 °C, B3 = Pouring height in mm:50
	Properties genes, C	C1 = Ultimate tensile strength in MPa:360, C2 = Elongation in (0.5-1)%, C3 = Modulus of elasticity in kN/mm^2:73, C4 = Brinell Hardness:100-150 , C5 = Engineering stress in N/mm^2:200.
Application chromosomes, X	Dimensions genes, D	D1 = Height in mm:80, D2 = Diameter in mm:80, D3 = Crown side thickness in mm:5, D4 = Skirt side thickness in mm:3, D5 = Shrinkage allowance of aluminium: 1.3%, D6 = Accuracy ± 0.05mm
	Internal soundness genes, E	E1 = Microstructure fine dendrite grain structure, E2 = Voxels size in μm^3:0.3, E3 = Porosity: less
	Surface condition Genes, F	F1 = RMS value:6µm, F2 = Volume:145146(±0.05%) mm^3, F3 = Surface area:±25132 mm^2, F4 = Volume/surface area: 4.2 (±.05%) mm
Process chromosomes, Y	Moulding type & preparation genes, G	G1 = Die temperature:288^{0}C, G2 = Clamping force in kN:900, G3 = Die cycle time in s:25, G4 = Shot size in mm^3: 193044,G5 = Clamping force in kN: 900, G6 = Platent size in mm:480× 640, G7 = Cavity pressure in MN/m^2: 48
	Core type & preparation Genes, H	H1 = Preheated cavity:200^{0}C, H11 = Diameter in mm: 79.5, H12 = Height in mm:100
	Metal filling genes, I	I1 = Metal pouring temperature 680-700^{0}C, I2 = Pouring time:2.1s
Geometry chromosomes, Z	Weight & size of the casting genes, J	J1 = Casting weight in g: 392, J2 = Pouring weight in g:521
	Thickness (minimum, maximum, average) of casting genes, K	K1 = Crown thickness in mm:5, K2 = Skirt thickness in mm: 3
	Holes and cavities genes, L	L1 = Piston inside dia in mm:74±0.1, L2 = Piston gudgon pin hole dia in mm :18 ±0.05

The 73rd World Foundry Congress
„Creative Foundry"
23 - 27 September 2018, Krakow, Poland

INVESTIGATION OF THE IMPACT OF ROUGHNESS ON ADHESION FORCES AND FILTRATION EFFICIENCY IN A WATER-BASED MODEL SYSTEM WITH CERAMIC FOAM FILTERS

Hoppach Daniel *, Ditscherlein Lisa *, Peuker Urs A.

Technische Universität Bergakademie Freiberg/ Institute for Mechanical Engineering and Mineral Processing, Agricolastrasse 1, 09599 Freiberg, Germany,
**Corresponding address: e-mail: daniel.hoppach@mvtat.tu-freiberg.de; lisa.ditscherlein@mvtat.tu-freiberg.de*

Key words: filtration, atomic force microscopy, nanobubbles, microbubbles, roughness

Metal melt filtration plays an important role to improve cast product purity due to the removal of endogenous and exogenous inclusions like oxides. During filtration, the inclusions can attach on the filter surface if they come close to the surface and adhesion forces surpass viscous and inertial forces [1]. A better understanding of the interaction mechanisms and thus an optimization of the filter itself are the aims of the subproject B01 in the CRC920. Therefore AFM and a water-based filtration plant was used.

Because it is nearly impossible to investigate filtration and adhesion mechanisms in the real process, a water-based model system was developed [2]. Metal melts are characterized by very high surface tensions and wet the filters poorly. To mimic this wetting behavior in the model system, ceramic foam filters and particles that represent the impurities are coated with a silane [3]. Both consist of alumina. Fritzsche et al. have shown that adhesion forces on poor wetted rough surfaces are enlarged compared to wetted surfaces not only by a structuring of the liquid, but mainly because of the formation of small cap-shaped bubbles [4-6]. These so-called nanobubbles are generated during immersion, when a particle perturbs the surface or if a temperature gradient occurs. They act as capillary bridges between two solid surfaces and increase adhesion. For metal melt filtration, the presence of nanobubbles increases the filtration efficiency and hence the metal purity significantly [3]. Nanobubbles have to stay stable on the filter surface whilst the filtration process takes place. This may be realized by increasing roughness of filter surface as larger contact force is needed to let them detach [7].

In this study, the impact of filter roughness on adhesion forces and filtration efficiency is investigated using a water-based filtration pilot plant and an atomic force microscope (AFM). It can be seen for both, filtration and AFM experiments, that higher roughness of the filter material leads to stronger adhesive forces and hence better filtration efficiencies (Figure 1). This was also seen in another study with other foam filter material [8].

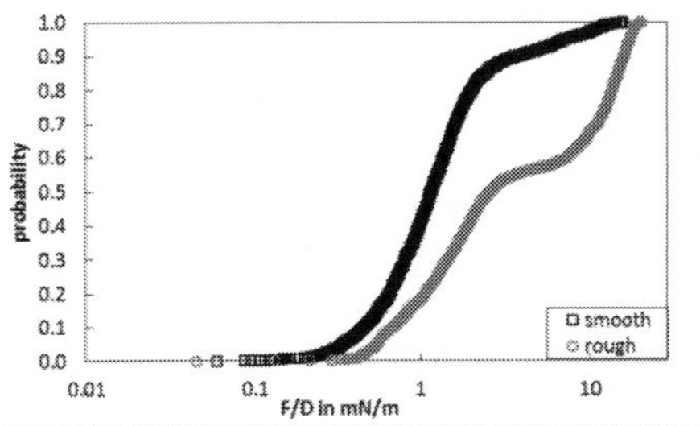

Fig. 1. ***Adhesion force distributions for hydrophobic Al_2O_3 foam filter webs with different roughness, no temperature gradient, only immersion***

This investigation distinguishes not only between filter morphology (smooth, rough), but also between the methods of bubble generation: Both, filtration and adhesion experiments were done with using filter samples covered with nanobubbles that are generated only during immersion with water and via an additional temperature gradient that decreases the gas solubility and leads to bubble growth. Furthermore, the influence of microbubbles on the filter surface is examined in detail to bridge the gap between nano- and microbubbles and their influence on the filtration efficiency.

Acknowledgments
The authors would like to thank the German Research Foundation (DFG) for supporting the studies in subproject B01 which is part of the Collaborative Research Center CRC920.

References

[1] Fabian Heuzeroth, Joerg Fritzsche, Eric Werzner, Miguel A.A. Mendes, Subashis Ray, Dimosthenus Trimis & Urs A. Peuker. (2015). Viscous force – An important parameter for the modeling of deep bed filtration in liquid media. Powder Technology. 283, 190-198.

[2] Fabian Heuzeroth, Joerg Fritzsche & Urs A. Peuker. (2014). Rasterkraftmikroskopie zur Evaluation der Abscheideeffizienz bei der Metallschmelzefiltration. Chemie Ingenieur Technik. 86 (6), 874-882.

[3] Fabian Heuzeroth, Joerg Fritzsche & Urs A. Peuker. (2015). Wetting and its influence on the filtration ability of ceramic foam filters. Particuology. 18, 50-57.

[4] Joerg Fritzsche & Urs A. Peuker. (2014). Particle adhesion on highly rough surfaces: The distribution of interaction mechanisms. Colloids and Surfaces A: Physicochemical and Engineering Aspects. 459, 166-171.

[5] Joerg Fritzsche & Urs A. Peuker. (2015). Wetting and adhesive forces on rough surfaces – An experimental and theoretical study. Procedia Engineering. 102, 45-53.

[6] Joerg Fritzsche & Urs A. Peuker. (2016). Modeling adhesive force distributions on highly rough surfaces. Powder Technology. 289, 88-94.

[7] Lisa Ditscherlein, Joerg Fritzsche & Urs A. Peuker. (2016). Study of nanobubbles on hydrophilic and hydrophobic alumina surfaces. Colloids and Surfaces A: Physicochemical and Engineering Aspects. 497, 242-250.

[7] Lisa Ditscherlein, Anne Schmidt, Enrico Storti, Christos G. Aneziris & Urs A. Peuker. (2017). Impact of the Roughness of Alumina and Al2O3-C Substrates on the Adhesion Mechanisms in a Model System. Advanced Engineering Materials. 19 (9), 1-11 (1700088).

The 73rd World Foundry Congress
„Creative Foundry"
23 - 27 September 2018, Krakow, Poland

COMPARATIVE STUDY OF WASTE FOUNDRY SAND RECLAMATION TECHNIQUES

M. Moiz Khan[a], G. N Jadhav[b], S. M. Mahajani[a*], D. Murthy[c], R. Vishwakarma[a], D. Gosavi[a], S. Mandre[d]

[a]Department of Chemical Engineering, Indian Institute of Technology Bombay, Mumbai-400076, India
[b]Department of Earth Science, Indian Institute of Technology Bombay, Mumbai-400076, India
[c]Department of Civil Engineering, Indian Institute of Technology Bombay, Mumbai-400076, India
[d]Department of Metallurgy, Government Polytechnic, Kolhapur-410064, India
sanjaym@iitb.ac.in

Key words: Waste foundry sand, Binder, Total clay content, Grain fineness number, Reclamation techniques.

Abstract
Binders (Bentonite, Phenolic resin etc.) are used for making molds and cores in the foundries. The binder present in the foundry sand becomes dead upon application of high molten metal temperature (~1400 ^{0}C). As a result, these binders forms a brittle coating on the surface of sand grains leaving them unsuitable for further use. In order to remove the dead binder coating from waste foundry sand (WFS), two types of techniques have been used namely, mechanical and thermal reclamation. This WFS after being reclaimed, mechanically and thermally, has been later compared with each other using the various tests. These tests include namely, Compressive and shear strength, Scanning Electron microscopy (SEM) and/or Optical microscopy, Total clay content, Grain fineness number (GFN), Particle size distribution, and finally loss on ignition (LOI). The above mentioned tests have been used to characterize the quality of foundry sand used in foundry industry at Kolhapur, Maharashtra, India and to test the suitability of reclaimed foundry sand for application in mold and core making in those foundry industries.

1. Introduction
According to Kolhapur, Maharashtra, India foundry survey, it has been observed that around 60 % green sand and 40 % chemical bonded sand are being used in the foundries [1]. In most of the foundries, green sand is used with the chemical bonded sand. Waste sand of these foundries generally contains mixture of both, green sand and chemically bonded sand [2]. This mixing of sand occurs during knockout process. Therefore, to reclaim mixed sand we need to perform both thermal as well as mechanical treatment. Thermal reclamation technique is best for core sand as mentioned in the literature [3]. The typical total clay content of Waste foundry sand (WFS) is 10-14 % [4]. The aim of this study is to bring down the total clay content to less than 0.2 %. This reclaimed sand (total clay content <0. 2%) is suitable for both molds and cores [4]. Suitability of reclaimed sand can be checked by performing above mentioned standard foundry tests.

2. Experiments
2.1 Mechanical Treatment
Mechanical reclamation is based on the principle of attrition. A special ball mill based prototype called two stage attrition device was used for attrition followed by sieving to remove the coating of the dead binder from waste foundry sand particles as shown in figure 1. Waste foundry sand (5 kg) was fed into the device measuring 0.5 m in length and 0.2 m in diameter, it first passes through attrition offered by agate balls (3 kg). The attrition between agate (crypto-crystalline silica) balls and sand particles, and between sand particles and wall of the device, disintegrates the dead coating present on the surface of WFS. The operating speed of ball mill was 30 rpm over 25 minutes. The waste foundry sand was thus, reclaimed mechanically and the reclaimed samples were collected for further analysis.

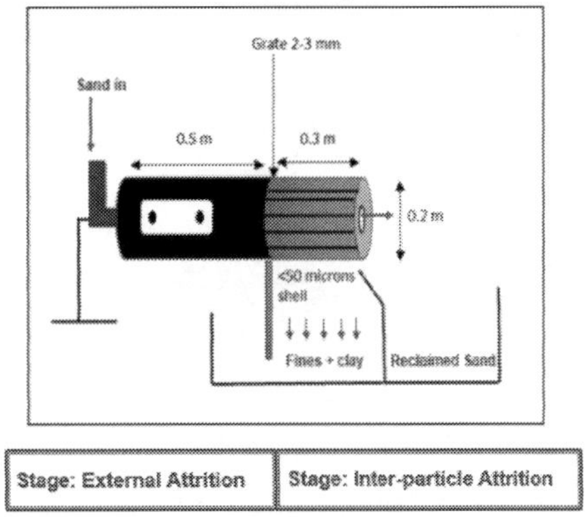

Fig.1 Two stage mechanical attrition device

2.2 Thermal treatment
A lab scale thermal fluidized bed reactor has been prepared which is 0.25 m in diameter 0.60 m high as shown in figure 2. The air compressor is connected for providing air for fluidization and combustion of sand particles. LPG and air mixture is fed into the reactor with optimized flow rate. An LPG torch is mounted at the top of the reactor to ignite the LPG, mixed with air. In Thermal reclamation process temperature was raised up to 700 ^{0}C. Thermal reclamation process has been used for both used green sand and used core sand in this study.

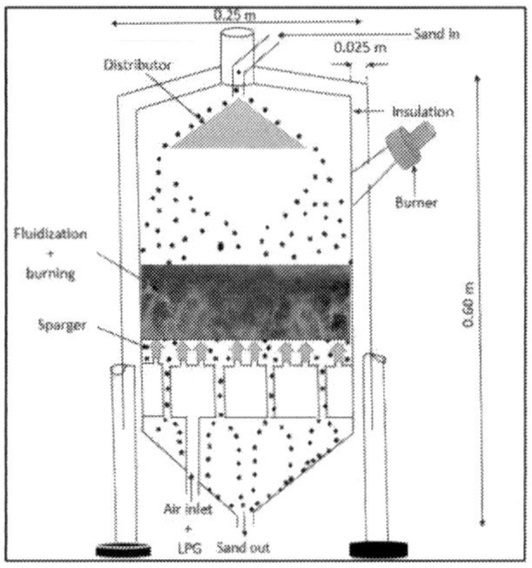

Fig.2 A lab sacle thermal fluidized bed reactor

3. Process Sequence of Reclamation of waste foundry sand

To remove dead binder coating from the sand grain, there is a need to develop a hybrid unit having a combination of both thermal and mechanical reclamation for treating all types of sand. The following sequence of reclamation process was proposed:

1. Mechanical Reclamation
2. Thermal Reclamation
3. Thermal + Mechanical Reclamation
4. Mechanical + Thermal Reclamation

4. Results and Discussion

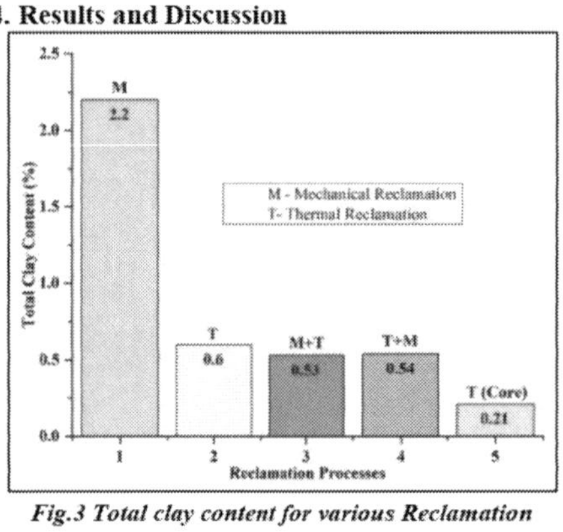

Fig.3 Total clay content for various Reclamation processes

In mechanical reclamation, a ball mill with sieve was used and the clay content was brought down from 10% to 2.2 % as shown in figure 3. In this technique, only attrition took place between, sand and sand (autogenous), sand and ball and sand and wall. A hard-shell coating is deposited on the surface and within the sand grains because, phase changes in the sand occur at the high temperature during the casting process. Mechanical reclamation alone, is therefore not sufficient to bring down clay content to less than 2%.

In thermal reclamation, the clay content was brought down significantly from 10% to 0.6%, which is quite remarkable as shown in figure 3. This happened because, upon heating the used green sand at 700 ^{0}C, binders such as Bentonite etc. becomes dead and forms a brittle coating on the sand grains. This brittle coating was removed due to particle-particle attrition and particle-wall attrition during fluidization. Furthermore, mechanical reclamation was used before and after thermal reclamation to examine the effect and figure 3 shows clearly that the clay content was brought down to only 0.53% by Mechanical attrition before or after thermal reclamation. This is because sand was heated for 30 min. at 700 ^{0}C in the fluidization chamber and same sand had been heated several times in the foundries during casting. Hence, the hard-shell coating is formed on the sand grains and is not removed effectively by mechanical attrition even after thermal reclamation. Total clay content of thermally reclaimed core sand was also determined and found to be 0.21 % as shown in figure 3. In the preparation of core sand, clay was not used as a binder. Therefore after thermal reclamation of core sand it becomes good as good as fresh sand.

5. Conclusions

The dead clay content of pure used green sand was brought down from 10% to 2.2 % in mechanical reclamation. This process works well for used green sand but cannot be used for chemically bonded or mixed type of sand. Mechanical Reclamation alone is not sufficient for reclaiming all types of sand. Thermal treatment brought down total clay content significantly from 10% to 0.5 - 0.6%. Still, the thermally reclaimed green sand is not suitable for cores due to high clay content but can be used for molds. This thermally reclaimed sand can be used for core making by mixing with fresh sand in appropriate proportion.

Acknowledgement

The financial support received from Tata Centre for Technology and Design, IIT Bombay is gratefully acknowledged.

References

[1] Khan, M., M., Singh M., Mahajani S., Jadhav G. N., Mandre. S. (2018). Reclamation of used green sand in small scale foundries. Journal of Materials Processing Technology. (In press, 10.1016/j.jmatprotec.2018.01.005)

[2] Cruz, N., Briens, C., Berruti, F. (2009). Green sand reclamation using a fluidized bed with an attrition nozzle. Resources, Conservation and Recycling 54, 45–52.

[3] Lucarz, M. (2015). Thermal reclamation of the used moulding sands. Metalurgija 54, 109–112.

[4] Khan, M., M., Singh M., Mahajani S., Jadhav G., N. (2017). A method and a device for reclaiming green waste foundry sand, Indian Patent No. 201721023693.

The 73rd World Foundry Congress
„Creative Foundry"
23 - 27 September 2018, Krakow, Poland

SPAGHETTI FILTERS – A NOVEL APPROACH FOR STEEL MELT FILTRATION BASED ON ALGINATE GEL-CASTING

Tony Wetzig [1]*, Christian Ode [1], Steffen Dudczig [1], Christos G. Aneziris [1]

[1]*Technische Universität Bergakademie Freiberg, Institute of Ceramics, Glass and Construction Materials, Agricolastraße 17, 09599 Freiberg, Germany,*
**Corresponding address: e-mail: tony.wetzig@ikgb.tu-freiberg.de*

Key words: filtering, steel casting simulator, gel-casting, carbon-bonded alumina, non-metallic inclusions

1. Introduction

Filtration is a modern approach for foundry industry to remove non-metallic inclusions from the melt and to influence the melt flow in a positive manner. Due to the overall severe conditions in the case of steel casting, cellular refractory materials with both excellent thermomechanical behavior and large surface area as well as high porosity are necessary in order to avoid filter breakage or blockage. The products of choice usually are refractory ceramic foam filters (CFFs) produced by the replication of polyurethane foams. This type of filter usually represents a good compromise regarding efficiency and cost [1, 2].

However, there also are some major limitations to CFFs. The strength of the component is restricted due to the hollow struts formed by the burnout of the inner polymer skeleton [3]. For materials such as carbon-bonded alumina, which possess comparatively low strength at room temperature, this results in problems during transport, handling and installation. Furthermore, the final pore structure is determined by the used polymer structure [4]. That means that the specific adjustment of the melt flow is not possible with this technique [5].

Therefore, a new approach has been investigated to address these problems. By alginate gel-casting, continuous strands of carbon-bonded alumina with filled cross-section were created. These can be used to form cellular components with either random or periodic pattern. Tests in contact with molten steel showed the thermal shock resistance and reactivity of the filters contributing to steel cleanliness.

2. Materials and Methods

For the filter production, the same base material was used as for carbon-bonded alumina CFFs within the studies of the Collaborative Research Center (CRC) 920 [6]. It consists of 66 % fine-grained alumina, 20 % modified pitch binder, 7.7 % graphite and 6.3 % carbon black. The main binder is the modified mesophase pitch that forms a carbonaceous matrix with a carbon yield of about 80 %.

The additive system is different from that used for the replica slurries. Water, dispersing agent and antifoam agent and poly vinyl alcohol as temporary binding agent were added to the dry material. As gelling agent for enabling the actual gel-casting process, 0.65 % of sodium alginate were used. Due to its special structural properties, it is able to conduct a sol-gel transition by exchanging sodium counter-ions with alkaline earth ions [7]. Therefore, a 1 m% solution of calcium chloride was used for the in-situ gelation.

A thin stream of the mixed slurry with about 59 m% or 32 vol% is cast or pumped into the solution. This results in immediate diffusion of sodium and calcium ions and the in-situ triggering of the gelation.

The actual filter layers are formed either by computer-assisted movement of the pumping tube or the random packing resulting from the cast slurry stream.

The gelled component has to be dried at room temperature or by freeze-drying. The latter prevents excessive dry shrinkage of the material. In a final step, the filter is dip-coated with a conventional slip-casting slurry in order to interconnect the strand's points of contact. In this case, the before-mentioned original carbon-bonded alumina slurry of the CRC 920 was used [6]. The once again dried filters then were fired in reducing atmosphere at 800°C for 3 hours.

Different sample geometries were prepared for the investigations, including periodic "finger" samples with dimensions of 25 x 25 x 110 mm³ and 2 mm spacing between the filter struts as well as flat filters with 80 x 80 x 20 mm³ and 4 mm spacing.

These sample types were tested in contact with molten 42CrMo4 steel in a steel casting simulator. Thereby, two modes were tested. The "finger" samples were immersed into the inductively melted steel at 1585°C. The flat filter was placed in a casting system and 100 kg of 1650°C was cast through the filter into a tundish vessel distributing the steel into two copper molds. Both tests were conducted under argon atmosphere.

After the test, the filters were checked for potential damage due to thermal shock. Furthermore, the filter surface was investigated by digital light microscopy, scanning electron microscopy (SEM), energy-dispersive X-ray spectroscopy (EDS) and electron backscatter diffraction (EBSD) regarding collected inclusions.

3. Results and Discussion

In both tests, the filter structure was still intact after contact with molten steel. Single cracks occurred due to the thermal stresses, however, they did not result in component failure or loss of material.

Due to the lower spacing in the case of "finger" samples, a higher ferrostatic pressure of the melt was necessary for the priming. The impregnation of the whole filter occurred at an infiltration depth of over 25 mm. Such problems were not observed for flat filters with 4 mm spacing or common 10 ppi CFFs in previous test. That implies that the pore width of 2 mm is too low for proper priming and through-flow and may lead to excessive pressure drop or blockage during casting.

In both cases, a change in appearance was observed after the melt contact. The filter surface exhibited a bright color. The origins of this phenomenon are (i) the partial decarburization of the material due to carbon solubility and (ii) the formation of alumina-based in-situ layers. This effect was well described in previous studies of the CRC 920 for carbon-bonded alumina CFFs [6]. However, in the case of gel-casted filters, the surface seemed more reactive. More pronounced layers are visible in narrow sleeves, near points of contact and cracks as well as in dead end pores. In certain areas, there were also large inclusion formations in the range of several hundred micrometers. Besides translucent colorless inclusions similar to typically observed alumina crystals, also a new type of vermicular deposit was observed.

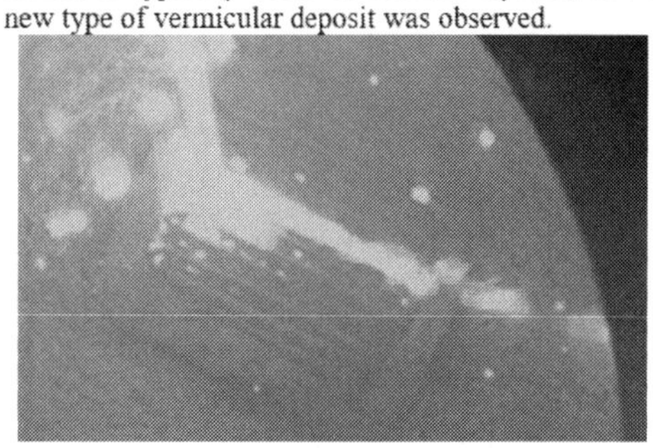

Fig. 1. Cast-on carbon-bonded alumina spaghetti filter.

With the aid of SEM, EDS and EBSD, the found formations were investigated regarding their qualitative phase and chemical composition. In the case of translucent structures, it was found that the material was mainly based on aluminum, oxygen and calcium. Furthermore, EBSD measurements indicated that the structures were at least partially crystalline. These results indicate that these formations consist of alumina-based inclusions together with calcium of the alginate residuals. Vermicular structures also mainly were composed of aluminum, calcium and oxygen. However, an increased iron and chlorine content was observed compared to the base material. Furthermore, the EBSD measurement showed no signal of crystallinity. As a result, it is possible that amorphous phases of the Al_2O_3-CaO-FeO system were formed. Furthermore, the formation of spinel type crystals such as hercynite cannot be fully excluded.

Besides the special structure, the main difference to foam filters was the presence of calcium and chlorine, which led to new chemical processes during filtration. It is not clear yet, if these mechanisms are just an insignificant byproducts or have an effect on the purification efficiency of the steel melt.

In order to quantify this in comparison to other filter types, flat filters were produced for actual castings in the steel casting simulator (Fig. 1). Once more, these filters showed the same change in appearance without critical damage due to the steel impact or thermal shock. In the future, the cast steel will be investigated regarding its purity compared to various reference tests.

4. Conclusions

Carbon-bonded alumina filters were successfully produced by sodium alginate gel-casting. It was shown that the material is thermal shock resistant and strong enough for the casting of molten steel. The spacing between the gel-casted struts should exceed 2 mm in order to ensure proper priming and flow of the melt.

Regarding in-situ layer formation, the filters acted similarly compared to carbon-bonded foam filters. However, new phases and increased reactivity were observed due to the special geometry and the calcium content of the filter.

The actual effect of the structure and chemistry on the filtration will be investigated in future tests in comparison to conventional filter systems.

Acknowledgments

The Authors would like to thank the German Research Foundation for financially supporting the subproject T01 in the framework of the Collaborative Research Center 920.

References

[1] Antsiferov V. N. (2003). Foam ceramic filters for molten metals: Reality and prospects. Powder Metallurgy and Metal Ceramics. 42 (9/10), 474–476.

[2] Kondrat'ev A. S., Popov V. N., Aksel'rod L. M., Baranovskii M. R., Suvorov S. A., Tebuev N. B. (1990). Possibilities of filter-refining of metals and the required characteristics of the filter elements: A review. Refractories. 31 (7/8), 384–391.

[3] Colombo P. (2006). Conventional and novel processing methods for cellular ceramics. Philosophical Transactions. Series A, Mathematical, Physical, and Engineering Sciences. 364 (1838), 109–124.

[4] Dawson J. R., Shortall J. B. (1982). The microstructure of rigid polyurethane foams. Journal of Materials Science. 17 (1), 220–224.

[5] Werzner E., Mendes M. A. A., Ray S., Trimis D. (2013). Numerical investigation on the depth filtration of liquid metals: Influence of process conditions and inclusion properties. Advanced Engineering Materials. 15 (12), 1307–1314.

[6] Storti E., Emmel M., Dudczig S., Colombo P., Aneziris C. G. (2015). Development of multi-walled carbon nanotubes-based coatings on carbon-bonded alumina filters for steel melt filtration. Journal of the European Ceramic Society. 35 (5), 1569–1580.

[7] Oppelt M., Wenzel C., Aneziris C. G., Berek H. (2014). Processing and characterization of mmc beads based on zirconia and trip steel. Metallurgical and Materials Transactions B. 45 (6), 2000–2008.

The 73rd World Foundry Congress
„Creative Foundry"
23 - 27 September 2018, Krakow, Poland

EXCHANGEABLE CARBON-BONDED ALUMINA FOAM FILTER SYSTEMS FOR THE CONTINUOUS CASTING OF STEEL – DEVELOPMENT AND TESTING

Tony Wetzig [1]*, Jana Hubálková [1], Steffen Dudczig [1], Andreas Baaske[2], Sven Karrasch [2], Christos G. Aneziris [1]

[1]*Technische Universität Bergakademie Freiberg, Institute of Ceramics, Glass and Construction Materials, Agricolastraße 17, 09599 Freiberg, Germany,*
[2]*Thyssenkrupp Steel Europe AG, Kaiser-Wilhelm-Straße 100, 47166 Duisburg, Germany*
**Corresponding address: e-mail: tony.wetzig@ikgb.tu-freiberg.de*

Key words: filtering, continuous casting, steel casting simulator, ceramic foam filter, carbon-bonded alumina

1. Introduction

Not only exogenous macro-inclusions but also agglomerating micro-inclusions have considerable negative effects on the material properties of cast metals and cause excessive tool wear as well as high reject rates [1]. Therefore, melt filtration became a viable approach for the foundry industry that also has been considered and tested for continuous casting of steel [2]. Thereby, research and industry mostly focused on the implementation of filter walls and dams in the tundish of the continous casting plant [3]. The main problem for highly efficient foam filters is the large quantity of steel processed within a continuous casting sequence. The filter capacity depends on many parameters and is exhauted rather quickly, especially in the case of inclusion-rich batches [4]. Therefore, the exchange of the applied filter systems during the casting sequence has to be enabled.

In this study, cylindrical large-scale ceramic foam filters were produced by the replication of polyurethane foams. The prepared filters were tested for their mechnical strength, porosity and structural properties with regard to different coating techniques. The filter capacity and the mechanical stresses occuring in the tundish can be estimated with the aid of the found parameters.

Based on the results, one filter was chosen for a test in contact with molten steel in a steel casting simulator in order to ensure the save application prior to industrial tests.

2. Materials and Methods

The filters were prepared by a modified replication technique. Thereby, cylindrical polyurethane foams with 10 pores per inch and a height as well as a diameter of 200 mm served as a basis. Furthermore, the filters had nine channels with diameters of 40 mm for fixing, flow conditioning and easier firing.

The used refractory material consisted of 66% fine-grained alumina, 20% modified pitch binder, 7.7% graphite and 6.3% carbon black. During firing, the pitch forms a mesophase associated with the release of about 20 mass% volatile constituents, resulting in a carbonaceous binding matrix. Since the prepared slurry is water-based, lignin sulfonate was used as wetting agent for the carbonaceous raw materials and as temporary binder. Furthermore, a dispersing agent and an antifoam agent were added. The solid content of the slurry was adjusted regarding the particular coating method.

The polymer foams were coated using combinations of three different techniques: primary impregnation-centrifugation (C, 81% slurry solid content) secondary dip-coating (D, 60% slurry solid content) and secondary spray-coating (S, 70% slurry solid content). In order to investigate the influence of secondary coatings, three different coating procedures were applied: (i) C, (ii) CS and (iii) CDS with three filter samples each. After each coating step, the component was dried for 24 hours at room temperature. The final filters then were embedded in pet coke and fired at 800°C for 3 hours.

The prepared filters were tested for their porosity and density using the Archimedes' principle. Based on these results the buoyancy force in molten steel can be calculated and the filter capacity can be estimated according to Raiber et al. [4].

Afterwards, the filters were tested in mechanical investigations for their maximum load under compression.

One additional CDS filter was prepared for a test in contact with molten steel. Therefore, 100 kg of 42CrMo4 steel were melted in a steel casting simulator at 1650°C and cast on top of the filter under argon atmosphere. The steel formed a melt bath in the tundish vessel of the device before entering the two copper molds. Afterwards, the filter was investigated by scanning electron microscopy and computer tomography.

3. Results and Discussion

The overall porosity of the filters is directly correlated with the component mass resulting in lower values for triple-coated CDS filters. Thereby, the share of porosity accessible for the steel melt is crucial for the filtration. Impacts of the secondary dip-coating on the closed porosity due to inhibited drainage of the slurry were negligible (see Table 1). With the aid of the filter pore volume and the material parameters of alumina inclusions, the filter capacity can be estimated using calculations as published by Raiber et al. [4]. Thereby, the two main unknown parameters are (i) the inclusion content of the steel melt and (ii) the maximum possible utilization of the pore volume.

Table 1. Open porosity, closed porosity and maximum compressive load of different filter types

Filter type	Open porosity [%]	Closed porosity [%]	Max. compressive load [MPa]
C	84.2 ± 0.5	6.2 ± 0.2	0.06 ± 0.00
CS	80.6 ± 0.7	7.5 ± 0.4	0.08 ± 0.01
CDS	77.4 ± 0.5	7.8 ± 0.3	0.23 ± 0.02

Assuming a conventional filter pore volume utilization of 20 %, a filter with about 4 dm^3 volume and 77.4 % open porosity can remove up to 0.6 dm^3 of alumina inclusions. This is equivalent to lowering the total oxygen content (TOC) of 56.2 tons steel by 20 ppm for example. If the pollution is higher, the filtration kinetics is enhanced resulting in higher inclusion removal and thus lower filtration capacity. For example, doubled TOC reduction of 40 ppm results in halved filtration capacity of only 28.1 tons steel. This illustrates that the filter capacity is heavily influenced by the filtered melt and shows the necessity of exchangeable filters since the amount of treatable steel heavily depends on the specific batch.

The maximum load that can be borne by the CDS filters was 0.23 MPa (see table 1). Excluding the macro-channels from the calculation, this value increases to 0.35 MPa and can be seen as strength of the general carbon-bonded alumina foam structure. This number is agreement with investigations of Storti et al. for smaller filters of the same material [5]. Solarek et al. showed that for increasing temperature there is an increase in strength of carbon-bonded filters. However, at over 1400 °C creep and deformation may occur [6].

Considering the immersion of the filter in the steel melt of the tundish with a cardboard protection cover, the compressive load caused by buoyancy is about 0.01 MPa. That means that the filter's reliability is ensured. Furthermore, the occurring load is far below the deformation onset observed by Solarek et al. [6].

In the steel casting simulator test, the filter was soused with 100 kg of molten steel. The bottom of the filter was longer in contact with the melt due to the formation of a bath level. As expected, the filter showed no critical thermal shock, damage or deformation. However, a radial shrinkage of about 2% was observed because of the continuing coking of the material that was treated at 800°C beforehand. Computer tomographic investigations revealed that no freezing occurred. With the aid of scanning electron microscopy, the formation of oxide layers was observed on the filter surface. This is in agreement with findings of previous investigations on carbon-bonded alumina filters [5]. In the present study, the formed layers were less pronounced in regions with reduced flow velocity, which is favorable considering generally low flow rates in future industrial tundish applications.

In a first trial, a filter was immersed in an industrial tundish for about 45 min. surviving without damage. Dense clogging layers and inclusion clusters on the surface of the filter struts indicate similar mechanisms to findings reported by Storti et al. [7].

4. Conclusions

In this study, the technology for the production of carbon-bonded alumina filters was successfully adopted for an upscaling by triple-coating replication towards industrial tests in continuous casting.

Calculations showed that the filter definitely is able to treat high amounts of steel in the range of industrial batches under average conditions. The lab-scale steel casting test showed that the components are resistant to stresses in contact with steel. Furthermore, the proper priming and melt flow are ensured. First industrial tests confirmed these assumptions.

Acknowledgments

The Authors would like to thank the German Research Foundation for financially supporting the subproject T01 in the framework of the Collaborative Research Center 920. Special thanks go to Thyssenkrupp Steel Europe AG for the excellent cooperation.

References

[1] Zhang L., Thomas B. G. (2003). Inclusions in continuous casting of steel. In proceedings of xxiv national steelmaking symposium, Nov 26-28 (138–183). Morelia, Mich, Mexico.

[2] Kondrat'ev A. S., Popov V. N., Aksel'rod L. M., Baranovskii M. R., Suvorov S. A., Tebuev N. B. (1990). Possibilities of filter-refining of metals and the required characteristics of the filter elements: A review. Refractories. 31 (7/8), 384–391.

[3] Janiszewski K., Panic B. (2014). Industrial investigations of the liquid steel filtration. Metalurgija. 53 (3), 339–342.

[4] Raiber K., Hammerschmid P., Janke D. (1995). Experimental studies on Al2O3 inclusion removal from steel melts using ceramic filters. ISIJ International. 35 (4), 380–388.

[5] Storti E., Emmel M., Dudczig S., Colombo P., Aneziris C. G. (2015). Development of multi-walled carbon nanotubes-based coatings on carbon-bonded alumina filters for steel melt filtration. Journal of the European Ceramic Society. 35 (5), 1569–1580.

[6] Solarek J., Himcinschi C., Klemm Y., Aneziris C. G., Biermann H. (2017). Ductile behavior of fine-grained, carbon-bonded materials at elevated temperatures. Carbon. 122, 141–149.

[7] Storti E., Dudczig S., Schmidt A., Schmidt G., Aneziris C. G. (2017). Filter functionalization with carbon nanotubes and alumina nanosheets for advanced steel filtration. steel research international. 88 (10), 1700142.

The 73rd World Foundry Congress
„Creative Foundry"
23 - 27 September 2018, Krakow, Poland

SIMULATION OF SAND FILLING PROCESS AND ANALYSIS OF FLOW BEHAVIOR WITH DIFFERENT BINDER RATIO

Lele Tong [1], Jianxin Zhou [1]*, Yajun Yin [1], Xiaoyuan Ji[1], Xu Shen[1]

Huazhong University of Science and Technology, State Key Laboratory of Materials Processing and Die &Mould Technology, Wuhan 430074, Hubei Province, People's Republic of China
**Corresponding address: e-mail: zhoujianxin@hust.edu.cn*

Key words: sand core shooting; two fluid model; flow behavior; pressure distribution; binder ratio

Abstract

Core shooting process has a great influence on the quality of sand cores. In present, there are two computational fluid dynamics(CFD) models applied to simulate the hydrodynamics of core shooting process. In this paper, a comparison between two-fluid model(TFM) and a discrete element model(DEM) was made by dynamic and force analysis. Both models are able to get a reasonable flow behavior of sand particles and the distribution of gas can be well predicted. The binder ratio has a significant effect on sand hydrodynamic behavior and the distribution of residual gas pressure.

1. Introduction

With the development of near net shape casting, gas-cured cold box binder systems are most widely used to make sand cores of metal castings[1]. The core shooting process includes two stages: filling and compressing. At the filling stage, sand-binder mixture is shot at high velocity into the core box through blow tubes. At the compressing stage, the air escapes through vents and then the sand compresses adequately[2-3].

In the last two decades, two main methods of numerical modeling are applied in sand core shooting simulation. The two-fluid models(TFM) which is based on Eulerian-Eulerian theory, treats both particles and gas phases as interpenetrating continua[4-5]. The discrete particle model(DPM) has been used for a wide range of application with determining trajectories of particles by integrating the Newtonian equation of motion[6].

In this paper, we made a comparison between both two models on hydrodynamic behavior of core shooting process and predict the distribution of gas. Then a comparison between different percentage of binders was made.

2. Models

2.1 TFM model

In Two-Fluid-Model(TFM), the continuity equation for phase q is

$$\frac{\partial}{\partial}(\alpha_q \rho_q) + \nabla(\alpha_q \rho_q \upsilon_q) = 0 \tag{1}$$

The conservation of momentum for the phase q is

$$\frac{\partial}{\partial t}(\alpha_q \rho_q \vec{\upsilon}_q) + \nabla \bullet (\alpha_q \rho_q \vec{\upsilon}_q \vec{\upsilon}_q) = -\alpha_q \nabla p + \nabla \bullet \tau_q + \sum_{p=1}^{n} K_{pq}(\vec{\upsilon}_p - \vec{\upsilon}_q) + \alpha_q \rho_q (\vec{F}_q + \vec{F}_{lift,q} + \vec{F}_{Vm,q}) \tag{2}$$

Where τ_q is the fluid stress-strain tensor, F_q is external volume force, $F_{lift,q}$ is lift force, $F_{vm,q}$ is virtual mass force.

The phase interaction between two phase is introduced by the fluid-solid momentum exchange coefficient K_{pq}, for granular and gas phase, in our study, we used a model proposed by Gidaspow which is a combination of the Ergun and Wen-Yu models.It is given as follows:

$$K_{gs} = \begin{cases} \frac{3}{4} C_D \frac{\alpha_s \alpha_g \rho_g |\vec{\upsilon}_s - \vec{\upsilon}_g|}{d_s} \alpha_g^{-2.65} & \alpha_g > 0.8 \\ 150 \frac{\alpha_s^2 \mu_g}{\alpha_g d_s^2} + 1.75 \frac{\rho_g \alpha_s |\vec{\upsilon}_g - \vec{\upsilon}_s|}{d_s} & \alpha_g \le 0.8 \end{cases} \tag{3}$$

In order to solve particle -particle collisions and thermal motion of molecules of gas, a transport equation was introduced from kinetic theory takes the form:

$$\frac{3}{2}\left[\frac{\partial}{\partial t}(\rho_s \alpha_s \Theta_s) + \nabla \bullet (\rho_s \alpha_s \vec{\upsilon}_s \Theta_s)\right] = (-p_s I + \tau_s) : \nabla \vec{\upsilon}_s + \nabla \bullet (k_{\Theta_s} \nabla \Theta_s) - \gamma \Theta_s + \varphi_{gs} \tag{4}$$

2.2 DPM model

In discrete particle model(DPM), it is assumed each individual particle with mass m_p and volume V_p and the model is given by as follows:

$$m_p \frac{d\upsilon_p}{dt} = \frac{V_p}{1-\alpha_c}(\upsilon_c - \upsilon_p) + m_p g\left(1 - \frac{\rho_c}{\rho_p}\right) + F_c \tag{5}$$

The angular momentum of the particle is computed with:

$$I_p \frac{d\omega_p}{dt} = T_p \tag{6}$$

2.3 boundary conditions

At the wall and outlet, the nomal velocity of each phases is set to zero. The granular temperature is set by Gidaspow assumption. And the energy flux at the wall is set to zero.At the inlet,the velocity and volume fractions of each phase is set experiments.

The governing equations of TFM models and DPM models were discretized into linear equations by FLUENT and EDEM respectively. Table 1 gives the sand shooting parameters used in simulations.

Table1 Simulation parameters

Property	value
Sand diametet(mm)	0.3
Restitution coefficient	0.001,0.3,0.5,0.8,0.9
Max volume fraction	0.8
Sand density(kg/m3)	2600
Particle Static friction	0.8
Particle rolling friction	0.1

3. Results and discussions

3.1 Flow behavior of shooting process

Figure 1(a)~(b) shows the simulation of sand particles velocities at time t=150 ms by TFM and DPM models respectively.

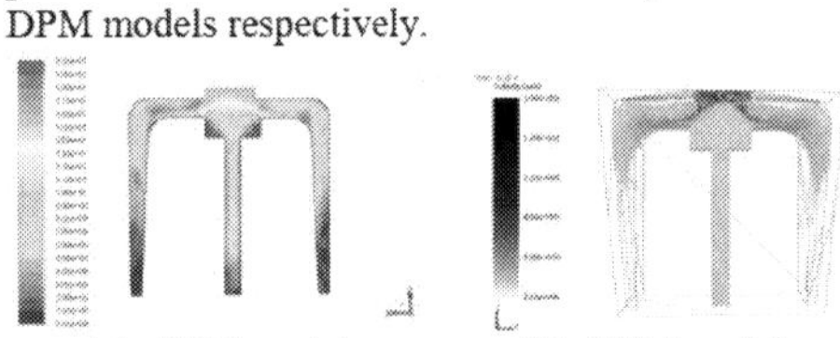

(a) TFM model (b) DPM model

Figure 1 Sand particles velocities at t=150 ms

It is observed that both models have a similar filling process of solid phase. At t=150s, The velocity of sand particles has a maximum value near the nozzle and an arch-bridge is formed at the top of the center branch.

Figure 2 shows the variation of sand particles velocity and compressive force in the sand box. The velocity increases sharply before t=10s and increasing until reach the bottom in the center branch of the sand box. Then the velocity reduces to a steady condition before t=80 ms. At last, with the friction of particle-particle and particle-geometry, the increasing of velocity is far less than the beginning stage.

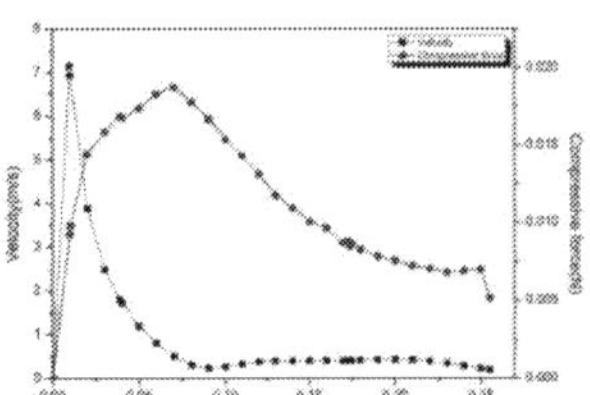

Figure 2 Variation of velocity and compressive force

3.2 Gas distribution of shooting process

Figure 3 is the TFM simulation results of gas distribution when core shooting process finishes. The residual gas is located in inwall of side branch. With comparison of DPM simulation results in Figure 4, inwall of side branch is the last reached part. With combination of these models, it is able to predict that the location of residual gas which has a bad influence on quality of shooting process when it can't exhaust completely.

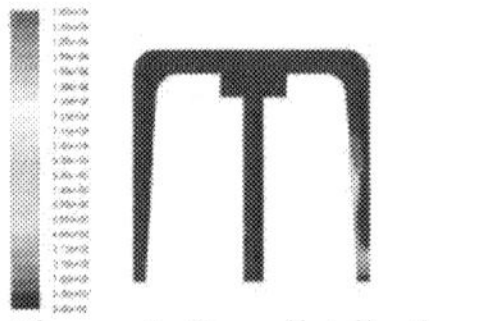

Figure 3 Gas distribution by TFM

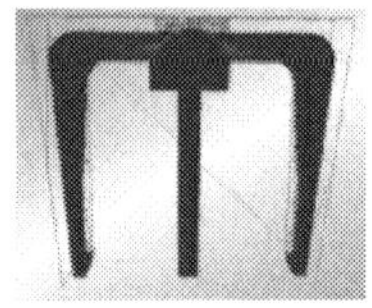

Figure 4 Sand filling simulation by DPM

3.3 Effect of binder ratio

The percentage of binder in core shooting process is 0.5%~2%. The content of binder has a great effect on flow behavior by influencing the restitution coefficient of particle-particle interaction. Figure 5 is the sand velocity along x-axis at the same location of core box.

Figure 5 Sand velocity along x-axis for different values of restitution coefficient

4. Conclusions

1. Both TFM and DPM models can get a reasonable result of sand hydrodynamic behavior in sand shooting process.
2. With combination of these models, it is able to predict distribution of gas precisely.
3. With the increasing percentage of binder, the sand velocity has a significant increase.

Acknowledgments

This research was financially supported by the National Nature Science Fund Projects, China (no. 51775205), and State Key Laboratory of Materials Processing and Die & Mould Technology Research Project (2015-2017).

References

[1] Bakhtiyarov S I, Overfelt R A. 2003. CFD modeling and experimental study of resin-bonded sand/air two phase flow in sand core making process. Powder Technology. 103:68-78.

[2] Winartomo B, Vroomen U, Hrig-Polaczek A B, et al.2005.Mulitiphase modeling of core shooting process. International Journal of Cast Metals Research.18:13-20.

[3] Ni C, Lu G, Zhang Q,et al. 2016. Influence of core box vents distribution on flow dynamics of core shooting process based on experiment and numerical simulation. China Foundry. 13(1): 22-29.

[4] Ni C, Guo E Y, Zhang Q.2016.Frictional-kinetic modeling and numerical simulation of core shooting process. International Journal of Cast Metals Research. 29:214-221.

[5] GidaspowD.1994. Multiphase flow and fluidization: continuum and kinetic theory descriptions. Academic Press. New York.

[6] Linmin Li,Baokuan Li, Zhongqiu Liu.2017. Modeling of spout-fluidized beds and investigation of drag closures using OpenFOAM. Powder Technology.305:364-376.

The 73rd World Foundry Congress
„Creative Foundry"
23 - 27 September 2018, Krakow, Poland

OPTIMAL GATING SYSTEM DESIGN OF STEEL CASTING BY FRUIT FLY OPTIMIZATION ALGORITHM BASED ON CASTING SIMULATION TECHNOLOGY

Tong Wang [1], Xu Shen [1], Jianxin Zhou [1*], Yajun Yin [1], Xiaoyuan Ji [1], Qin Zhou [2]

[1] *Huazhong University of Science and Technology, State Key Laboratory of Materials Processing and Die & Mould Technology, Wuhan 430074, China,*

[2] *Huazhong University of Science and Technology, Engineering Training Center, Wuhan 430074, China,*

**Corresponding address: e-mail: zhoujianxin@hust.edu.cn*

Key words: Gating System Design, Steel Casting, Fruit Fly Optimization Algorithm (FOA), InteCAST, CAD

Abstract: This paper presents an optimal method for gating system design by fruit fly optimization algorithm based on casting simulation technology. Taking an upper center plate casting for example, the results indicate that the proposed optimal method could provide practical and useful suggestions for designer to obtain higher quality and lower resource cost designs in the gating system design process.

1. Introduction

Gating system plays an important role in casting process for high quality casting production. And it refers to those channels through which the metal flows from the ladle to the mold cavity [1-2].

2. Geometry mathematical model of gating system

The optimization goal is to minimize the filling time of mold cavity in sand casting. So the objective function is as follows (Eq. (1)):[3]

$$\tau_f = \int_0^{h_c} \frac{A(h)}{\sum_{j=1}^{N} A_{ingate\_j} \times v_{ingate\_j}} dh \tag{1}$$

Where τ_f is the filling time of mold cavity, $A(h)$ is the instantaneous cross sectional area of casting layer being filled, A_{ingate_j} is the cross sectional area of ingate, v_{ingate_j} is the instantaneous velocity of molten metal in the mold cavity, N is the number of ingates, h_c is the height of mold cavity, h is the instantaneous height of casting part.

The gating system model of upper center plate casting is shown in Fig.1 The cross section shape of ingate is rectangular. There are two ingates with the same cross section. So we have Eq. (2).

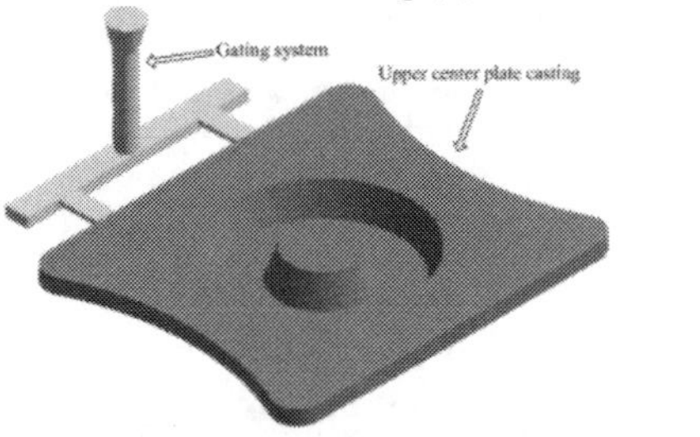

Fig. 1. The gating system model of upper center plate casting

$$\tau_f = \frac{1}{2ab\sqrt{2g}} \int_0^{h_c} \frac{A(h)}{\sqrt{h_m - h}} dh \tag{2}$$

Where a and b are the side lengths of ingate, respectively, g is the gravity acceleration, h_m is the distance between the bottom of upper center plate casting and the level of molten metal in the pouring cup.

The design of gating system must meet the following constraints.

(1) Reynolds number constraint [2]

In order to avoid high turbulence flow, the mathematical expression is shown as Eq. (3).

$$R_e = \frac{\rho_m \times v_{ingate} \times d}{\mu_m} \le 20000 \tag{3}$$

Where R_e is the Reynolds number, ρ_m is the density of molten metal, d is the characteristic length, μ_m is the dynamic viscosity of molten metal, v_{ingate} is the velocity of molten metal at ingate. The characteristic length d of the ingate can be expressed as Eq. (4). The Reynolds number is expressed as Eq. (5).

$$d = \frac{4A_{ingate}}{P_{ingate}} = \frac{2ab}{a+b} \tag{4}$$

$$R_e = \frac{2ab\rho_m\sqrt{2gh_m}}{\mu_m(a+b)} \le 20000 \tag{5}$$

Where A_{ingate} is the cross sectional area of ingate, P_{ingate} is the wetted perimeter of ingate section.

(2) Ingate modulus constraint

The ingate modulus should be less than or equal to the casting modulus [4] (Eq. (6)).

$$M_{ingate} \le M_{casting} \Rightarrow \frac{ab}{a+b} \le \frac{V_{casting}}{A_{casting}} \tag{6}$$

Where M_{ingate} is the ingate modulus, $M_{casting}$ is the casting modulus, $V_{casting}$ is the volume of casting, $A_{casting}$ is the cooling area of casting.

3. Gating system optimization strategy

The gating system optimization stragy is shown in Fig.2

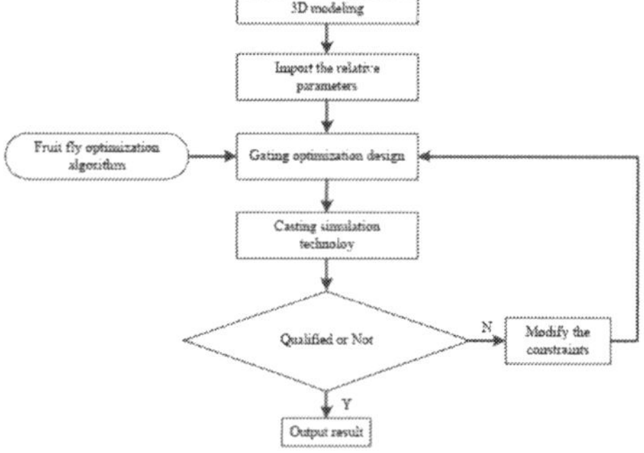

Fig. 2. ***The gating system optimization strategy***

4. Optimization results and analysis

$V_{casting}$ =6.06*10^{6}mm^3, $A_{casting}$ =5.36*10^{5}mm^2. The input parameter values are: ρ_m =7600kg/m^3, μ_m =0.007kg/(m.s), g=9.8m/s^2, Population size=100, Maximum number of iteration=5000, Multiplier=100. The minimal filling time of mold cavity is τ_f =10.58s. The optimized results of geometric dimension parameters are as follow:a=29.9611mm, b=6.0694mm and h_m=168.3857mm. Fig.3 shows the minimal value curves of iterative search the filling time of mold cavity. With the increase of iterative times, the filling time also gradually converges and tends to be stable.

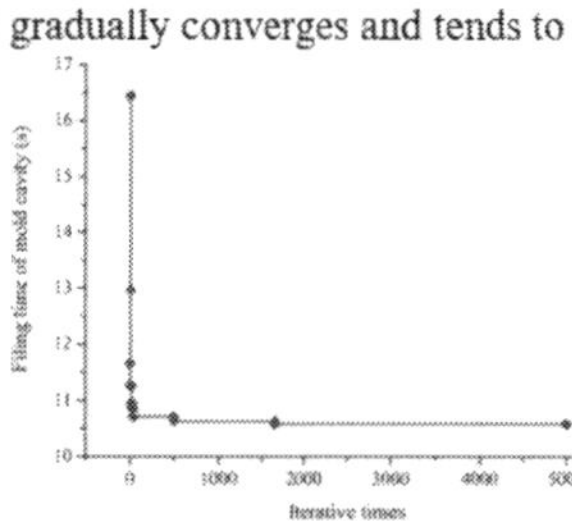

Fig. 3. ***The minimal value curves of iterative search the filling time of mold cavity***

The area ratio of sprue, runner and ingate is 1:0.8:0.6. Based on the ingate area, the sprue area is 606.1530 mm^2 and the runner area is 484.9224mm^2. The sprue is circular in cross section and the diameter is 28mm. The cross section of runner is rectangular and the two side lengths are 31mm and 16mm.

InteCAST software package [6] is used to simulate the above design. The total filling time is 28.37s and is consistent with the optimized filling time of mold cavity due to the total filling time includes casting part, gating system and risers. Fig.4 represents the velocity distribution of filling fluid at different filling time. Fig.5 represents the flow streamline of filling fluid at different filling time. These filling results show that the filling process is uniform and smooth without high turbulence. The highest velocity is 1.619m/s and located in the ingate during the filling process. Therefore, the optimal method can be used to the gating system design of steel casting and the optimal results provide practical and useful suggestions for designer to obtain higher quality and lower resource cost designs.

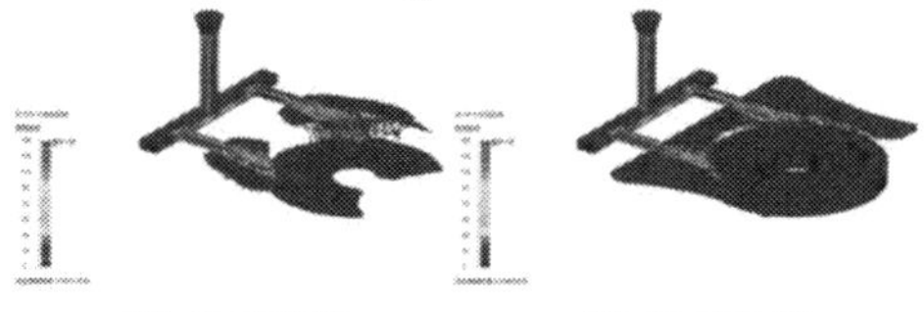

(a) The filling time is 1.97s (b) The filling time is 6.61s

Fig. 4. ***The velocity distribution of filling fluid at different filling time***

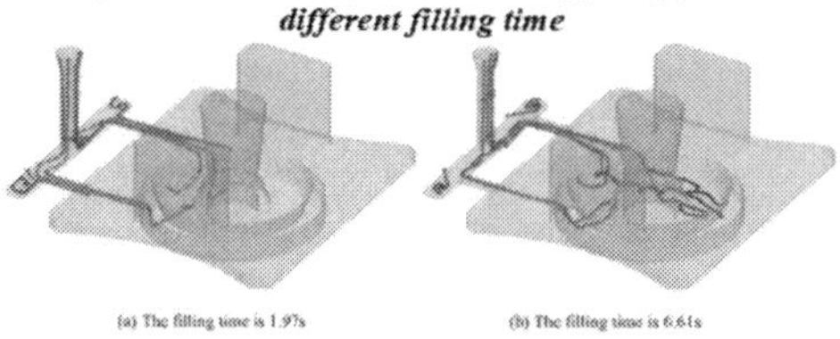

(a) The filling time is 1.97s (b) The filling time is 6.61s

Fig. 5. ***The flow streamline of filling fluid at different filling time***

5. Conclusions

In this paper, an optimal method for gating system design by FOA based on casting simulation technology is proposed. Taking a upper center plate casting for example, the gating system geometry model is established and the model is solved by FOA. Based the optimization results, each dimension of gating system is determined. The simulation results based on InteCAST confirms the correctness of the designed gating system based on this optimal method. Therefore, the optimal method can be used to the gating system design of steel casting and the optimal results provide practical and useful suggestions for designer to obtain higher quality and lower resource cost designs.

Acknowledgments

This research was financially supported by the National Nature Science Fund Projects, China (no.51605174), and State Key Laboratory of Materials Processing and Die & Mould Technology Research Project (2015-2017).

References

[1] Lin X Y. (2011). Casting manual-Volume 5- Casting process. Beijing: China Machines Press.

[2] Lampman S. (2009). Casting design and performance. Ohio: ASM International.

[3] Wukovich N, Metevelis G. (1989). Gating: the Foundryman's Dilemma or Fifty Years of Data and Still Asking How?. AFS Transactions. 97, 285-302.

[4] Dučić N, Žarko Ćojbašić & Manasijević S, et al. (2017). Optimization of the gating system for sand casting using genetic algorithm. International Journal of Metalcasting. 11(2), 255-265.

[5] Pan W T. (2012) A new fruit fly optimization algorithm: taking the financial distress model as an example. Knowledge-Based Systems. 26, 69-74.

[6] http://www.intecast.com/

The 73rd World Foundry Congress
„Creative Foundry"
23 - 27 September 2018, Krakow, Poland

BAYESIAN INFERENCE BASED OPTIMIZATION OF PROCESS PARAMETERS FOR CHEMICALLY BONDED MOLDING SYSTEM

Himanshu Khandelwal[1*], Amit Sata[2], B. Ravi[3]

[1] *Research Scientist, MPM Infosoft Pvt. Ltd., India,*
[2] *Professor, Marwadi Education Foundation, Rajkot., India,*
[3] *Institute Chair Professor*, IIT Bombay, Mumbai, India,
**Corresponding address: e-mail: khandelwal.iit@gmail.com*

Key words: Sand Casting, Chemically bonded molding, No Bake mold, Bayesian inference, Optimization

Introduction of chemically bonded process is considered to be one of the key improvements in molding of sand casting process. This further leads to improved surface finish and better dimensional accuracy in castings. However, quality of chemically bonded mold is primarily driven by different process parameters including Grain Fineness Number (GFN), Binder (%) and curing time. These process parameters affect different mold characteristics (mold compressive strength, shear strength, mold shrinkage and core hardness). In this work, the effect of different process parameters on the characteristics of the mold has been studied. Three-part binder system (one of chemically bonded molding system) has been taken for the study, as it is widely accepted in most metal casting foundries. The process parameters mentioned above varied in three levels, and experiments were performed. The experimental results were then optimized using Bayesian inference based tool. The Bayesian inference method is essentially based on probabilistic approach. This tool is developed in such a way that it takes input in the form of a spreadsheet, and provides an optimized range of values for each input parameter. It was observed that GFN (no.) with 40-48; Binder (%) with 2.20-2.40; curing time with 2-2.4 (hr) are optimized values to achieve higher strength while optimum range of values for minimum core shrinkage are found to be GFN: 70-80, Binder (%): 1.80-2.00; Curing Time: 1.75-2.50. Bayesian inference based tool is proven to be very effective in the optimization of different characteristics of chemically bonded mold.

1. Introduction

Sand casting is one of the most preferred manufacturing processes for producing metal parts. Increasing demand of dimensionally accurate near net castings has shifted researcher's interest from classical Green Sand Molding process to newer molding process, which uses resin binders [1]. Chemically bonded process are considered more stable and reliable process because of its ability to produce dimensionally accurate castings with good surface finish [2]. Though it is well accepted that the quality of the cast part primarily depends on the quality of mold, still this is least explored for the chemically bonded system [3]. In this work, an attempt was made to optimize the molding parameters of chemically bonded molds in order to produce dimensionally accurate and mechanically stable molds. For the optimization, Bayesian-based optimization technique was explored.

The Bayesian approach can be considered an important tool for optimization when data of various parameters associated with the process is available. One of the earliest applications of this approach was in medicine [4], followed by diagnosis of sensor failure [5], causal analysis for software industries [6], root cause analysis in statistical process control [7], and fault diagnosis for semiconductor manufacturing equipment [8]. Recently, Bayesian inference approach has been explored in the area of metal casting [9]. However, this approach is relatively unexplored for this domain, and yet to be explored.

2. Experimentation

To explore the effect of molding parameters, oil urethane three-part no-bake binder system is used in mold making. The trial experiment was performed with varying input parameters (sand grain fineness no., binder amount and curing time) in three step. The process variable for nine experimental trial and corresponding response parameters data are reported in table 1. Mold compression strength, shear strength, core hardness and core shrinkage are measured for all the experiments. The detailed information about the binder system and experiment process can be obtained from previous research [10].

Table 1. Experimental trial (L9 OA) and corresponding mold properties [10]

No.	Process Variables			Response Parameters			
	GFN (no.)	Wt. % of sand	Hour (hr)	Compression Strength (Kg/cm^2)	Shear Strength (Kg/cm^2)	Core Hardness (Number)	Core Shrinkage (microns)
1	40	1.6	1	4.08	9.10	33	15.2
2	60	2	2	7.56	13.10	57	33.0
3	60	2.4	4	16.16	17.20	83	45.5
4	60	1.6	2	5.12	12.40	42	30.1
5	60	2	4	10.31	13.64	64	63.5
6	60	2.4	1	5.21	9.64	46	53.2
7	80	1.6	4	5.30	11.61	40	52.1
8	80	2	4	4.20	9.40	37	57.1
9	80	2.4	2	7.21	11.20	51	80.6

3. Bayesian Inference

The Bayesian inference is basically a probabilistic model of multivariate analysis, in which multiple outputs can be optimized. It relies on the Bayes' rule [7] that estimates the probability of a hypothesis (H) based on the evidences (e). The Bayes' rule is mathematically represented by following equation

$$P\left(\frac{H}{e}\right) = \frac{P\left(\frac{e}{H}\right)P(H)}{P(e)}$$

where, P(H/e) = conditional probability; P(e/H) = joint probability; P(H) = local probability of hypothesis; P(e) = local probability of evidence

The conditional probability estimates the relationship of the hypothesis (H) to the occurrence of evidence (e) while the joint probability estimates simultaneous occurrence of evidence (e) and hypothesis (H). This concept can be mapped for optimization of input parameters. The input data is processed using Bayesian inference (programmed in MS Excel) to identify optimized output parameters and their specific range of input parameters. It essentially involves computing of posterior probabilities of all input parameters and their range of values.

The total range of each input parameter as well as output parameters are divided into four (i = 1 to 4), and the observed value of each input is classified in one of the four ranges. The input parameter addend (PIP) and output parameter addend (OPMP) are calculated. The values of the posterior probability for the upper range of output parameters (RMP4) are generally computed for all input parameters from an input data. This, in turn, requires computing the values of local probability (LP), prior odd (Oprior), joint probability (JP), conditional probability (CP), likelihood ratio (LR), and posterior odd (Opost) based on the input data. These values can be found out using equations mentioned in table 2. The explanation of above-mentioned terms is briefly explained herewith (Sata, 2015).

Table 2. Computation of posterior probabilities

Parameters	Equation
LPIPRi	NIPRi/NIP
Oprior(IPRi)	LPIPRi/1- LPIPRi
JPOPRi/ IPRi	NIPRi/OPRi/NOPRi
CPIPRi/OPRi	POPRi* JPOPRi/ IPRi/ LPIPRi
LROPRi/ IPRi	CPIPRi/OPRi/1- CPIPRi/OPRi
Opost(OPRi/(IPRi)	Oprior(IPRi) * LROPRi/ RIPi
PRIPRi	Opost(OPRi/(IPRi)/1 + Opost(OPRi/(IPRi)

The values of posterior probabilities further used to identify optimum input parameters. The input parameters and their specific ranges of values are considered to be critical if its values of posterior probabilities are higher.

4. Results

The optimal molding parameter corresponding to maximum mechanical strength of the mold have been determined using Bayesian inference based approach. The optimized range of values for achieving maximum compressive strength, shear strength, as well as core hardness, are GFN: 60-70 (no.), Binder (%): 2.20-2.40; Curing Time: 3.25-4 (hr). Similarly, the optimum range of values for minimum core shrinkage are GFN: 70-80, Binder (%): 1.80-2.00; Curing Time: 1.75-2.50. These range of values are also tuned, and experiments were conducted to achieve maximum strength. It was observed that maximum strength can be achieved by tuning above mentioned values.

5. Conclusion

It can be concluded that Bayesian inference based approach can be employed to optimize the outputs. It mainly computes values of posterior probabilities of input parameters. These values signify the importance of particular input parameter and their specific range of value on output parameters. This approach is considered to be very useful as it has been programmed in MS Excel. This makes it very generic in nature and can be employed to optimize manufacturing process. However, the application of Bayesian inference in the domain of other manufacturing process is yet to be explored, and can be considered for further research in future.

References

[1] H. Khandelwal and B. Ravi, "Effect of Binder Composition on the Shrinkage of Chemically Bonded Sand Cores," *Mater. Manuf. Process.*, vol. 30, pp. 1465–1470, 2015.

[2] J. Thiel, "Thermal Expansion of Chemically Bonded Silica Sands," *AFS Trans.*, vol. 116, pp. 369–378, 2011.

[3] W. A. Ayoola, S. O. Adeosun, O. S. Sanni, and A. Oyetunji, "Effect of Casting Mould on Mechanical Properties of 6063 Aluminum Alloy," *J. Eng. Sci. Technol.*, vol. 7, no. 1, pp. 89–96, 2012.

[4] J. Kershaw, B. a Ardekani, and I. Kanno, "Application of Bayesian inference to fMRI data analysis.," *IEEE Trans. Med. Imaging*, vol. 18, no. 12, pp. 1138–1153, 1999.

[5] O. J. Mengshoel, U. Riacs, and M. Field, "Sensor Validation using Bayesian Networks," *Heal. (San Fr.*, no. ii, 2008.

[6] M. Kalinowski, E. Mendes, and D. Card, "Product-Focused Software Process Improvement," in *In Product-Focused Software Process Improvement*, 2010, vol. 6156, no. June, pp. 92–106.

[7] A. Alaeddini and I. Dogan, "Using Bayesian networks for root cause analysis in statistical process control," *Expert Syst. Appl.*, vol. 38, no. 9, pp. 11230–11243, 2011.

[8] B. Li, T. Han, and F. Kang, "Fault diagnosis expert system of semiconductor manufacturing equipment using a Bayesian network," *Int. J. Comput. Integr. Manuf.*, vol. 26, no. 12, pp. 1161–1171, Dec. 2013.

[9] A. Sata and B. Ravi, "Bayesian inference-based investment-casting defect analysis system for industrial application," *Int. J. Adv. Manuf. Technol.*, vol. 90, no. 9, pp. 3301–3315, Jun. 2017

[10] H. Khandelwal and B. Ravi, "Effect of Molding Parameters on Chemically Bonded Sand Mold Properties," *J. Manuf. Process.*, vol. 22, pp. 127–133, 2016.

[11] Amit Sata:PhD Thesis: IIT Bombay (2015).

The 73rd World Foundry Congress
„Creative Foundry"
23 - 27 September 2018, Krakow, Poland

COUPLED METHODS OF THERMAL ANALYSIS IN THE STUDY OF MATERIALS USED IN THE FOUNDRY INDUSTRY

Agnieszka Roczniak[1], Mariusz Holtzer[1*], Aleksandra Benko[3], Angelika Kmita[2]

[1]AGH University of Science and Technology, Faculty of Foundry Engineering, Mickiewicza 30 Ave , 30-059 Krakow, Poland.
[2]AGH University of Science and Technology, Academic Centre for Materials and Nanotechnology, A. Mickiewicza 30 Ave., 30-059 Krakow, Poland.
[3]AGH University of Science and Technology, Faculty of Materials Science and Ceramics, A. Mickiewicza 30 Ave., 30-059 Krakow, Poland.
**Corresponding address: e-mail: arocznia@agh.edu.pl*

Key words: coupled methods, evolved gases, chemically cured binder, environmental

1. Introduction

Foundry plants emit VOCs and hazardous air pollutants (HAPs), many of which pose a threat to the environment and human health. Over 40 compounds from the HAPs list were identified in the gases emitted from the foundry plants. One of the main sources of VOC and HAPs emissions in foundries are binders used in moulding and core sands. The release of harmful gases takes place at the stage of: forming moulding sands and castings knocking out but the biggest emission occurs during the moulds pouring (as a result of **"slow"** and **"flash"** pyrolysis). It should also be noted, that organic binders in the mould are heated under quite different conditions - in various places of the mould - than during pyrolysis (e.g heating rate, maximum temperature). Differences in heating rates result in the emission volume and composition of gases. Therefore, the results of pyrolysis obtained in the laboratory can not be uncritically used for modeling emissions under real conditions of a foundry [1-9]. The identification and quantification of the pyrolysis products of foundry binders is very important because:

- gases emitted can be harmful for foundry workers and local residents;
- these substances may be adsorbed or may condense in the waste sands, and then leach into the environment when they are used as soil components or for the construction of embankments on garbage dumps.

The assessment of the harmfulness of moulding sands subjected to a high temperature impact in the laboratory tests, can be divided into two stages:

- stage 1: release of gases from the test sample,
- stage 2: determination of the quantity and composition of the emitted gases.

In order to realize both stages in one measurement, we can use the so-called coupled methods, i.e. a combination of thermal decomposition techniques with techniques used to identify gaseous components [10-13].

2. Coupled methods applied for the analysis of materials used in the foundry industry

Coupled methods in the case of "**flash**" pyrolysis are: e.g. Py/GC/MS (pyrolyzer/gas chromatograph / mass spectrometer (Fig.1).

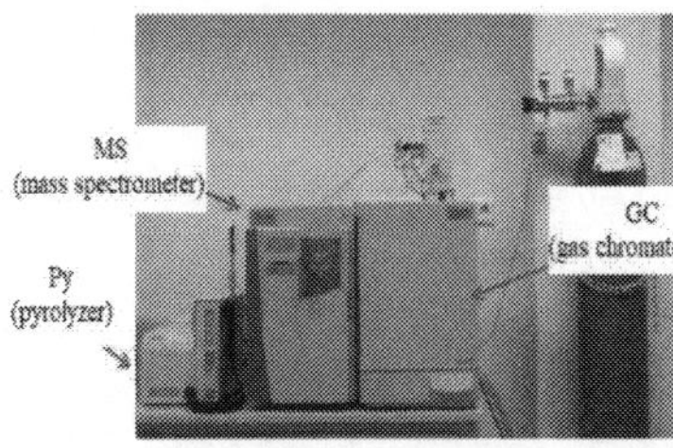

Fig. 1 Measuring system of Py/GC/MS pyrolyzer/ gas chromatograph/mass spectrometer

An exemplary chromatogram and mass spectrum is presented in Fig. 2 a-b.

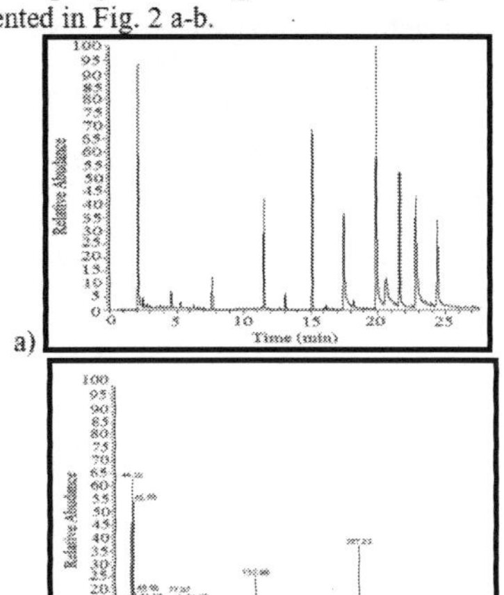

Fig. 2 Exemplary chromatogram of pyrolyzate obtained from hardened resin (a) and mass spectrum (b)

When we want to simulate **"slow"** pyrolysis, we use the thermal analysis coupled with methods used to identify emission components e.g.:

TGA – FTIR, TGA – MS; TGA – GC – MS;

TGA – FTIR – MS; STA – MS, STA – FTIR;

STA – GC –MS; STA – FTIR – MS;

STA - FTIR – GC – MS. ([1])

Figure 3 shows examples of results obtained by the TGA-FTIR method. With this method we obtain results in the form of thermograms and infrared spectra.

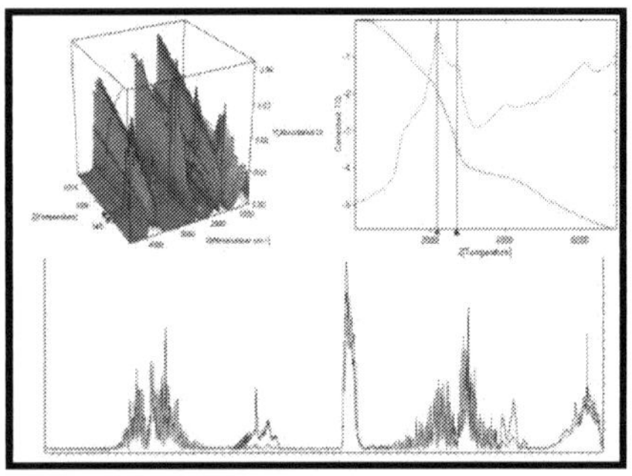

Fig. 3 Sample results obtained by TGA-FTIR method

Simultaneous thermal analysis (STA) is a method of simultaneous application of two research techniques: thermogravimetry (TG) and differential scanning calorimetry (DSC) in relation to one sample. The conditions prevailing during the measurements are identical for TG and DSC signals, i.e. the same atmosphere, pressure, heating rate, thermal contact of the sample with crucible and sensor, radiation effects, etc. This condition can not be met when two separate instruments would be used.

The TGA and DSC curves recorded during the "slow" pyrolysis of the cured resin are shown in Figure 4. Characteristic inflections of the curve, when the sand reduction rate was decreased, is shown in the DTG curve).

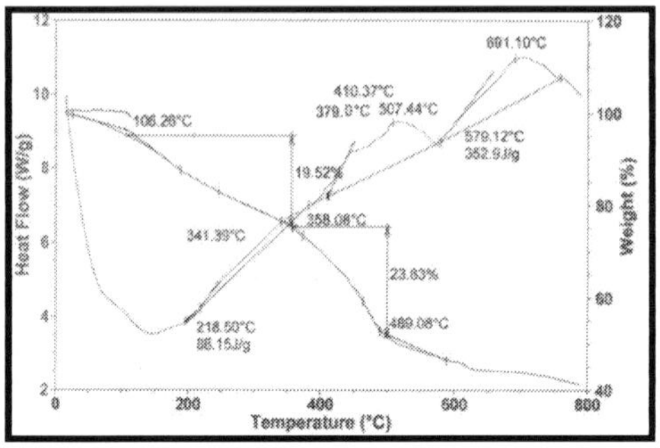

Fig. 4 TGA and DSC curves of the cured resin, heated up to 800°C

Acknowledgments

This work was supported by the grant (statutory work) realized in 2018, at the Faculty of Foundry Engineering AGH (principal investigator: prof. Mariusz Holtzer)

References:

[1] B. Strzemiecka, J. Zięba-Palus, A. Voelkel, T. Lachowicz & E. Socha. (2016). Examination of the chemical changes in cured phenol-formaldehyde resins during storage. J. Chromatogr. A,1441,106-115.

[2] Y. Chen, Z. Chen, S. Xiao & H. Liu. (2008). A novel thermal degradation mechanism of phenol-formaldehyde type resins. Thermochim. Acta 476, 39-43.

[3] A. Kmita, G. Piwowarski, A. Roczniak & M. Holtzer. The activation energy of the thermal degradation of the cured alkaline resole resin. (in press).

[4] M.V. Alonso, M. Oliet, J.C. Dominguez, E. Rojo & F. Rodriguez. (2010). Thermal degradation of lignin–phenol–formaldehyde and phenol–formaldehyde resol resins. J. Therm. Anal. Calorim. 105, 349-359.

[5] B. Howell, C. Yumin & B. Duane. (2003). Assessment of the thermal degradation characteristics of isomeric poly(styrene)s using TG/MS and TG/GC/MS. Thermochim. Acta 396, 167-177.

[6] Authors' own research unpublished.

[7] J.J. Sunol, J. Saurina, F. Carrillo & X. Colon. (2003). Comparison of the thermal behavior of three cellulosen fibres mercerized or submitted to solar degradation. J. Therm. Anal. Calorim. 72, 753-758.

[8] H.E. Kissinger. (1956). Variation of peak temperature with heating rate in differential thermal analysis. J. Res. Natl. Bur. Stand. 57, 217-221.

[9] R. Lee. (2007). Phenolic resin proposed mechanism chemistry and for thermal decomposition. NASA Marshall Space Flight Center. Carton Cloth Phenolic Composite for Abiative Rocket Nozzle Liners.

[10]J.H. Flynn. (1983). The isoconversional method for determination of energy activation at constant heating rates. J. Thermal. Anal. 27, 95-102.

[11]C.A. Lytle, W. Bertsch & M. McKinley. (1998). Determination of novolac resin thermal decomposition products by pyrolysis-gas chromatography-mass spectrometry. J. Anal. Appl. Pyrolysis, 45, 121-131.

[12] R.H. Huang, B.P. Zhang & Y.J. Tang. (2016). Application conditions for ester cured alkaline phenolic resin sand, China Foundry, 13, 231-237.

[13] T. Ozawa. (1970). Kinetic analysis of derivative curves in thermal analysis, J. Thermal. Anal. 2 (1970) 301-324.

---

[1] STA/TG-DSC - simultaneous thermal analysis,
TGA - thermogravimetric analysis,
DSC - differential scanning calorimetry,
FTIR - infrared spectroscopy with Fourier transformation,
GC - gas chromatography,
MS - mass spectrometry.

GATING SYSTEM DESIGN OPTIMIZATION FOR INVESTMENT CASTED PART

Marek Bruna [1]*, Danka Bolibruchová [2], Richard Pastirčák[3], Augustín Sládek[4]

[1]NUniversity of Žilina, Faculty of Mechanical Engineering, Department of Technological Engineering, Univerzitná 8215/1, 010 26 Žilina, Slovakia
*Corresponding address: e-mail: rapture.marko@gmail.com

Key words: Investment casting, microstructure, numerical simulation, shrinkage, optimization

1. Introduction

The paper deals with elimination of critical shrinkage occurring in the investment casted part – link of a track for heavy vehicles. The casting is made of the 20ChGSNM steel described by the Russian GOST standard. Shell was composed of 6 layers with thickness 0.5 mm. Pouring temperature was 1600 ± 5 °C and and the pouring height 500 mm. After pouring, air cooling with exothermic topping material took place. The position of the shrinkage is unacceptable due to the functional significance of the casted part. The occurrence of shrinkage is caused mainly because of hot-spot area and poor gating system design. By failure to ensure directional solidification, the area with an excessive volume of molten metal is isolated and formation of closed or open cavities with oxidized surface occurs. The location of the critical areas are shown in fig.1.

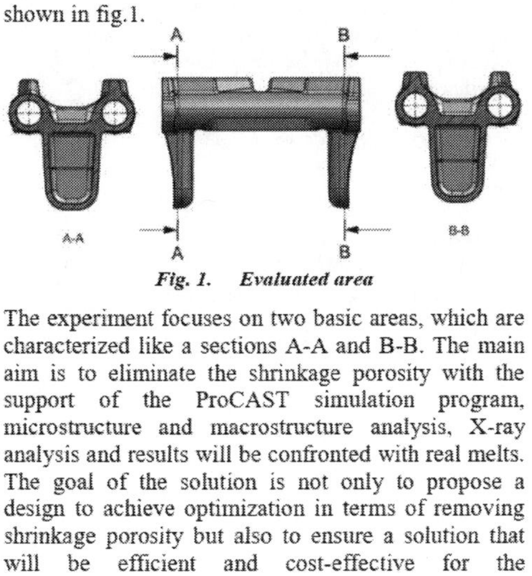

Fig. 1. Evaluated area

The experiment focuses on two basic areas, which are characterized like a sections A-A and B-B. The main aim is to eliminate the shrinkage porosity with the support of the ProCAST simulation program, microstructure and macrostructure analysis, X-ray analysis and results will be confronted with real melts. The goal of the solution is not only to propose a design to achieve optimization in terms of removing shrinkage porosity but also to ensure a solution that will be efficient and cost-effective for the manufacturer.

2. Initial gating system

The initial solution consisted of gluing the casting wax model to the „H-type" inlet system (Figure 2). Dimensions of gates are 35x6x10 (length, width and height). During analysis by the numerical simulation (fig. 3), it was found that there is a high probability of shrinkage porosity occurrence in evaluated areas. The results from the simulations correspond to the results from real melts as can be seen on macrostructure evaluation (Fig. 4).

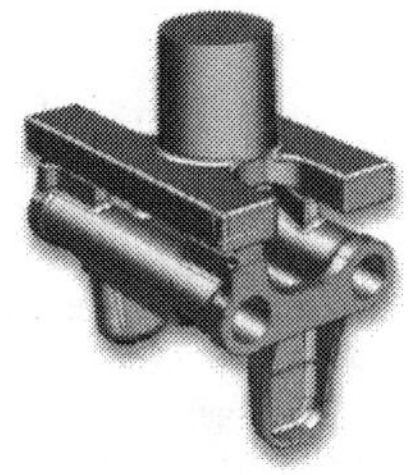

Fig. 2. Initial design

In addition to macrostructure (fig. 4) and microstructure (fig. 6) evaluation, also x-ray analysis was carried out (fig. 5). When looking at the microstructure at the cuts A-A and B-B, it is possible to see in particular the oxide films. Their occurrence is induced by secondary oxidation (reoxidation) on the surface of the melt during melting, casting and filling of the mold. Their entrainment into the casting cavity is caused by the formation of a turbulent flow. The occurrence of shrinkage is also evident. Shrinkage is primarily associated with an inadequately designed gating system and uneven wall thicknesses in the cast. In this case, shrinkage is caused mainly due to the limited feeding of the metal.

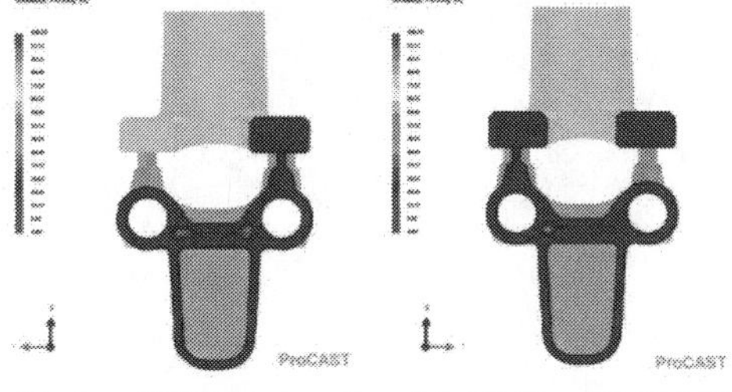

Fig. 3. Shrinkage porosity-cut A-A (left), cut B-B(right)

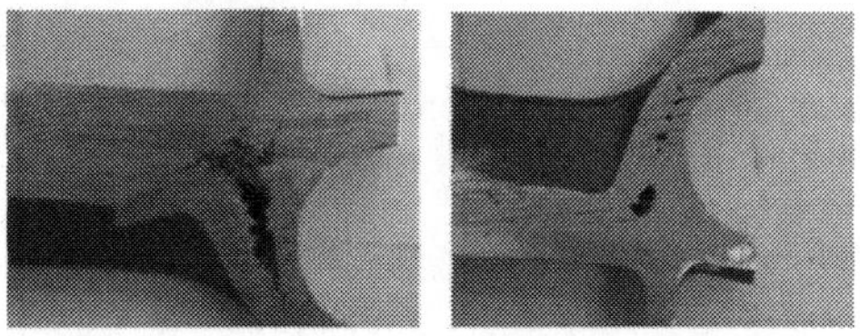

Fig. 4. Macrostructure – cut A-A(left), cut B-B (right)

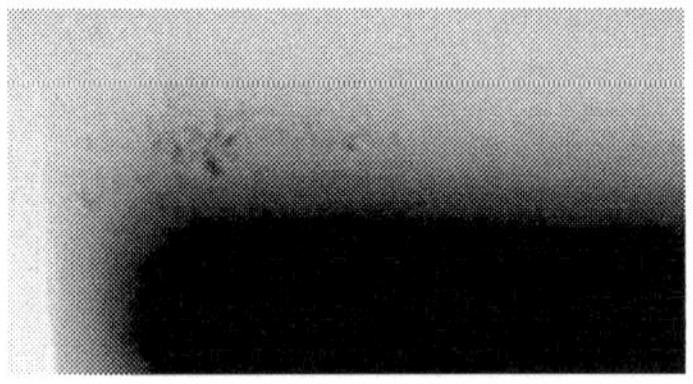

Fig. 5. X-ray – cut A-A

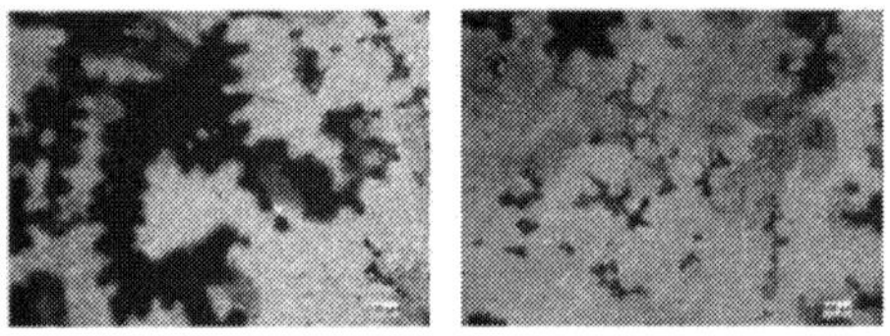

Fig. 6. Microstructure - cut A-A (left), cut B-B(right)

3. Modified gating system

Variant 2 was aimed at changing the overall gating system function and to ensure proper filling character. The „H-type" gating system has been replaced by a "Rod" gating system (Fig. 7), with basic functional dimensions of 40x40x290 mm. The connection between the gating system and the casting was ensured by creating two gates. The gates had a gradually decreasing cross-section to ensure a directional solidification from the casting to the gating system. The gates were created using the FDM rapid prototyping method.

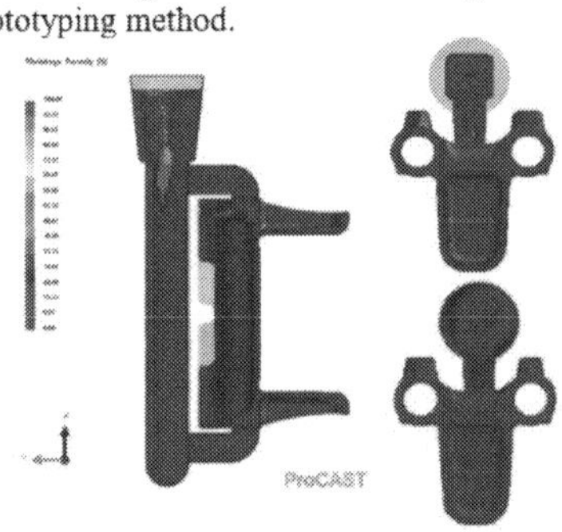

Fig. 7. Modified design (left), cut A-A (up right), cut B-B (bottom right)

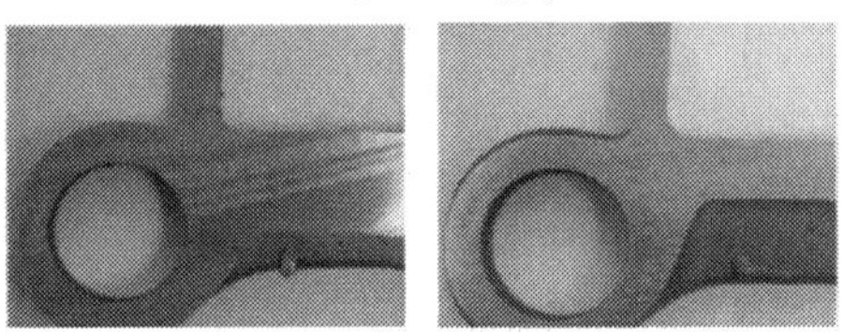

Fig. 8. Macrostructure – cut A-A(left), cut B-B (right)

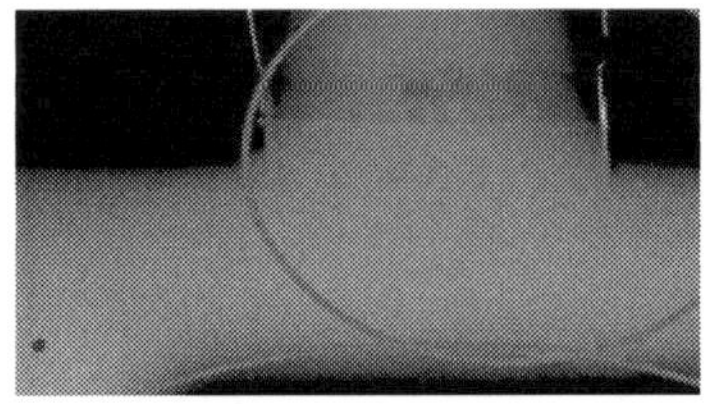

Fig. 9. X-ray – cut A-A

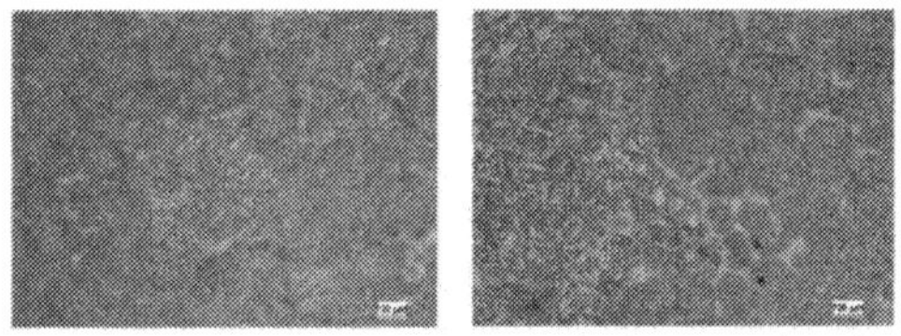

Fig. 10. Microstructure - cut A-A (left), cut B-B(right)

The bottom gate was deliberately shifted from the bottom of the sprue to prevent the cooler metal from entering to the casting cavity and to partially refine the metal. By performing the x-ray test (Fig. 9), the occurrence of the shrinkage was only detected in the upper gate. Transfer of the shrinkage porosity into a gate is a positive result. Also, numerical simulation before the casting process did not identify the occurrence of shrinkage porosity in critical areas (fig. 7). In order to define the overall quality of the controlled areas, the microstructure assessment (fig. 10) and macrostructure analysis (fig. 8) in areas A-A and B-B was also performed. The oxide films have been completely removed, which can be attributed mainly to the ensuring of a uniform metallostatic pressure. The shrinkage and microporosity occurred, but their decrease compared to the primary design was approximately 90%. This phenomenon may be related to a fact, that new design provides bottom filling. The upper gate initially acts as an exhaust, allowing the removal of gases from the casting cavity.

4. Conclusion

The second variant ensures the removal of shrinkage in the evaluated areas and does not represent a major interference in production because it is based on the concept already used by the manufacturer's. Experimental work also pointed to the significant benefit of numerical simulations during optimization of production processes.

The 73rd World Foundry Congress
„Creative Foundry"
23 - 27 September 2018, Krakow, Poland

DEVELOPMENT OF A METHOD FOR MEASURING THE EJECTOR FORCES IN THE DIE CASTING PROCESS

Sebastian Krischke [1]*, Sebastian Müller [1], Torsten Schuchardt[1] and Klaus Dilger [1]

[1]Institute of Joining and Welding, TU Braunschweig, Braunschweig, Germany
**Corresponding e-mail: s.krischke@tu-braunschweig.de*

Key words: die casting, ejector forces, measurement, sensors, casting simulation

The ejector system of the die casting tool has a high influence on the technological and economic performance of the die casting process as it influences the production reliability as well as the quality of the die cast components. Despite its high importance, the layout and the dimensioning of the ejector system is usually based only on the experience of the designer. Especially in the case of new designs this can lead to insufficient part ejection and thus to component defects and production interruptions. The aim of this work was therefore to develop a measuring system for the experimental determination of ejector forces in the casting tool. These investigations provide the basis for the identification of main influencing parameters and for the development of models to simulate ejector forces and component distortion.

1. Introduction

The literature provides only limited information on the knowledge-based dimensioning of ejector systems. The most comprehensive information with regard to this objective can be taken from [1] and [2]. There is also few experimental data available for the transient measurement of ejector forces [3–5]. On the one hand, these sources are relatively old, which is regarded as critical with respect to the use of current release agents, materials, production methods and tolerances. On the other hand, systematic results cannot be derived from the available data, as the data volume is limited and numerous boundary conditions are not documented. An approach for the simulation of ejector forces has already been successfully demonstrated [6]. To improve the calculation results, however, experimental data on the coefficient of adhesion are still required.

2. Scientific approach

The measuring system was developed for an existing die casting tool used to manufacture a structural cast component (Fig. 1). In order to understand the behavior of the casting with regard to local component stress and ejector forces, casting process simulations were carried out. Since none of the simulation programs currently available are capable of calculating the ejector forces, the von-Mises stress in the cast component and the contact pressure between part and die surface were used as the initial indicator for the evaluation of ejector forces. Based on the results of the simulation, influencing parameters that can be attributed to the temperature dependent material behavior of the casting and the tool can be described. The developed measuring system is then to be tested by means of casting tests under parameter variation. For an initial evaluation, the ejector forces measured in the tests were compared with the contact pressures calculated in the simulations.

3. Casting process simulation

Casting simulation software MAGMASOFT and ProCAST were used for the casting process simulation. The simulations were initially performed with standard casting parameters in order to better understand the behavior of the casting. After the first simulation runs, variations of different parameters were carried out. The parameters varied included the closing time, the temperature of the melt, the alloy as well as the duration and the heat transfer coefficient of spraying process. The simulations were evaluated on the basis of the von Mises stresses and the level of the contact pressure right before part ejection. Fig. 1 shows the von Mises stress in the cast component for the two alloys AlSi9Cu3 and AlSi10Mg. Considering the fact the identical casting boundary conditions were applied, Fig. 1 visualizes the extent to which the calculation results depend on the selected material properties. In particular, the faster solidification and the different gradients of the yield strength were considered as main causes.

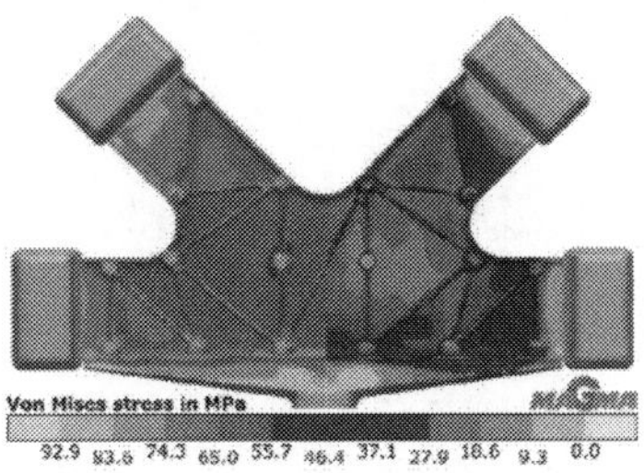

Fig. 1. Simulated von Mises stress depending on the alloy type (left: AlSi9Cu3, right: AlSi10Mg)

The influence of the cast component temperature on ejection forces has already been extensively investigated and described [1-3]. Nevertheless, the depicted approach also enables to study different influencing variables and to quantify the main effects, taking into account the superposition of the individual influencing variables. In this context, Fig. 2 shows the curves for the maximum of the contact pressure between casting part and the surface of the die depending on the closing time as well as the temperature of the melt. As depicted in Fig. 2, the contact pressure increases with higher closing times and declines with higher melt temperatures. The temperature of the part at the time of ejection is assumed to be the main driving force, as it is affected by both of the parameters.

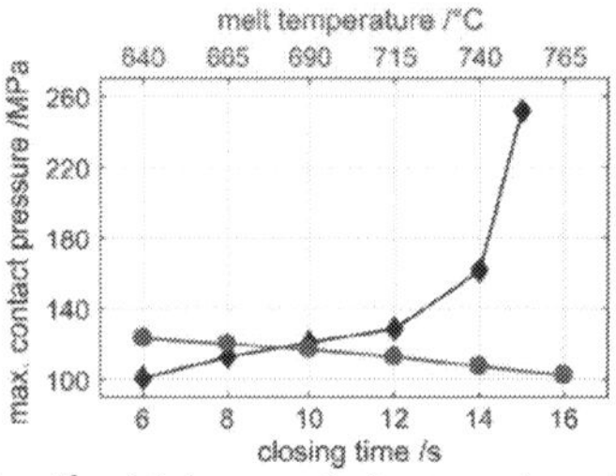

Fig. 2. Simulated max. contact pressure depending on the closing time and the melt temperature

4. Development of the measurement system

For the experimental investigations, an existing die casting tool for the production of a structural casting component was modified. The existing ejector plate was fitted with eight piezoelectric force sensors from KISTLER with a maximum sample rate of 62.5 kHz. Fig. 3 illustrates the schematic representation of the installation position. In order to gain a comprehensive understanding of the ejector force distribution, in total 33 positions can be instrumented. An additional steel cylinder serves to distribute the pressure of the ejectors homogeneously over the surface of the sensors. Next to the force sensors, the tool was also equipped with four thermocouples in order to validate calculated temperatures at the surface of the die casting tool.

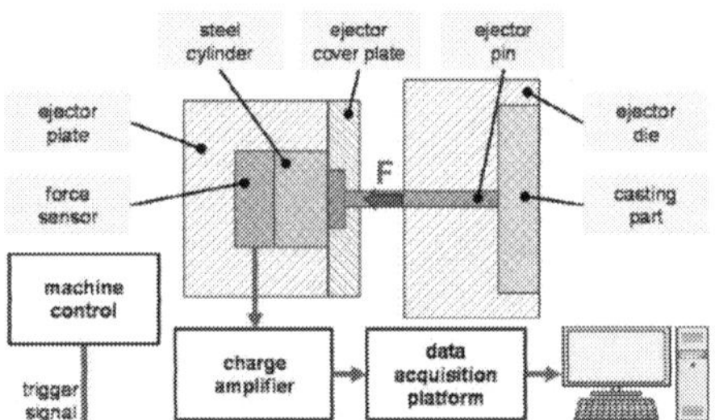

Fig. 3. Sensor configuration for the force measurements

The charge of the sensors was processed with two KISTLER charge amplifiers; the signal was transmitted through a data acquisition platform (DAQ) from NATIONAL INSTRUMENTS to a computer. Due to the high sampling rate and number of sensors, numerous data are already generated for one casting cycle. In order to reduce the amount of data needed for the acquisition of large casting series, a trigger signal was configured in order to activate the data acquisition only during the part ejection process. The testing of the measurement system was done by casting experiments performed on a BÜHLER 53D cold chamber die casting machine with AlSi10Mg as casting material. With the described configuration, it was possible to measure the forces on the ejector pins, as shown in Fig. 4. In addition to the compressive forces of the ejection itself, it was also possible to record the compressive forces during filling and intensification as well as the tensile forces during the opening of the die.

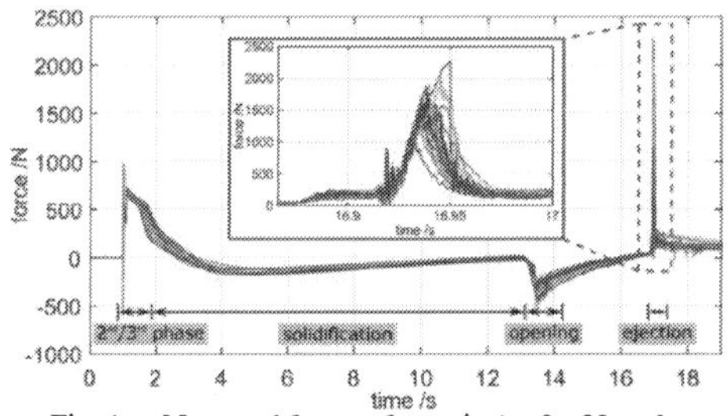

Fig. 4. Measured forces of one ejector for 22 cycles

5. Summary and outlook

Within the framework of this work, an automated system for the measurement of ejector forces in the die casting process was developed. Casting process simulation was used to study the behavior of the already existing die casting part and to identify parameters for casting experiments. Based on the conducted casting series it could be demonstrated that a measurement of highly dynamic ejector forces is possible. Further casting experiments are planned to collect extensive data on the behavior of the ejector forces. This is particularly important for relevant influencing variables that cannot yet be described by the simulation, for example the application of different release agents and the condition of the tool surface. Based on this data, models and methods for calculating and simulating the ejector forces are to be developed in further investigations.

Acknowledgments

The IGF-research project 19.258N of the Research Association Foundry Technology e. V. (FVG) Hansaallee 203, 40549 Düsseldorf, is funded by the AiF within the program to promote Joint Industrial Research (IGF) by the Federal Ministry for Economic Affairs and Energy, following a decision by the German Bundestag.

References

[1] Nogowizin, B. (2008). Entformen des Gusstücks aus dem Formhohlraum beim Druckgießverfahren. DRUCKGUSS, no. 3, pp. 125–136

[2] Nogowizin, B. (2011). Theorie und Praxis des Druckgusses. Berlin: Schiele & Schön Verlag

[3] Tosa, H.; Urakami, A. (1973). Einflussgrößen auf die Druckguss-Auswerfung. Gießerei-Praxis, no. 5, pp. 79–84

[4] Klein, F.; Wüst, J. (1990). Klebeneigung der Legierung GD-AlSi9Cu3 beim Gießen von Druckgussteilen. GIESSEREI 77, no. 10, pp. 344–351

[5] German, A. (1999). Auftrag und Wirkung wassermischbarer Formtrennstoffe beim Druckgießen. Dissertation. Otto-von-Guericke-Universität Magdeburg

[6] Köser, O.; Rückert, J.; Ubl, P. (2015). Modeling and Optimization of Part Ejection in Magnesium High Pressure Die Casting. Proceedings of EMC 2015

The 73rd World Foundry Congress
„Creative Foundry"
23 - 27 September 2018, Krakow, Poland

TEACHING-AND-PLAYBACK APPROACH BASED ON POURING FLOW RATE IN TILTING-LADLE-TYPE POURING MACHINE

Taito Yajima [1*], **Yoshiyuki Noda** [2]

[1] *University of Yamanashi Department of Mechanical Engineering / 4-3-11, Takeda, Kofu, Yamanashi, Japan,*
[2] *University of Yamanashi Department of Mechanical Engineering / 4-3-11, Takeda, Kofu, Yamanashi, Japan,*
**Corresponding address: e-mail: g17tm017@yamanashi.ac.jp*

Key words: tilting-ladle type pouring machine, teaching-and-playback approach, flow rate control

1. Introduction

The casting industry often uses a pouring process in which molten metal is poured from a ladle into a mold by tilting the ladle. Since the process involves molten metal, it is inherently dangerous for workers. In recent years, an automatic pouring machine has been used for improving the working environment of the pouring process and stabilizing the production quality[1]. The automated pouring process has been required that molten metal is quickly and precisely poured into the mold. In the present tilting-ladle type automatic pouring machine, a teaching-and-playback approach has mainly been applied to satisfy such requirements. In the teaching-and-playback approach, the operator teaches the pouring pattern to the pouring machine by manipulating the angular velocity of tilting ladle on the remote-site. Then, the pouring machine pours automatically the molten metal by the playback of the taught pouring pattern[2]. However, since the operator manipulates the angular velocity of tilting ladle on the remote-site, it is difficult to pour precisely the molten metal into the mold. Moreover, even though the molten metal is poured in the same pattern of the angular velocity of tilting ladle, the pouring flow rate can be varied depending on the tilting angle of the ladle. Thus, in the playback mode, it is difficult to pour precisely the molten metal by the pouring machine with the teaching data of angular velocity.

In order to solve these problems, we propose the automatic pouring control system with the teaching-and-playback approach based on the pouring flow rate as shown in Fig. 1. In the proposed approach, the operator can manipulate directly the pouring flow rate, and the flow rate data can be memorized as the teaching data. In the playback mode, the molten metal is poured precisely by the flow rate control with the teaching data of the flow rate.

2. Teaching-and-playback Approach Based on Pouring Flow Rate

In order to enable the operator to manipulate directly the pouring flow rate in teaching mode and pour precisely the molten metal according to teaching data in playback mode, teaching-and-playback approach based on the pouring flow rate is proposed in this study. The block diagram of the proposed approach is shown in Fig. 2. In the teaching mode, the operator makes the joystick rotate for manipulating the flow rate. By using the flow rate control[2], the direct flow rate manipulation can be realized. The teaching data based on the pouring flow rate can be created by the flow rate estimation. The flow rate estimation consists of model simulation of the pouring process and parameters optimization. The model simulation of the pouring process is shown as

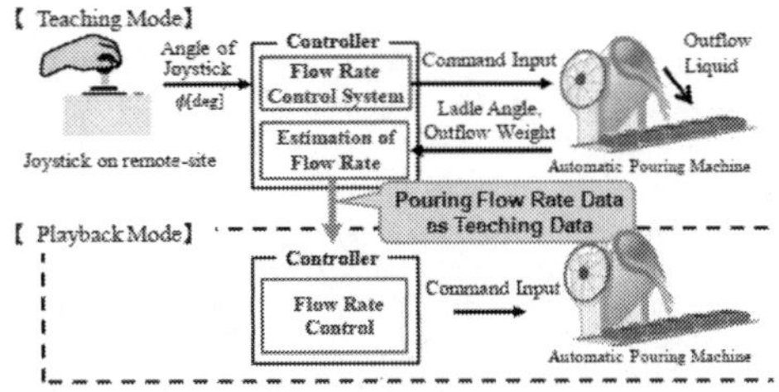

Fig. 1. ***Teaching-and-playback Approach of Pouring Flow Rate***

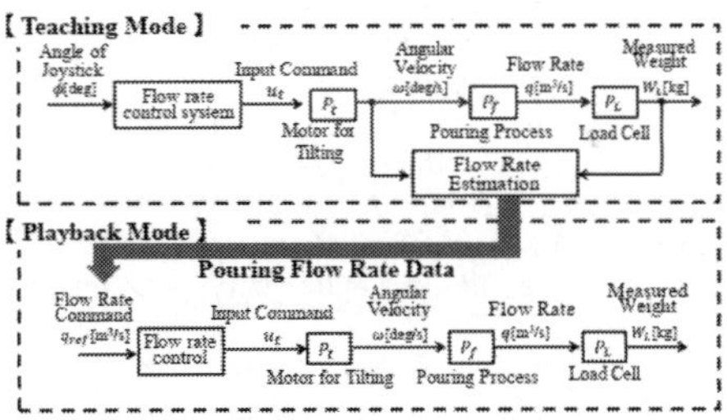

Fig. 2. ***Block Diagram of Teaching-and-playback Approach Based on Pouring Flow Rate***

$$\begin{bmatrix} \dot{\omega} \\ \dot{\theta} \\ \dot{h} \\ \dot{W} \\ \dot{W}_L \end{bmatrix} = \begin{bmatrix} -\frac{1}{T_{mt}}\omega + \frac{K_{mt}}{T_{mt}}u_t \\ \omega \\ -\frac{q(h)}{A(\theta)} - \left(\frac{h}{A(\theta)}\frac{\partial A(\theta)}{\partial \theta} + \frac{1}{A(\theta)}\frac{\partial V_s(\theta)}{\partial \theta}\right)\omega \\ \rho q(h) \\ -\frac{1}{T_L}W_L + \frac{1}{T_L}W \end{bmatrix}, \quad (1)$$

$$q(h) = c\int_0^h L_f(h_s)\sqrt{2gh_s}\,dh_s,$$

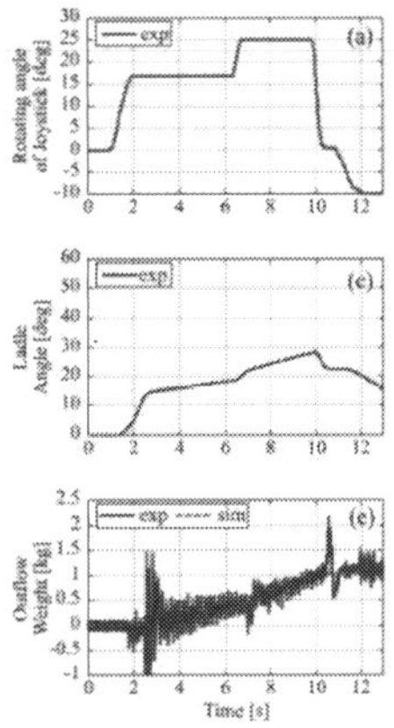

Fig. 3. Experimental Results in Teaching Mode

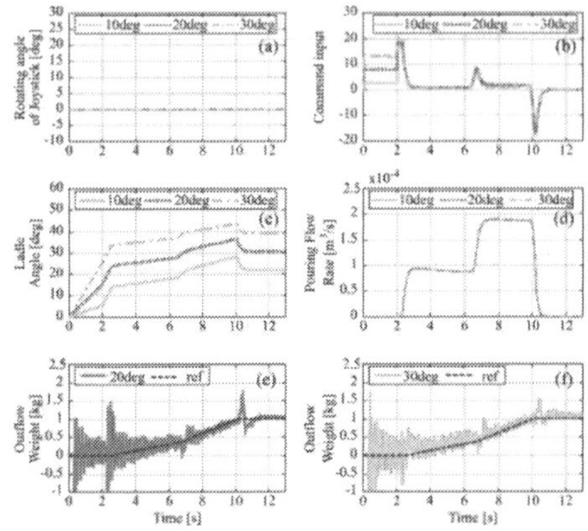

Fig. 4. Experimental Results in Playback Mode

where ω, θ, and u_t show the angular velocity of tilting ladle, tilting angle, and the input command added to the motor, respectively. T_m and K_m are the time constant and the gain of the motor for tilting ladle, respectively. h, q, A, and V_s are the liquid height on the pouring mouth, the pouring flow rate, the upper surface area of the liquid in the ladle, and the liquid volume under the tip of the pouring mouth, respectively. W and ρ are the weight of the outflow liquid and the liquid density, respectively. W_L and T_L are the weigth of the outflow liquid measured by the load cell, and the time constant of the load cell, respectively. c, L_f, and g are the flow rate coefficient, width of the pouring mouth, and a gravity acceleration, respectively. In the model of the pouring process, the flow rate coefficient c and the tilting angle θ_s at begining of the liquid outflow are the unknown model parameters. Therefore, the model parameters are identified by the parameters optimization with the cost function shown as

$$[c, \theta_s] = arg\ min \int (W_{Lact} - W_{Lsim})^2 dt, \quad (2)$$

where W_{Lact} and W_{Lsim} are the weight of the outflow liquid measured actually by the load cell, and the outflow liquid simulated by the model.

In the playback mode, the flow rate control[3] is implemented, and the teaching data of the pouring flow rate is applied to the flow rate control. The teaching data of the pouring flow rate is derived from the model simulation with the optimization parameters. The pouring process can be represented precisely by the model simulation with the optimized parameters. The model parameters optimization is performed by a Gauss-Newton method, and can be solved within 1[s] [3]. By the flow rate control, the pouring flow rate can track precisely the teaching data, even if the tilting angle of ladle is different. Therefore, the molten metal can be poured precisely in the repetitive pouring motion of the pouring machine in the playback mode.

3. Experiment Verification

Fig. 3 show the experimental results in the teaching mode. In Fig.3, (a), (b), (c), (d), and (e) show the rotating angle of the joystick, command input added to the motor, the tilting angle of the ladle, the pouring flow rate, and the weight of the outflow liquid from the ladle, respectively. In Fig. 3 (e), the solid line is the experimental data, and the dashed line is the simulation data derived from the model in Eq. (1). As seen from the results, the operator can manipulate the pouring flow rate easily.

Fig.4 shows the experimental results in the playback mode. In Fig. 4, the graphs from (a) to (f) show the rotating angle of the joystick, the command input, the tilting angle of the ladle, the pouring flow rate, the outflow weight in case of the tilting angle at beginning of the outflow liquid, $\theta_s = 20$[deg], and the outflow weight in case of $\theta_s = 30$[deg], respectively. In Fig. 4(e) and (f), the line styles are the same manner as Fig. 3(e). As seen from the results, the taught pouring flow rate was reproduced precisely in the playback mode with the proposed approach.

4. Conclusions

In this study, we proposed the flow rate based teaching-and-playback approach which has the flow rate control and estimation. In the teaching mode, the operator can manipulate the pouring flow rate directly, and the estimated pouring flow rate is memorized as the teaching data. The taught pouring flow rate can be reproduced precisely in the playback mode. The effectiveness of the proposed approach was verified by the experiments.

References

[1] W. Lindsay (1983). Foundry Trade. Automatic Pouring and Metal Distribution System. Journal, February, no. 6. 151-176.

[2] Y. Noda, K. Terashima. (2007). Modeling and Feedforward Flow Rate Control of Automatic Pouring System with Real Ladle, Journal of Robotics and Mechatronics, Vol.19. No. 2. 205-211.

[3] Y. Noda, M. Suzuki, K. Ohta. (2016). High-precision Automatic Pouring Machine with Adaptive Control Parameters. Proceedings of World Foundry Congress. O151.

The 73rd World Foundry Congress
„Creative Foundry"
23 - 27 September 2018, Krakow, Poland

OPTIMIZATION OF THE RUNNER SHAPE CONSIDERING THE MOLTEN METAL FLOW OF PRODUCT IN DIE CASTING

Tokiya Okuno[1], Keigo Somiya[1], Yuto Takagi[1], Ken'ichi Yano[1], Ryusei Kawatani[2], Jun'ichi Ogura[2]

[1]Mie University, Faculty of Eng., Dept. of Mech. Eng., 1577 Kurimamachiya-cho, Tsu, Mie, Japan,
[2] Yamaha Motor Co., Ltd., Iwata, Japan
**Corresponding address: e-mail: yanolab@robot.mach.mie-u.ac.jp*

Key words: die casting, CFD, optimum design, runner, air entrainment

1. Introduction

The die casting method has a number of advantages and is used for the production of various industrial products such as automotive parts. Gating system design in die casting is a key to realizing satisfactory molten metal flow and suppressing product defects. Particular care must be taken in designing the runner. A large amount of air entrainment occurs when the molten metal passes through the runner interior, so that an internal defect such as a blow hole can occur in the runner. Runners and gates are now being specifically designed to prevent air entrainment defects [1-3]. The generation of air entrainment defects depends greatly on the shape of the product. If optimization is only carried out without considering the molten metal flow in the product, the optimization process might not be sufficient.

In this research, in order to reduce air entrainment defects in die casting, we optimized the runner shape with consideration of the molten metal flow in the product.

2. Formulation of optimization problem

The products targeted in this research are asymmetrical, with left and right cage shapes. The runner shape is trifurcated and the gate position cannot be changed.

Optimization of the product shape was first carried out without considering the optimization of the gating system or the flow of molten metal. When optimization of gating system design was performed without considering the product, it was confirmed that air entrainment defects were still occurring in the left circular part. In order to reduce air entrainment defects, it is necessary to change just the flow in the product interior.

Optimization of the runner shape was based on the results of computational fluid dynamics (CFD) simulation analysis. In order to evaluate air entrainment defects in the product, the following equation was defined to model the optimization problem.

$$\text{Minimize}\, J = \frac{1}{|\Omega_{\text{mold}}|} \sum_{i \in \Omega_{\text{mold}}} V_{\text{air,i}} \tag{1}$$

Here, V_{air} is the entrained-air volume ratio in each mesh cell in the CFD simulator, Ω_{mold} is the set of mesh cells in the product section in Fig. 1, excluding the exhaust system such as overflow and vent. The smaller the function value J, the fewer the air entrainment defects in the product. This optimization problem is solved using a nonparametric shape optimization method based on a genetic algorithm (GA).

3. Nonparametric shape optimization method

In this method, a curve on the xy plane, which is a component of the shape, is an individual in GA, and this curve is expressed as a vector of a polynomial function of parameter t ($0 \le t \le 1$) as shown in equation [4]:

$$\mathbf{S}(t) = \sum_{i=0}^{n} \begin{bmatrix} a_i^x \\ a_i^y \end{bmatrix} t^i \tag{2}$$

In equation (2), n is the order of the polynomial, and a_i^x and a_i^y are the coefficients. Also, a genetic operation crossover X is defined as the following mathematical synthesis of the parents $\mathbf{S}_p(t)$, $\mathbf{S}_q(t)$:

$$\begin{aligned} \mathbf{S}_c(t) &= X\left(t, r_0, r_1, \mathbf{S}_p(t), \mathbf{S}_q(t)\right) \\ &= f(t, r_0, r_1)\mathbf{S}_p(t) \\ &\quad + (1 - f(t, r_0, r_1))\mathbf{S}_q(t) \\ &\quad + (r_1 - r_0)t(1 - t) \\ &\times \left\{\left(\mathbf{S}_p(1) - \mathbf{S}_q(1)\right)t - \left(\mathbf{S}_p(0) - \mathbf{S}_q(0)\right)(1 - t)\right\} \end{aligned} \tag{3}$$

In equation (3), $\mathbf{S}_c(t)$ is a newly generated child, while r_0 and r_1 are standard normal random numbers. Also, the function f is a random linear fuction with respect to t defined as

$$f(t, r_0, r_1) = \frac{1}{2}\{(r_0 + 1)(1 - t) + (r_1 + 1)t\} \tag{4}$$

4. Setting of CFD simulation

The mesh cell in the CFD simulator was uniformly set at 3 mm, and it was set at 1.5 mm at extremely narrow places such as the gate part. In addition, analysis was made in consideration of gravity, surface tension, viscosity and heat conduction. The molten metal was assumed to be aluminum alloy ADC 12, which is commonly used in die casting. In this study, FLOW-3D, a general-purpose CFD software manufactured by Flow Science Co., Ltd., was used for CFD simulation [5].

5. Optimization result and analysis

The individual distribution map in optimization is shown in Fig. 1; the shape derived by optimization is shown in Fig. 2.

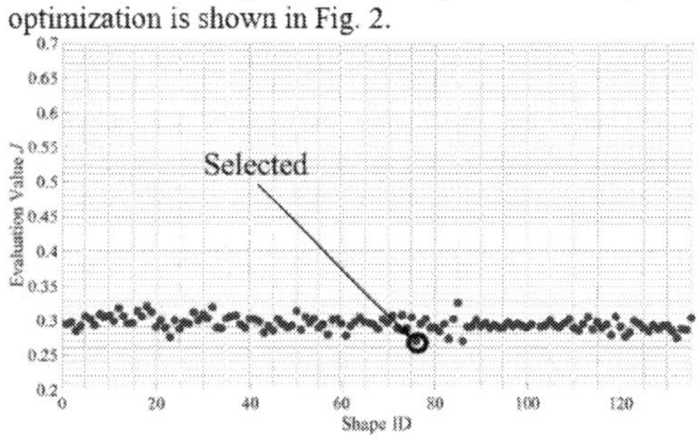

Fig. 1. Distribution diagram generated by optimization

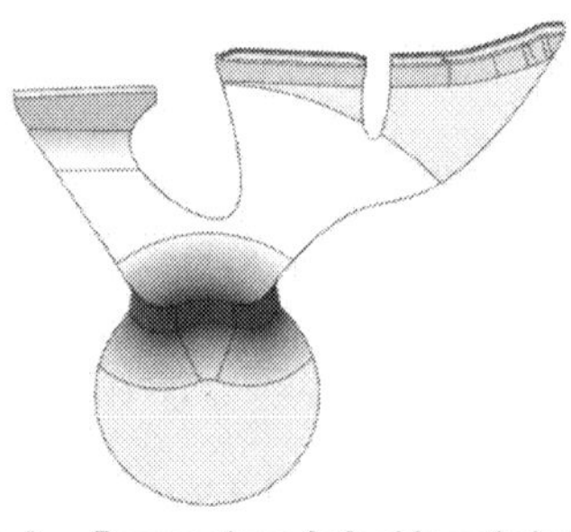

Fig. 2. Runner shape derived by optimization considering molten metal flow of product

Evaluation value J is the entrained-air volume ratio calculated using the air entrainment model in the CFD simulator. This model is a numerical representation of the degree of rampage of flow. For this reason, the entrained-air volume ratio is a value representing relative probability of air entrainment. In the experiment of the actual machine carried out beforehand, there was a correlation between the locations of air entrainment defects in the product and the distribution the of entrained-air volume ratios. This result confirmed that the lower the numerical value of J, the lower the probability of air entrainment defects.

In the gating system design derived by optimization, the value of the entrained-air volume ratio at the circular part on the left side of the product confirms that the case taking into account the molten metal flow is smaller than that not taking into account the molten metal flow. In the shape shown in Fig. 2, it is clearly shown that the width of the left runner is widened. As the width of the runner increases, the inflow speed of the molten metal slows down. As a result, the width of the runner is considered to be a major factor to prevent air entrainment in the molten metal. Comparing the runners on the center and the right side, it was confirmed that the shape was very similar.

Also, the average value of the entrained-air volume ratio in the product, which is the function value J in the optimization problem, was compared in each gating system design. The results are shown in Table 1.

Table 1. Comparison of average value of entrained-air volume ratio in product

gating system design	Evaluation value J
Conventional	0.3651
No consideration of the molten metal flow	0.3013
Consideration of the molten metal flow	0.2695

As is shown from this result, the shape derived by optimization considering the product shows the smallest value. In sum, it is possible to improve the flow of the molten metal in the product part and to realize a gating system design that can reduce the occurrence of air entrainment defects.

6. Conclusions

In this study, we optimized the design shape considering the molten metal flow of the production die casting. First, we carried out an optimum design analysis without considering molten metal flow in the product; then, we conducted actual experiments using the gating system design derived by optimization, which revealed the importance of considering the molten metal flow in the product. Next, we adopted the average value of the entrained-air volume ratio calculated by the CFD simulator as the evaluation value, thereby deriving a scheme to prevent air entrainment defects by optimization. The optimization process performed with consideration of the part's molten metal flow showed good results for air entrainment defects.

References

[1] B. H. Hu, K. K. Tong, X. P. Niu and I. Pinwill. (2000). J. Mater. Process. Technol., 105. 128-133.
[2] B. D. Lee, U. H. Baek and J. W. Han. (2012). J. Mater. Eng. Perform. 21. 1893-1903.
[3] K. Kanazawa, K. Yano, R. Kawatani, J. Ogura and Y. Nemoto. (2016). Proc. The 72nd World Foundry Congress, paper O-146.
[4] K. Kanazawa, K. Yano, J. Ogura and Y. Nemoto. (2014). Proc. ASME Int. Mech. Eng. Congr. & Expo. paper No. IMECE2014-37419.
[5] Flow Science, Inc., http://www.flow3d.com.

The 73rd World Foundry Congress
„Creative Foundry"
23 - 27 September 2018, Krakow, Poland

MULTI-SCALE TOPOLOGICALLY OPTIMIZED COMPONENTS MADE BY CASTING AND ADDITIVE MANUFACTURING

Kalle Jalava [1]*, Mika Salmi [1], Kirsi Kukko [1], Juhani Orkas [1]

[1]*Department of Mechanical Engineering, Aalto University, Otakaari 4, 00076 Aalto, Finland*
**Corresponding address: e-mail: kalle.jalava@aalto.fi*

Key words: Additive manufacturing, topological optimization, manufacturability, 3D-printing, design of castings

1. Introduction

The onset of additive manufacturing has enabled production of very complex component geometries, even straight to metal. Strictly taken, casting methods with traditional style patterns, moulds, and other types of permanent tooling are not capable of feasibly producing these type of geometries, while different direct manufacturing methods have their own challenges [1]. E.g. regarding component size, batch size, speed etc., these production processes are on complete different ends of the spectrum. However, taking a more wholesome approach where different methods are used complementary rather than exclusively, one could feasibly manufacture components that might have seemed unnecessarily complex, or even contrived, earlier.

Topological optimization is one method where such highly complex, almost organic looking geometries arise. Topological optimization is a method where material placement is optimized within a specific design space and is based on specific loads and boundary conditions [2]. In most cases, topological optimization is more about studying a concept to be refined rather than a tool to produce final designs. This is mostly due to the method itself not having anything to do with manufacturability by default. Advances in different manufacturing technologies and simulation capabilities might very well change all that. Constraints and penalization can be used to promote more processable geometries [3]. To demonstrate some possibilities of mixing manufacturing processes, three cases of topology-optimized components manufactured through applying casting and additive manufacturing routes are presented in this work.

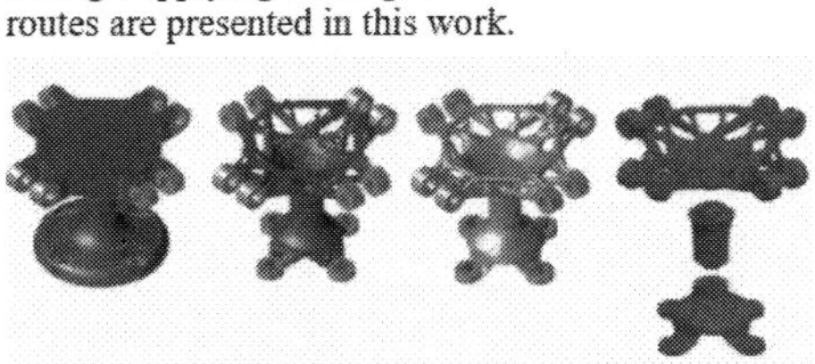

Fig. 1. From left to right, propeller hub design space, raw data, smoothed geometry and final design split into 3 pieces for 3D-printing

2. Experimental work

Investment casting methods lend themselves nicely to the production of intricate shapes, as the process is based on lost patterns. 3D printable waxes and plastics are easily integrated into contemporary industrial processes without the need for major changes. Fig. 1. shows process stages of a hypothetical propeller hub geometry optimization and design to be produced with investment casting. In this case, the nature of the process meant that the final design was to be a positive shape. The final 3-part design was 3D printed in wax and used as a pattern in conventional investment casting process and cast in aluminium.

Fig. 2. A-arm design space, raw data and smoothed geometry

Fig. 2. shows process stages of a hypothetical automotive A-arm optimization process, designed to be manufactured with sand casting. Sand casting is one of the most traditional industrial production methods, capable of producing small to massive components. The needed negative mould shapes are often produced through permanent patterns and core-boxes. Thus, manufacturing many raw topologically optimized shapes would require almost absurd use of cores. Methods like sand binder jetting can alleviate those problems, as there is not as much need to contemplate design parameters like undercuts, drafts etc.

Fig. 3. Multiple piece mould construction

For the A-arm, the process required a negative geometry, a mould cavity to be designed for sand printing. To substantiate the claim that parting lines and other basic casting design viewpoints are not as important as in a traditional process, the cavity geometry was simply split in multiple parts. The heavy splitting mostly helps in cleaning unbound sand from complex cavities. Fig. 3. illustrates the used multi-part mould assembly. The final component size was roughly 60x40x10 cm (LxWxH). Fig. 4. shows the cast component.

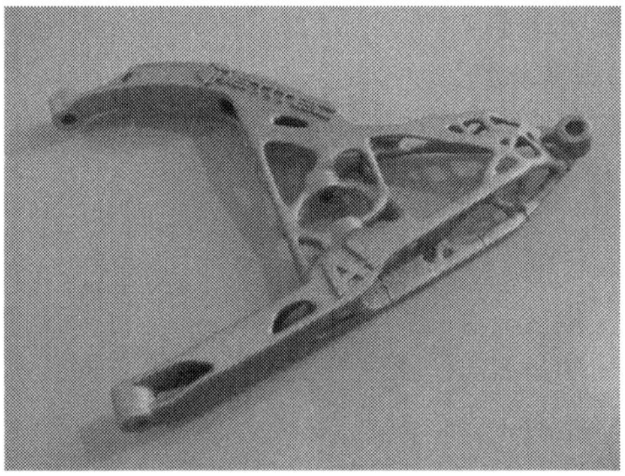

Fig. 4. Final cast a-arm

If a pick-the-best-of-everything style of design thinking is taken into the extreme, additional possibilities come into play. Mixing and matching materials produced with different processes is possible. Demonstrating this approach, the propeller hub geometry was reused. The more intricate top part was 3D printed in wax, and investment cast, the middle shaft machined, and the base additively manufactured with laser sintering. These separate parts were made in stainless steel, namely AISI 316, and later laser welded to complete the assembly. The resulting component is shown in Fig. 5.

Fig. 5. Multi-part assembly pictured on the left and a straight cast design shown on the right

3. Discussion

Approaches like topological optimization are often associated to processes like metal 3D printing, as manufacturing complex geometries lend themselves nicely to many additive processes. There has been relatively little focus on optimized components made as castings, although the processes fit nicely together if even traditionally mouldable manufacturing constraints are considered [4]. However, there are many modern technologies to be taken advantage of in both additive manufacturing, and metal-casting fields. Straight to metal printing generally still has challenges like size and material restrictions compared to castings, although the situation is constantly changing in a developing market.

Using additive manufacturing as a tool alongside conventional casting processes can spawn a best-of-both-worlds kind of a situation. The three different cases presented in this work are examples of such possibilities. The small-scale propeller hub is easily producible through casting, metal printing and even a big mix of many processes. The bigger A-arm is big enough to present challenges to metal, wax or plastic printing but not as a sand mould print. Although the concept of topological optimization itself is nothing new, a lot current research is being made partly influenced by all the new additive manufacturing methods. As with simulation approaches, one should always have a healthily questioning mind on the performance of algorithms, applied methods and inner workings of the software used [5].

4. Conclusion

Three distinctive styles of component manufacturing were studied in this work. The following conclusions can be made from the findings

- The use of wax patterns in a conventional investment casting process is a viable way to produce topologically optimized complex components, as the process uses lost patterns
- Sand mould printing is one conceivable way to achieve complex geometries in much larger scales
- Mixing of different production methods such as casting, machining, welding and additive manufacturing of metal parts is possible using suitable materials

References

[1] Thomas-Seale, L.E.J., Kirkman-Brown, J.C, et al, (2018). The barriers to the progression of additive manufacture, Int. J. Prod. Econ., 198, 104-118.

[2] Bendsøe, M.P., Sigmund, O., (2004). Topology Optimization, Springer.

[3] Gardan, N., Schneider, A., Topological optimization of internal patterns and support in additive manufacturing, (2015), J. Manuf. Syst., 37(1), 417-425.

[4] Li, Q., Chen, W., et al, (2018) Topology optimization design of cast parts based on virtual temperature method, Computer-Aided Design, 94, 28-40.

[5] Rozvany, G., (2009). A critical review of established methods of structural topology optimization., Struct. Multidiscip. Optim., 37(3), 217–23

The 73rd World Foundry Congress
„Creative Foundry"
23 - 27 September 2018, Krakow, Poland

PRODUCT DESIGN AND MANUFACTURING IN FOUNDRY INDUSTRY USING AUGMENTED VIRTUAL REALITY TECHNOLOGY

Yashesh Pandya [1]*, Jozef Suchy [1], Nagaraj S. V. [2], Jagadeesh Kannan R. [2]

[1]AGH University of Science & Technology, Faculty of Foundry Engineering, D-8 Reymonta23, Kraków, Poland,
[2] VIT University (Chennai Campus), School of Computing Science & Engineering, Chennai, India,
*Corresponding address: e-mail: ypandya@agh.edu.pl

Key words: Virtual Reality, Augmented Reality, Industry 4.0, Modern Foundry of Future, Internet of Things

1. Introduction

This paper deals with virtual reality in context of Industry 4.0 projects. Attributing to ample computing power, virtual reality permits visualisation of virtual objects in both professional and public spheres. With virtual reality, solving advanced problems has become very easy; particularly those based on industry 4.0 standards. Industry 4.0 requires association between all systems of modern machines (physical systems, embedded systems, sensors, actuators, electronic hardware, software system etc.) via integrated data chains and these links are used among different things, for monitoring operational conditions and share virtual models within the development cycle, both of which can be visualized via virtual reality.

2. Virtual Reality – Step in the Future

The virtual reality technologies can be divided into 2 groups: Semi-Immersive virtual reality and full-immersive virtual reality. Semi-Immersive virtual reality includes devices such as monocular headsets. Binocular headsets, Oculus Rift and HTC VIVE VR are full-immersive devices. The difference between these groups is in the rate of immersion to the virtual surroundings. Full-Immersive devices offer three-dimensional projection with a large field of vision to the user, compared to semi-immersive device.

a. Application of Virtual Reality in Machine Transformation

Virtual reality can be used to reduce time necessary for developing machine tools and thus reduce the manufacturing cost. Virtual handling and operation of machine, Fig. 1, will facilitate corroborative functionality of the designed solution and procure results of simulations (e.g. casting, kinematics, and dynamics) [1]. A virtual foundry industry may be a simulation model whereas its sub models represent production cells of an actual foundry machines. This simulation model is used for testing of manufacturing system design and control. [2] The model of virtual foundry industry can be used for analyzing operations at the workplace. The results of this simulation can be used to improve layout of machines within the workplace environment. Effective layout could cause easy operability and thus to allow proper placement of worker stations within the surroundings [3].

Fig 1. Operation of Machinery using Virtual Reality Platform [7]

Virtual environment can also be used for virtual machining, Fig. 1, during which the method of edge, turning, drilling and grinding is simulated. The results of this simulation provide information concerning factors affecting the quality and machining time [4]. The main advantage is eliminating the danger of use for each users and the machines [5]. As an example, it can be used for learning the operation of Kiln. The trainee program may also contain real-time material removal method simulation [6].

3. Industry 4.0 Initiative

In the current state of Foundry Industry, machines are designed and produced from parts created by corporations from various corners of the globe, manufactured with respect to approved standards in specific factories with chains of single purpose production cells optimized for one certain operation. With industry 4.0, a simpler approach was presented, transforming from single-purpose cells and unidirectional production into fully integrated cloud-based product flow. In this so-called Modern Foundry of Future, every production cell has access to Global Virtual Model (GVM) of manufacturing machine and based on the communication with the individual cells an automated operation with real-time self-adjustments is achieved. During the manufacturing process, user intervention is least and most of the machine-to-human interaction is related to monitoring and managing of the overall flow, Fig. 3. The pre-production phase also advantages from this transformation. With advanced visualisation techniques, such as virtual reality, global Virtual Model will function as a common base for sharing current state of product development, simulation results, predicted operating conditions and other information for effective collaboration. In the post-production phase, same approach can be used to monitor operating conditions (e.g. vibrations, fatigue, heat transfer, sound emission etc.) for diagnostics purposes and further optimisation of the product.

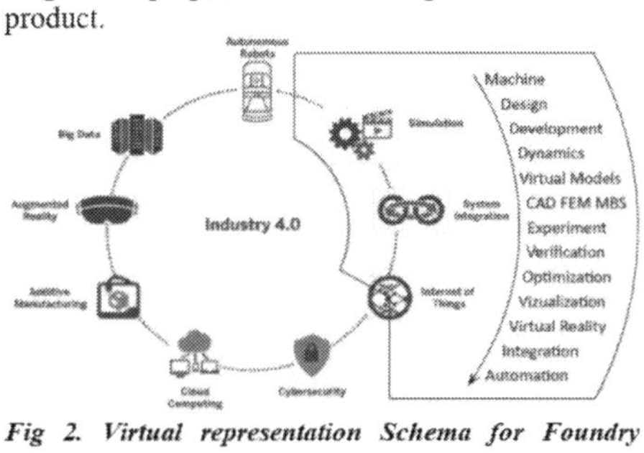

Fig 2. Virtual representation Schema for Foundry Industry 4.0

As mentioned above, sharing key info via GVM during development cycle is the base for effective collaboration of cooperating facilities. Downloading, uploading and updating status of developed product and simulation results means processing vast amount of information, with respect to industry 4.0 referred to as big data. Unfortunately, current state of virtual modeling and not powerful enough processing systems aren't prepared for this step. However, the need of modern Foundry Industry is to, among different things, increase efficiency, the time frame for designing, simulations and optimisation within the pre-production phase is very limited.

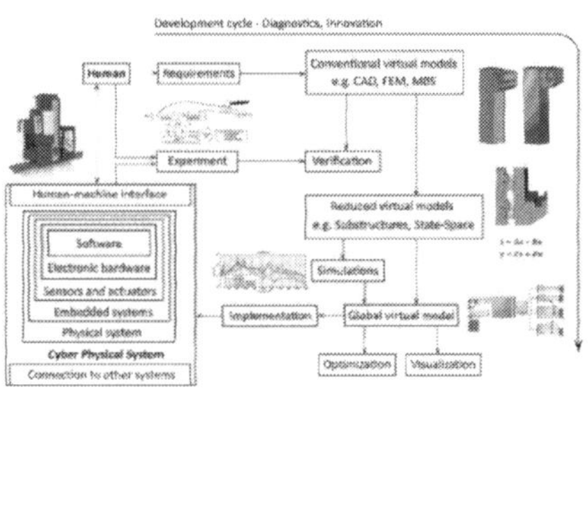

Fig 3. Representation of Virtual Model using CPS (Cyber Physical System) Ecosystem.

Besides, designing machines from only static point of view is no longer adequate and approach based on dynamic models is required. Therefore, the next logical step is to create ways and techniques for implementation of virtual models into additional manageable form, which can be effectively implemented into GVM and CPS, allowing us to perform real-time model updating and visualization.

Fig 4. Product Designing & Visualization using Virtual Reality Platform [7]

4. References

[1] Y. Altintas, C. Brecher, M. Weck, and S. Witt, "Virtual Machine Tool," CIRP Ann. - Manuf. Technol., vol. 54, no. 2, pp. 115–138, Jan. 2005.

[2] S. Jain, N. Fong Choong, K. Maung Aye, and M. Luo, "Virtual factory: an integrated approach to manufacturing systems modeling," Int. J. Oper. Prod. Manag., vol. 21, no. 5/6, pp. 594–608, May 2001.

[3] X. Yang, R. C. Malak, C. Lauer, C. Weidig, H. Hagen, B. Hamann, J. C. Aurich, and O. Kreylos, "Manufacturing System Design with Virtual Factory Tools," Int. J. Comput. Integr. Manuf., vol. 3052, no. March 2014, pp. 1–16, 2013.

[4] T. S. Mujber, T. Szecsi, and M. S. J. Hashmi, "Virtual reality applications in manufacturing process simulation," J. Mater. Process. Technol., vol. 155–156, no. 1–3, pp. 1834–1838, 2004.

[5] A. Sanz Lobera, I. González, A. J. Castejón, and J. L. Casado, "Using Virtual Reality in the Teaching of Manufacturing Processes with Material Removal in CNC Machine-Tools," Mater. Sci. Forum, vol. 692, pp. 112–119, 2011.

[6] T. Yu, X. Zhang, W. Liang, and W. Wang, "A web-based virtual system for turn-milling center," Int. J. Adv. Manuf. Technol., vol. 67, no. 9–12, pp. 2395–2409, 2013.

[7] J. Kovar, K. Mouralova, F. Ksica, J. Kroupa, O. Andrs and Z. Hadas, "Virtual reality in context of Industry 4.0 proposed projects at Brno University of Technology," 2016 17th International Conference on Mechatronics - Mechatronika (ME), Prague, 2016, pp. 1-7.

[8] K. Jansson, M. Höynälänmaa, I. Karvonen and M. Ollus, "Bridging virtual models and reality for lifecycle support in process industries," 2007 IEEE International Technology Management Conference (ICE), Sophia-Antipolis, 2007, pp. 1-8.

STUDY OF VISCOSITY OF WATER- AND ALCOHOL-BASED PROTECTIVE COATINGS

Jarosław Jakubski, Emilia Wildhirt*, Małgorzata Sapińska, Sylwia Sitko

AGH University of Science and technology, 30-059 Kraków al. Mickiewicza 30, Poland
**Corresponding address: emiliaw@agh.edu.pl*

Key words: protective coatings, viscosity, flow curves, thixotropic fluids

This article presents a study of viscosity which was conducted on the Rheotest 2 rheometer located at the laboratory of the Department of Moulding Materials, Moulding Technology and Cast Non-ferrous Metals at the Faculty of Foundry Engineering in Krakow. Measurements were taken of water- and alcohol-based protective coatings both in the initial form and after the addition of appropriate amounts of diluents in the form of water or ROKAfenol. During the study, flow curves were determined, and the impact of modification of the diluent on the viscosity of selected coatings was studied.

1. Introduction

For the study, one water-based coating and one alcohol-based coating were used. The water coating is based on a zirconia slurry with the addition of ceramic materials. It is used with steel castings. It is intended to prevent the occurrence of defects arising due to expansion of quartz sands and penetration of the molten metal.

The alcohol coating is a zirconia-graphite coating based on fire-resistant zirconium silicate with the addition of dark beige graphite. It is used in high-quality iron castings and spheroidal casting with variable wall thickness. Its most important feature is its very good coverage.

Study of viscosity was conducted on these coatings to establish the impact of amount of diluent and the addition of surfactants on the behaviour of the coating. Undiluted coatings and coatings diluted with water and rokafenol were used. The water-based coating was diluted by the addition of 2%, 5% and 10% of water, while the alcohol-based coatings was diluted by the addition of 2% and 5% rokafenol.

The Rheotest 2 rheometer, illustrated in figure 4, consists of two coaxial cylinders. The internal cylinder allows for easy regulation and temperature stabilization, and is linked to an external thermostatic system. A sample of the studied substance which is injected into the ring-shaped gap between the internal and external cylinders can be subjected to shear forces for any length of time. During rotation of the internal cylinder, laminar flow arises, based on which the shear stress, shear rate, and viscosity can be determined. The device is connected to a computer, thanks to which direct, graphic observation of all measured parameters is possible, as well as programming of the measurement process. Study using the rheometer allows data to be obtained which can be used to generate flow curves for the studied materials [1-3].

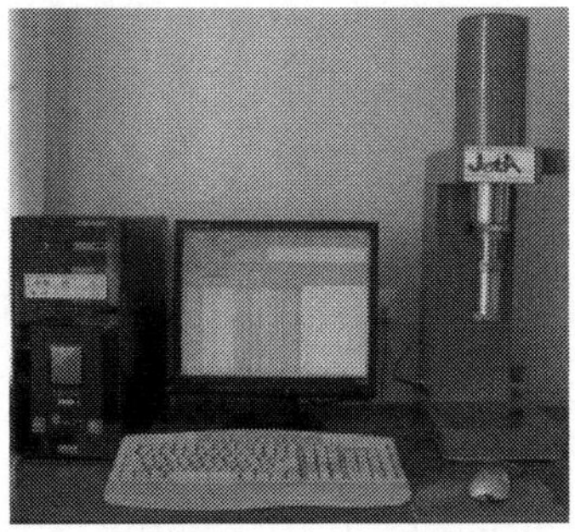

Fig. 1. Rheotest 2 rheometer [own photo]

2. Flow curves of protective coatings

The Rheotest 2 rheostat made it possible to generate flow curves for the studied protective coatings. Thanks to these curves, it is possible to compare viscosity and to observe the behaviour of the coatings after dilution with water or rokafenol. The studies were conducted at a temperature of approximately 27°C.

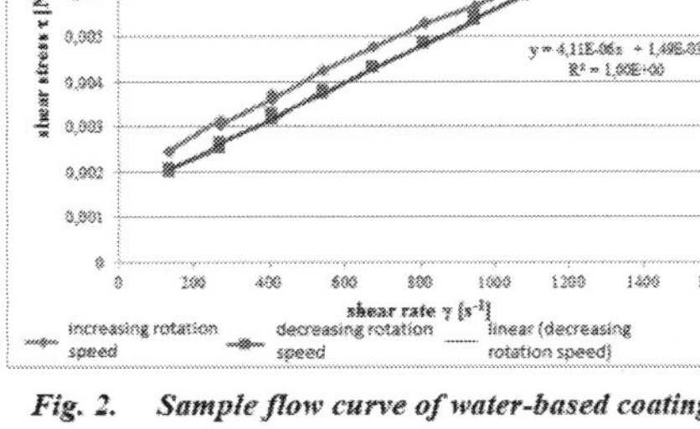

Fig. 2. Sample flow curve of water-based coating

In analysing the flow curve for a water-based coating, when water was added it could be observed that along with the increase in the amount of water added to the coating, the shear stress decreased. The addition of rokafenol caused in increase in shear stress; the more rokafenol was added, the greater was the shear stress.

The addition of rokafenol to the alcohol-based coating caused a slight decrease in shear stress along with increasing amounts.

The flow curves generated, both for water- and alcohol-based coatings, illustrate that these are thixotropic fluids. These are fluids which when applied to vertical surfaces of a mould prevent dripping of the coating [4]. Depending on the addition of diluent, the hysteresis loops change, leading to an observable change in viscosity.

3. Viscosity of coatings depending on the addition of diluent

Thanks to studies conducted on the Rheotest 2 rheometer, comparison graphs of the protective coatings can be generated showing their viscosity, dependent on the amount and type of diluent added.

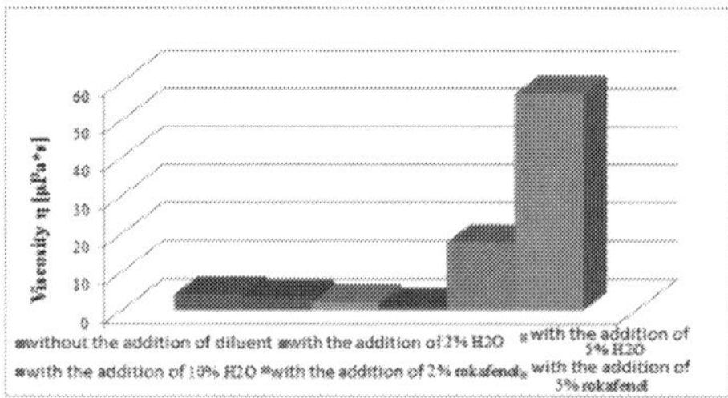

Fig. 3. Viscosity of water-based coating, dependent on the addition of diluent

The viscosity graph of the water-based coating without the addition of diluent, as well as with the addition of water and rokafenol, demonstrates that the addition of water causes a decrease in the viscosity of the coating. The more water is added, the more the viscosity decreases. In the case of addition of ROKAfenol, a significant increase in viscosity compared to the undiluted coating can be seen, as much as four times when 5% rokafenol was added.

By observing the graphs of viscosity of the water-based coatings, it can be seen that along with the increase in the amount of diluent added, the viscosity decreases. rokafenol causes a sharp increase in viscosity in amounts of 5%, yet this increase can also be observed when 2% diluent is added.

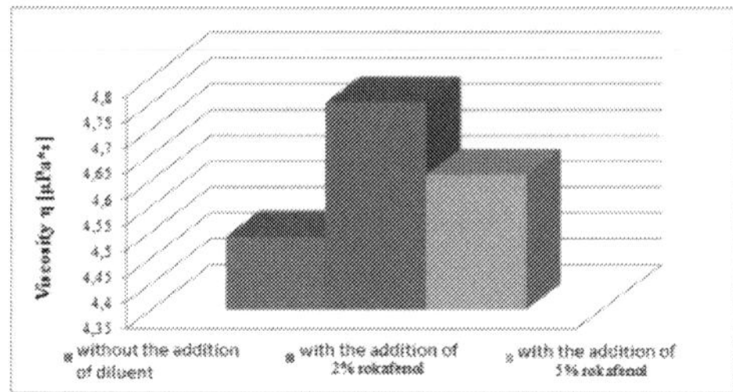

Fig. 4. Viscosity of alcohol-based coating dependent of the addition of diluent

In the case of the alcohol-based coating, to which rokafenol was added as a diluent, and increase in viscosity can be seen together with the addition of the diluent. The addition of 2% rokafenol caused a more than twofold increase in viscosity, while the addition of 5% decreased viscosity when compared to the addition of 2%. However, viscosity was still greater than in the case of the undiluted coating.

4. Conclusion

In analysing the flow curves and graphs illustrating the viscosity of coatings dependent on the addition of a particular diluent, it can be concluded that both water- and alcohol-based coatings are thixotropic fluids. Protective coatings possess a flow limit – they are viscoelastic fluids. Along with an increase in the amount of water added to the water-based coating, the viscosity decreases, while the addition of rokafenol dramatically increases the viscosity of the coatings.

The addition of Rrokafenol causes an increase in the viscosity of the alcohol-based coating, yet the addition of excessive amounts causes a decrease in viscosity.

References

[1] Lewandowski L. (1997). Materiały formierskie. Laboratorium. Kraków. Wydawnictwo AGH

[2] Hutera B. (2008). Znaczenie rozcieńczalnika w spoiwie dla przebiegu zjawisk powierzchniowych w układzie osnowa piaskowa- materiał wiążący. Kraków. Wydawnictwo Naukowe AKAPIT

[3] Ferguson J., Kembłowski Z. (1995). Reologia stosowana płynów. Łódź. Wydawnictwo MARCUS Sc

[4] Jeżowiecka-Kabsch K., Szewczyk H. (2001). Mechanika płynów. Wrocław. Oficyna Wydawnicza Politechniki Wrocławskiej.

INFLUENCE OF MANUFACTURING PROCESS PARAMETERS ON THE STABILITY OF LAYERS PRINTED FROM CERAMIC POWDERS USING ADDITIVE TECHNOLOGY

Adrianna Kania [1]*, Katarzyna Dębska [2], Katarzyna Berent [3]

[1]AGH University of Science and Technology, Faculty of Foundry Engineering, al. Mickiewicza 30, 30-059 Kraków, Poland
[2] AGH University of Science and Technology, Faculty of Materials Science and Ceramics, al. Mickiewicza 30, 30-059 Kraków, Poland
[3] AGH University of Science and Technology, Academic Centre for Materials and Nanotechnology, al. Mickiewicza 30, 30-059 Kraków, Poland
*Corresponding address: e-mail: kania.adrianna@gmail.com

Key words: robocasting, additive manufacturing, ceramic, nanoparticles, sintering

1. Introduction

Functional ceramic materials lead to the search for new, alternative methods of their production, as a result of increasing interest in their applications. The reduction of manufacturing costs by simplifying and automating technological processes has enabled the dynamic development of additive manufacturing techniques in recent years. Until now, ceramics has not been a widely used group of materials in incremental technologies. It constitutes approx. 10% of the entire 3D printing market [1].

Manufacturing of spatial solid structures from polymer composites with ceramic powders in additive technology is highly connected with appropriate rheological properties of raw material [2, 3], which are affected by the shape and size of the powder used and the proportion of the solid phase.

In the case of objects for which high strength requirements are imposed, powders with a diameter of fewer than 5 μm should be used [4]. The results presented in the publications and studies carried out so far show that depending on the polymer used (with Newtonian fluids [5]), the percentage of solid phase particles should be in the range of 40-60% by weight, to have the usage ability in robocasting technology. There are significant differences in rheological properties of suspensions made of polymers with pseudoplastic liquid properties and the use of nanometric powders for which the proportion of solid phase can be increased [6-7].

In addition to the composite components properties, an important factor affecting the properties of the final object made from these materials is the selection of appropriate process parameters, such as: extrusion speed (directly related to the shear rate, which allows thinning or thickening the suspension at a given shear rate), head moving speed, amount of extruded material per unit of time, height, width of paths and layers.

Determining the optimal conditions is a difficult task. However, thorough examination and optimization of process parameters ensure obtaining products in accordance with the quality and geometry requirements resulting from the 3D project and application. So far, there are few publications in literature related to the optimization of 3D printing processes.

The production of ceramic functional elements using robocasting technology has found application in casting filters. The use of ceramic filters supports the production of high-quality castings with a small number of defects. This applies mainly to high-quality castings needed in the automotive, machine and power hydraulics industries. Thanks to the use of liquid metal filtration in the casting mold, casts are obtained without non-metallic precipitates, which come from various sources - mainly slag, ashes, flux residues from the melting process, particles from furnace lining, ladle, molding sand (erosion from the gating systems), and oxidation products (magnesium oxides, metal sulfides, etc.)

Studies have shown that a simple, well-designed pouring system has a very limited ability to separate slag (about 10%). The main reason is the turbulent flow of metal and the impossibility of capturing the slags during the first seconds of pouring metal. Passing metal through the filters also reduces the amount of absorbed air and gases and reduces the number of microparticles, which results in better quality of final product. [8]

2. Aim of research

The aim of the research was to determine the impact of rheological properties of ceramic pastes (ie, viscosity, flow, shear rate) and 3D printing process parameters on the stability of printed layers of submicron Al_2O_3 ceramic powders, which can be used for special applications, working in high temperature. The polymer-ceramic paste properties are strongly dependent on the proportion of polymer and ceramic powder.

3. Methodology

The slurries were prepared using 50-60 %wt. of commercial alumina powder (ALCOA) and photosensitive acrylic resin. The suspension was mixed in a polyethylene container and placed in a high-speed mixer at 2000 rpm for 5 min. Layers were fabricated by robocasting (Material Extrusion) technology using an Additive Manufacturing Aerotech system.

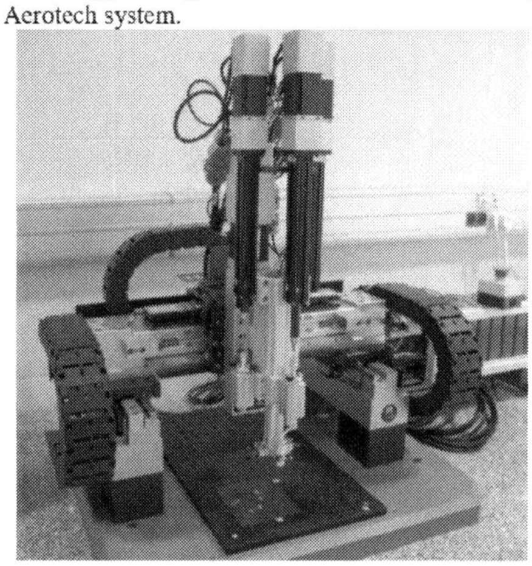

Fig. 1. Aerotech Additive Manufacturing Robocasting system

The microstructure of the sintered objects was studied using an FEI Versa 3D scanning electron microscopy (SEM). The elemental composition of the form was determined using an energy-dispersive X-ray (EDX) analysis.

4. Conclusions

The most important parameters affecting the final effect of both the created layers and the final sintering are primarily rheological properties, which change with the content of ceramic powder. The paste used should also have a uniform structure, without agglomeration and sedimentation, which should be achieved by proper mixing, drying, and proper placement input material in the additive system. For high precision of shape, the appropriate parameters of the speed of movement of the head and the piston should be adapted to the geometry. After the layers have been prepared, it is important to properly prepare the sintering process, which will allow to obtain a ready sample without cracking on the surface. Many factors influence the shrinkage of the created elements - an extended study is expected in the future scientific work of the authors.

Acknowledgments

This research has been financially supported by the Polish Ministry of Science and Higher Education under the project realized in 2018 at ACMiN AGH.

References

[1] Vitale M., Cotteleerv M. & Holdowsky J. (2016) An Overview of Additive Manufacturing, Defense AT& L. 11-12, 7-13.

[2] Feilden E. (2017) Additive Manufacturing of Ceramics and Ceramic Composites via Robocasting, Imperial College London.

[3] Cappi B et al. (2008). Direct inkjet printing of Si_3N_4 : Characterization of ink, green bodies and microstructure. Journal of the European Ceramic Society. 28, 2625–8.

[4] Richerson D.W. (2006). Modern ceramic engineering: properties, processing, and use in design. CRC Press, Taylor & Francis Group, Boca Raton, Florida.

[5] Sunthar P. (2010). Polymer rheology. Indian Institute of Technology, Bombay.

[6] Erwin P. et al. (2017) Robocasting of Dense Yttria-stabilized Zirconia Structures, National University of Singapore.

[7] Kranz S. (2013). Multinozzle printheads for 3D printing of viscoelastic inks. University of Illinois at Urbana-Champaign.

[8] John N. Stuecker, Joseph Cesarano III, James, E. Smay. (2001). Robocasting Periodic Lattices For Advanced Filtration. Solid Freeform Fabrication Symposium.6-8.8.2001.

The 73rd World Foundry Congress
„Creative Foundry"
23 - 27 September 2018, Krakow, Poland

FOUNDRY ENGINEERING AND ART. CASTING OF THE BRONZE STATUETTE INSPIRED BY SALVADOR DALI

Barbara Chachurska[1], **Aleksandra Chachurska**[2], **Damian Zięba**[1], **Janusz Kozana**[1], **Aldona Garbacz-Klempka**[1]

[1]*AGH – University of Science and Technology, Faculty of Foundry Engineering, "ARTEFAKT" - Students Science Club, Reymonta 23, 30-059 Kraków, Poland*
[2] *UMCS - Maria Curie-Skłodowska University, Faculty of Art, Kraśnicka 2a, 2b 20-718 Lublin, Poland*
**Corresponding address: e-mail: b.chachurska@gmail.com*

Key words: artistic casting , copper alloys, bronze, lost wax method, art

Abstract
Art is an inherent element of human life and accompanies us in its every aspect. We can divide it into plenty of categories such as music, painting, sculpture, architecture, fashion or drawing. Art is everywhere. We confront it virtually every day while strolling, shopping, commuting or watching posters on news boards. We don't have to be on an exhibition, in a museum or at a concert to see it, although in the case of this project, the idea formed while watching an exhibition of Salvador Dali's art (Fig.1).

Fig. 1. Inspiration: Salvador Dali Minotaur, 1981, bronze statue, lost wax cast [1]

Implementation
The Purpose of this project is making a sculpture inspired by the sculpture „Minotaur" by Salvador Dali. The sculpture is made of bronze using the lost wax method.

Copper alloys are popular in industry and artistic casting. The most commonly used casts are tin bronzes, silicon bronzes, and brass. Copper alloys have good strength properties and high abrasion resistance. They are also easy to form, resistant to corrosion, and high temperature. These properties had been valued for thousands of years. Initially people were using them to create simple tools, then jewellery, statues, and sculptures. Later on they found use for technology and industry. Nowadays bronze is used to make military casts, automotive elements, and electrical elements [2-3], using both sands forms and with the lost wax method. [4-5].

The lost wax method is one of the oldest methods of casting - this technology's purpose was already invented 5000 thousand years ago. It is used in singular and mass production. In the second case, models are combined into a form resembling a tree. Metal is poured into the form while rotating using centrifugal force. In consequence the form is precisely filled with metal. The lost wax method allows achieving high precision and accuracy in mapping shapes. [3].

The technology itself consists of a couple actions, namely:
- preparing a model
- preparing a ceramic form
- covering the model with subsequent layers of ceramic mass
- drying subsequent layers of mass
- disintegrating wax from the form
- annealing the form
- pouring liquid metal into the from
- fine-tuning

First a sketch inspired by French fashion magazines was made, giving the sculpture dynamism and keeping it fashion-like. One must pay special attention to details – a scorpion and a moth, which symbolise certain qualities. Moth symbolizes sensitivity, mystery, beauty, emotion, intemperate, disastrous desire, transformation, and intuition. Scorpion, on the other hand, means strength, toughness, devious, and deceitful. All of these character traits can be assigned to a woman, depending on her status, age, and circumstances.

Fig. 2. Sketch sculpture made by casting technique

As the drawing was made, a prototype model was sculpted in salt mass. After having been shaped, the model was protection-coated, preventing silicone from getting stuck to it. Next, the model was covered in silicone, forming a mould. The mould was then filled with wax, which after hardening, was fine-tuned to remove minor flaws. The model was divided into parts and moulded in oil mass and cast with bronze.

In the meantime a wax model of the figurine was made in silicone mould. The model was cleaned and fine-tuned, then cast with copper using the lost wax method.

After getting the cast out of the mould, finishing touches were applied and separately cast elements were combined together. To give an artistic effect, the surface of the cast was colour-treated by polishing, dying and covering certain elements – namely the moth and the scorpion - with smut. In the end, a copper-wire net was attached to the sculpture as an element of character's clothing.

Summary

Implementation of the project develops designing, prototyping, ceramic forming and artistic casting skills. In addition, it contributed to intercollegiate teamwork and integration. Moreover it enabled the art designer to design the product according to their artistic vision and directions given by them.

Acknowledgments

Project implemented within ARTEFAKT Students Science Club scheduled activities, in Faculty of Foundry Engineering of AGH UST.

References

[1] https://www.wroclaw.pl/walka-bykow-salvadora-dali-wideoprzewodnik-po-tauromachii-wideo, 29.03.2018, 21:18

[2] Rzadkosz, S. (2013). Casting of copper and copper alloys. Kraków: Akapit.

[3] Rzadkosz, S., Garbacz-Klempka, A., Kozana, J., Piękoś, M. & Kranc, M. (2014). Structure and Properties Research of Casts Made with Copper Alloys Matrix. Archives of Metallurgy and Materials. 59(2), 775-778. DOI:10.2478/amm-2014-0131.

[4] Rzadkosz, S., Kranc, M., Garbacz-Klempka, A., Kozana, J. & Piękoś, M. (2013). Investment Casting Technology Applied to Copper Alloys. Archives of Foundry Engineering. 13(special 3), 143-148.

[5] Rzadkosz, S., Zych, J., Garbacz-Klempka, A., Kranc, M., Kozana, J., Piękoś, M., Kolczyk, J., Jamrozowicz, Ł. & Stolarczyk, T. (2015). Copper alloys in investment casting technology. Metalurgija Metallurgy. 54(1), 293-296.

[6] Davey, C.J. (2009) The early history of lost-wax casting. J. Mei and Th. Rehren (eds), Metallurgy and Civilisation: Eurasia and Beyond Archetype, London.

[7] Garbacz-Klempka, A., Kozana, J., Piekoś, M., Kwak, Z., Długosz, P. & Stolarczyk., T. (2015). Ceramic moulds for precision casting in the bronze age and computer reconstruction of the casts. Archives of Foundry Engineering. 15(spec.1), 21-26.

[8] Allendorf H. (1960). Odlewnictwo precyzyjne za pomocą modeli wytapianych. Warszawa: PWT.

THE USE OF EXPERIMENTAL AND COMPUTER AIDED METHODS IN RECONSTRUCTION OF CASTING PROCESS OF METAL ARTIFACTS FROM THE BRONZE AGE

Daniel Gruszka[1]*, Aldona Garbacz-Klempka[1]*, Łukasz Karczmarek[2], Paweł Żak[1], Michał Szucki[1]

[1]AGH University of Science and Technology, Faculty of Foundry Engineering, Historical Layers Research Centre, Reymonta 23 Str., 30-059 Kraków, Poland,
[2] State Archaeological Museum, Długa Str.52 – Arsenal, 00-241 Warsaw, Poland.

**Corresponding address: e-mail: dgruszka@student.agh.edu.pl, agarbacz@agh.edu.pl*

Key words: Archaeometallurgy, Computer Modelling, Tin bronzes, Foundry Engineering, Bronze Age

Abstract

Main objective of modelling is reconstruction of casting process. To achieve this goal, non-destructive testing methods, computer aided methods and reverse engineering were used. Also research with radiographic tests and X-ray fluorescence to get information about artefacts were conducted.

In view of artefacts value only non-invasive composition and structure analysis of casts are acceptable, such as macroscopic and microscopic observations , X-ray fluorescence and radiographic methods. As a result material's chemical composition were determined as well as surface and internal defects. Results of researches are needed to reconstruct process of casting. Artefacts were facilitated by State Archeological Museum in Warsaw. During studies 3D models were developed in SolidWorks software and simulation in MAGMA5 of casting and solidification were carried out . This allow to apprise the level of metallurgy process advancement in bronze age.

Objective and research results

Hithrerto, 24 socketed axes -part of the treasure from Nowa Górna (Central Poland) were analysed. Most of the axes are Danubian type imports, the rest are local forms from Silesia and Greater Poland - western area of Lusatian urnfield culture. The find is preliminarily dated around 1000-900 B.C. (HB1). Studies are conducted by State Archaeological Museum in Warsaw and Faculty of Foundry Engineering at the University of Science and Technology in Cracow. One of the axes is shown at fig. 1.

Foundry men from the Bronze Age used variety of casting moulds – made of sand, clay, metal or stone [1]. In the cemetery of Lusatian urnfield culture well-preserved casting moulds were found, which presumably belonged to buried Lusatian foundry men [2].

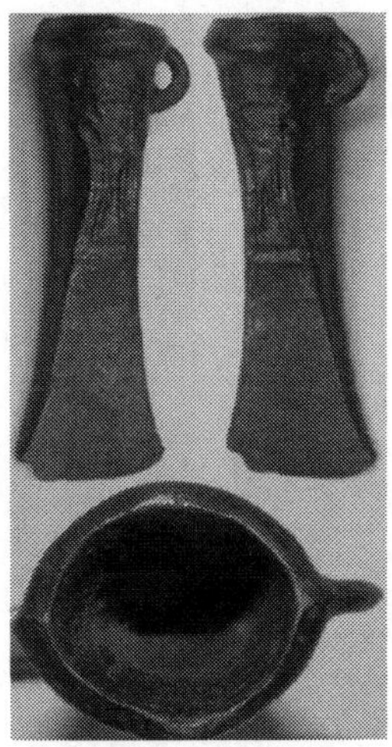

Fig. 1. ***Axe from Bronze Age found in Nowa Górna, (Central Poland). Archaeological Museum in Warsaw***

Chemical compound research (using XRF method) conducted on these moulds, confirmed their usage for bronze casting, because their composition consists elements such as copper, tin, lead, arsenic, antimony and nickel. Those compounds also were found in alloy used to cast analysed artefacts. Casting moulds were made with great accuracy. Inside were forged and polished to reproduced in details shapes of antic axes. Based on stone casting mould a 3D model was created (Fig.2)

The 3D models developed in SolidWorks software were next used for pouring and solidification simulations in MAGMA5 (Fig.3-4). This allow to apprise the level of metallurgy process advancement in bronze age.

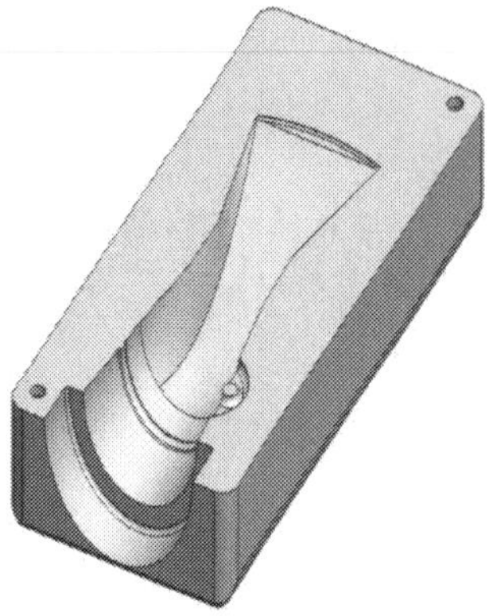

Fig. 2. Visualization of axe mould

Researches was conducted on CuSn10 bronze alloy, which has similar composition to analysed artefacts. [3]

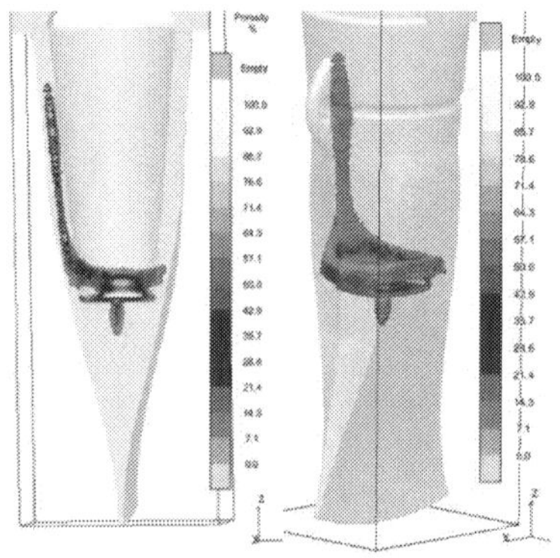

Fig. 3. Porosity created during simulation of solidification process, variant 1

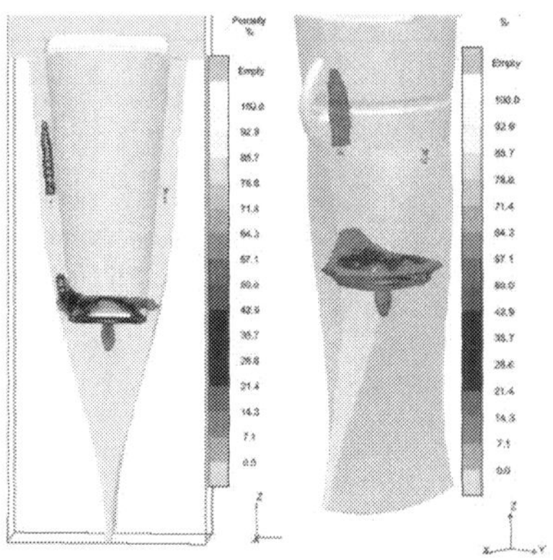

Fig. 4. Porosity created during simulation of solidification process, variant 2

Summary

Main objective of modelling is reconstruction of casting process. To achieve this goal, there were used non-destructive testing methods, computer aided methods and reverse engineering. Before that, artefacts were carefully analysed, to get necessary information about material, surface and internal defects, which occur during solidification process. For this purpose artefacts were submit to non-destructive tests. Results from macroscopic analysis showed that the most common surface defects were misruns and raptures. Radiographic examine indicated another faults (defects), such as porosity, blisters and shrinkage porosity. The final stage of tests was chemical composition determination by X-ray fluorescence, which defined that subjects of research were made of tin bronze which consist about 10% of tin.

Modelling launched out by gathering knowledge about comparable artefacts from a given epoch. Gathered information were used to reconstruct casting technology in SolidWorks software. 3D geometry obtained in this way was imported to MAGMA5 and where simulation of pouring and solidification were carried out.

Conducted simulations demonstrated conditions and technologies of making examined objects in the past. Radiological analysis showed, that artefacts might have numerous defects or do not have any, which indicate continuous improvement of skills by ancient foundry men. Only a combination of macroscopic, radiological and X-ray examines of chemical composition with computer aided methods enable gaining complete knowledge of ancient technologies.

Acknowledgments

The financial support of the State Committee for Scientific Research of Poland under the grant numbers 11.11.170.318 – 11 is acknowledged.

References

[1] Garbacz-Klempka A.,KozanaJ. Piękoś M., KwakZ., Długosz P. & StolarczykT. (2015) Ceramic moulds for precision casting in the bronze age and computer reconstruction of the casts. Archives of Foundry Engineering. 15(spec. iss. 1), 21–26.

[2] Garbacz-Klempka A., Kwak Z., Żak P.L., Szucki M.,Ścibior D., Stolarczyk T. & Nowak K. (2017). Reconstruction of the casting technology in the Bronze Age on the basis of investigations and visualisation of casting moulds. Archives of Foundry Engineering. 17(3), 184–190.

[3] Gruszka D. Modeling of archeological metal antiques with using computer aided methods (2018). Engineering Thesis under supervision of dr. inż. A. Garbacz-Klempka, AGH University of Science and Technology, Faculty of Foundry Engineering.

The 73rd World Foundry Congress
„Creative Foundry"
23 - 27 September 2018, Krakow, Poland

INFLUENCE OF ALLOY ADDITIVES AND CASTING PARAMETERS ON MICROSTRUCTURE AND MECHANICAL PROPERTIES OF SILICON BRONZES

D. Witasiak [1]*, A. Garbacz-Klempka [1], M. Papaj [2], P. Papaj [2]*

[1]AGH-University of Science and Technology, Faculty of Foundry Engineering, Reymonta 23, 30-059 Kraków, Poland,
[2] METAL-KOLOR, Ostrowiecka 5, 27-200 Starachowice, Poland
*Corresponding address: e-mail: dawid.witasiak@wp.eu

Key words: silicon bronzes, centrifugal casting, mechanical properties, microstructure

Abstract

Silicon bronzes are characterized by good mechanical properties and high corrosion resistance. During the research, the process of die-casting and gravity casting in sand molds was realized. The centrifugal casting of the sleeve in the machine with the horizontal axis of rotation of the mold was analyzed. The study evaluates the effect of the content of alloying elements and variable casting parameters on the microstructure and mechanical properties of alloys. The elements which concentration changed were nickel and zirconium. The research results were summarized on the basis of microstructure analyses and mechanical properties estimates. The aim of the experiment was to determine how the alloying additions and the change of pouring parameters affect the properties of the BK331 alloy.

1. Introduction

Bronze BK331 is characterized by a good plasticity and very high corrosion resistance [1]. The alloying elements contained in the studied bronze affect its properties. The addition of zinc makes these bronzes characterized by good fluidity, while additions of iron and manganese significantly improve their mechanical properties. For this reason, these bronzes have high mechanical properties both at room temperature and in elevated temperatures. Silicon bronze BK331 is characterized by high resistance to friction wear as well as to variable loads. BK331 alloy is used for the production of machine parts and accessories (bearings, pumps) that are exposed to corrosion, variable loads and bad lubrication. [2, 3]

2. Methodology

To determine how the addition of nickel and zirconium affects the microstructure and properties of bronze BK331 a gravity die casting and centrifugal casting was made of the analyzed alloy (Fig. 1).

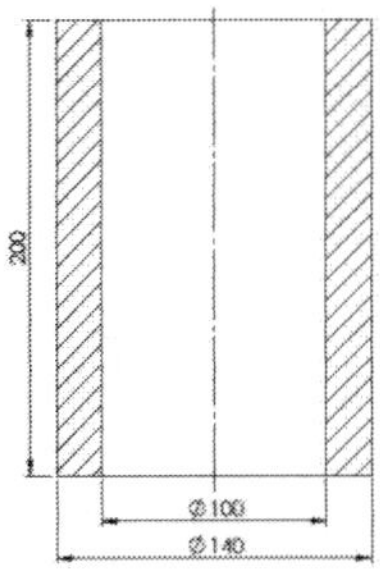

Fig. 1. ***Sleeve cast of silicon bronzes 100x140x200 mm***

The initial speed was selected based on the nomogram (Fig. 2), taking into account the optimal number of rotations depending on the diameter of the sleeve. The casting parameters were as follows: pouring temperature 1095°C, metal mold temperature 250 °C.

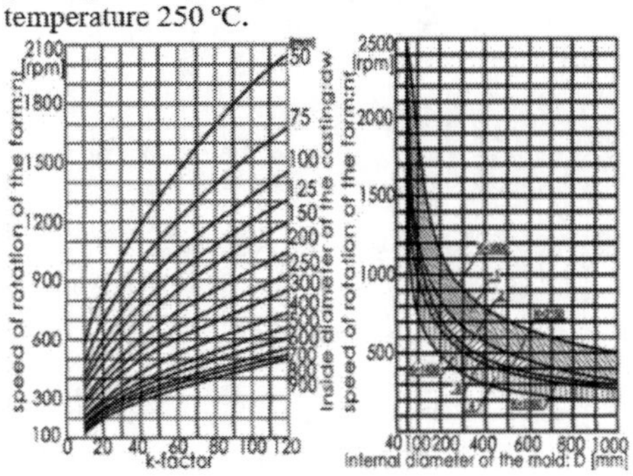

Fig. 2. ***Nomogram of rotational speed [4]***

3. Influence of alloy additions on microstructures

In order to perform macroscopic and microscopic analysis of the sleeve, samples were taken from the characteristic places on the cross-section of the sleeve (B), as well as on the outer (A) and inner (C) surface of the sleeve (Fig. 3).

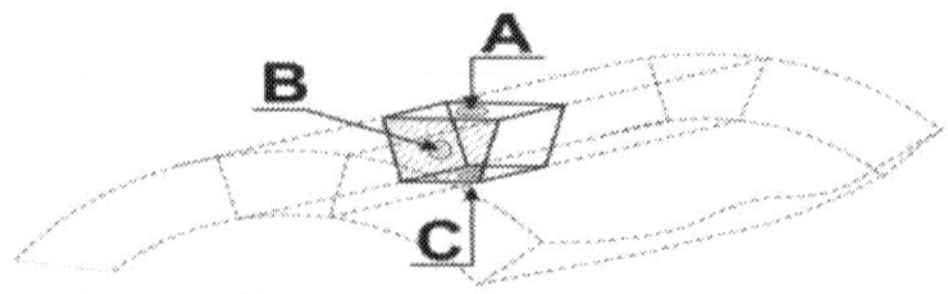

Fig. 3. Places of sampling for metallographic analysis [5]

Parameters of gravity casting are given in Table 1. The chemical composition of the cast was consistent with the composition of the normalized alloy BK331 (PN-91/ H-87026), containing: minimum 90% Cu, 3-5% Si, 3-5% Zn, 0.2% Pb, 0.2% Fe, 0.25% Mn.
Representative samples are shown in Fig. 4-5:
a) BK331,
b) BK331+0.9%Ni,
c) BK331+1.8%Ni
d) BK331+1.8%Ni+0.3%Zr

Fig. 4. The area of the microstructure cut from the fragment of the die casting

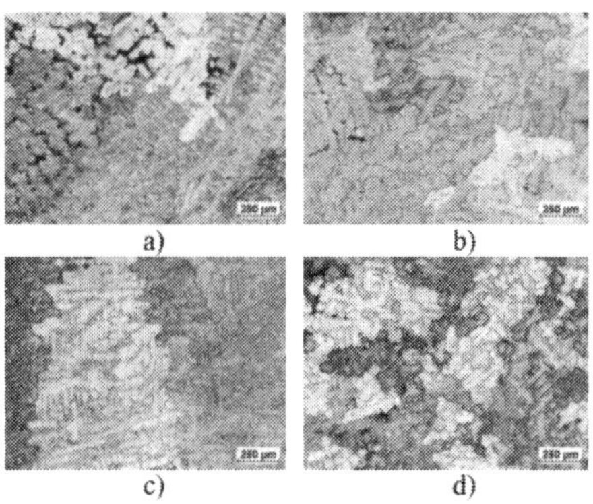

Fig. 5. The area of the microstructure cut from the fragment of the sand casting

Since these alloys are differing in grain sizes - they can be separated on the cross-section. These areas are the effect of mould variable heats and the presence of nickel and zirconium, leading in consequence to changes in the crystallization process.

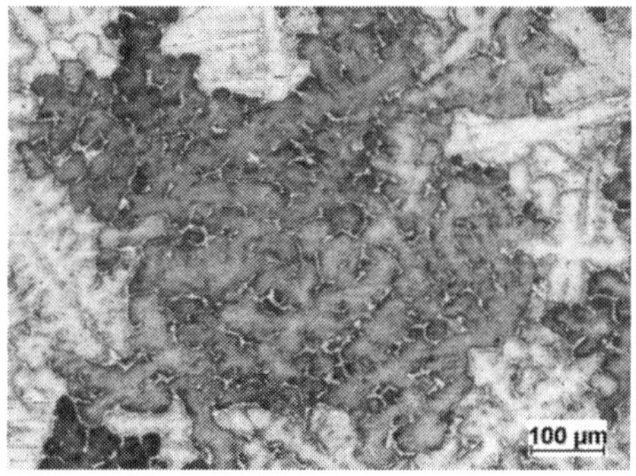

Fig. 6. The area of the macrostructure cut from the fragment of the sleeve

Figure 6 shows the microstructure of the BK331+1.8% Ni + 0.3% Zr centrifugal casting. Zirconium was introduced as a modifier, casting was made 35 minutes after the addition of zirconium.

Table 1. Casting parameters

Experiment	Metal temperature [°C]	Metal mould temperature [°C]
Stage **01** die casting BK331	1100	270
Stage **02** die casting BK331+0.9%Ni	1095	268
Stage **03** die casting BK331+1.8%Ni	1100	200
Stage **04** die casting BK331+1.8%Ni+0.3%Zr	1100	250

Acknowledgments
The financial support of the State Committee for Scientific Research of Poland under the grant numbers 11.11.170.318 – 11 is acknowledged.

References:

[1] Rzadkosz, S. (2013). *Foundry of copper and copper alloys*. Kraków: Akapit
[2] Najman, K., Muszyński, W. & Bydałek, A.W. (2004). The influence of refinement bronze BK331 on his abrasion resistance. *Archives of Foundry* 4(11), 29-34.
[3] Bydałek, A.W. & Najman, K. (2006). The Reduction Melting Conduction of Cu-Si Alloys. *Archives of Foundry* 6(22), 107-110.
[4] Górny, Z. (1966) *Casting in rotating forms*, Warszawa:WNT.
[5] Garbacz-Klempka, A., Kozana, J., Piękoś, M., Papaj, M., Papaj P., Perek-Nowak M. (2018). Influence of Modification in Centrifugal Casting on Microstructure and Mechanical Properties of Silicon Bronzes. *Archives of Foundry Engineering* (in print).

The 73rd World Foundry Congress
„Creative Foundry"
23 - 27 September 2018, Krakow, Poland

ANALYSIS AND EVALUATION OF THE EFFECT OF MANGANESE CONTENT ON THE PROPERTIES OF ALUMINUM ALLOY EN AC 46000

Martyna Pasternak [1*]

[1]*AGH University of Science and Technology, Faculty of Foundry Engineering , Poland,*
**corresponding address: e-mail: martpiw@gmail.com*

Key words: aluminium alloy, pressure casting, manganese, strength of material

1. Introduction

The article is the effect of industrial work and many survey in the one of the Polish foundry with specialization in aluminum alloys. The main goal of explorations is to evaluate manganese influence on Al-Si alloys properties.

In die-cast alloys, it is recommended to use larger amounts of iron to reduce the impact on the pressure chamber and the mould. Manganese neutralizes the harmful effect of ferrous by changing the morphology of the separated brittle ferrous phases. The unfavorable form of iron containing phases in the form of long needles changes into the forms of fine, compact crystals, which are less harmful to mechanical properties [1-3].

Nowadays the main clients require is to obtain right parameters in the best price. Specifically customer is automotive, where new technologies provides connection between high quality and low weight of cars parts.

The introduction of a completely new alloy involves often changes in casting process, purchase of new machinery or modernization of existing ones, training of employees, changes in occupational health and safety conditions. Not only do the casting parameters change, but sometimes also further processing. Attention should also be paid to the ratio of the price of alloying elements to the properties obtained by their use. These aspects lead to correction of existing processes, where sometimes a small change can have great, beneficial effects.

2. Experimental

One of the most important parameter is strengths of material, especially tensile strength. This parameter was the main indicator of quality improvement related to the increase of manganese content in the aluminum alloy. Additional, tests were carried out, such as: density index, chemical composition and x-ray examinations. X-ray scanning is a good and quick way to show the porosity of a casting. Porosity can affect on the strength of the castings. This parameter causes an increase in stress inside the casting, which is connected with exceeding the yield point of the material and permanent deformations. In the end, it can lead to the part crack.

Table 1. *Content of AC-46000 alloy alloying elements [6]*

	min.	max.
Al	-	-
Cr	-	0,15
Cu	2,0	4,0
Fe	-	1,3
Mg	0,05	0,55
Mn	-	0,55
Ni	-	0,55
Other	-	-
Other each	-	0,05
Other total	-	0,25
Pb	-	0,35
Si	8,0	11,0
Sn	-	0,15
Ti	-	0,25
Zn	-	1,2

Standard EN AC-46000 AC-AlSi9CU3(FE) provides for the maximum manganese content of 0,55% (tab. 1.). Initially, the spectrometric analysis indicated 0,29% of manganese content in the alloy. Higher content of this element was obtained by adding a charge with an increased content of manganese to the alloy. The main problem was required minimal content of metal in the furnace connected with die-casting machine. It was not possible to completely empty this furnace and load a new batch, therefore the chemical composition of the alloy was charged gradually by adding alloy with a higher manganese content to the furnace.

3. Conclusions

Conducting research under industrial conditions was a long-term process. Introduced changes could not have a negative impact on the demands placed on castings, therefore it had to be done with great attention. As mentioned earlier, changes in the chemical composition of the cast alloy was related to the technological parameters of the casting machine (minimum metal content of the furnace). During the introduction of new batches of metal into the machine it was noticed that there are problems with mixing of the filled alloy with the alloy which was already there before.
Strength tests have shown positive effect of increased content of manganese in the alloy, which confirmed the assumptions made at the beginning of the tests. In addition, the results of the density index were

improved, and most importantly, the number of production shortages decreased.

References

For books:

[1] Poniewierski Z. (1989). Crystallization, structure and properties of silumin. WNT Warszawa.

[2] Poniewierski Z. (1966). Silumin modification. Warszawa: Wydawnictwo Naukowo Techniczne. Modyfikacja siluminów.

[3] Górny Z., Sobczak J. (2005). Modern casting materials based on non-ferrous metals. Kraków: ZA-PIS.

For patents:

[6] EN AC-46000 AC-AlSi9CU3(FE) (2010)

The 73rd World Foundry Congress
„Creative Foundry"
23 - 27 September 2018, Krakow, Poland

ANALYSIS OF THE INFLUENCE OF THE DEGREE OF WEAR OF THE PRESSING PISTON IN COLD CHAMBER PRESSURE MACHINES ON THE QUALITY OF ALUMINIUM CASTINGS

Gabriela Piwowarczyk [1]*

[1]*AGH University of Science and Technology, Cracow, Poland*
**Corresponding address: e-mail: gabpiw@gmail.com*

Key words: Casting, high pressure, piston, cast defects, fin, flash

1. Introduction

High pressure die casting is used in high volume production. Due to the large number of advantages (mass production of components, casting time, high quality of castings), high-pressure casting occupies a high position especially in orders for the automotive industry. However, the customers from this branch of industry place very high demands on the quality of the obtained elements. This requires investment in high quality casting alloys and casting machines from leading companies in the market. The process of pressure casting under industrial conditions, however, poses many technological and technical problems, therefore, it is necessary to eliminate them and strive to stabilise the process in order to obtain high quality castings and, consequently, customer saftisfaction.

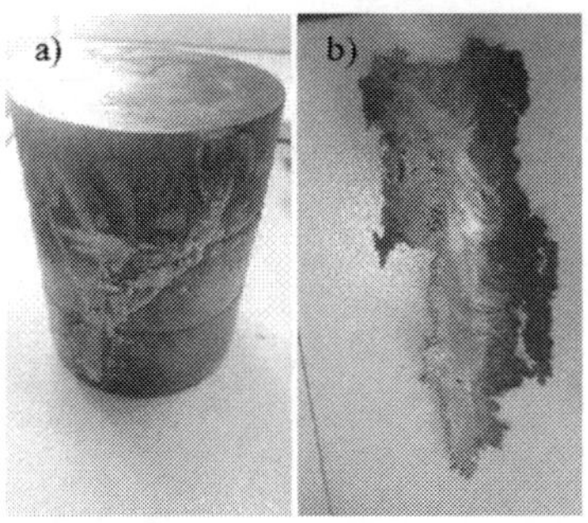

Fig. 1. a) Defective piston removed from the pressure machine, b) joint flash, also called as casting fin

2. Experiment

The aim of the study was to investigate the factors influencing the quality of aluminium castings obtained using the high-pressure method with the use of cold chamber machines in one of the Polish foundries. One of the most important factors influencing the occurrence of defects in the analysed foundry was the wear of the pressing piston in the pressing chamber, which generates a large number of defects such as porosity and shrinkage. The influence of piston operating time and its degree of wear on the type and number of defects occurring was analyzed and evaluated.

For this purpose, 2 attempts were made to optimize the piston replacement every 5 days and every 7 days. Since the piston was installed, 5 measurements have been taken at intervals of 4 hours (the number of measurements changed depending on the occurrence of a failure or other factors). Each injection was tested for casting defects and then assigned a class according to the defect catalogue. The paper presents a list of classes of subsequent castings for each of the 4 attempts at optimization. The process of changing the catalogue evaluation of samples in subsequent injections was also presented.

3. Conclusions

The analyses confirmed that the wear of the pressing piston and the chamber has a great influence on the quality of the castings obtained. This is due to the difficulty of maintaining a sufficiently high pressure in the third pressing phase, during which the casting is pressed. This results in defects such as greaseiness, gas porosity and shrinkage. However, due to frequent failures occurring during machine operation (e.g. piston jamming), it is difficult to determine precisely the impact of piston wear on the occurrence of defects in the performed parts.

References

For books:

[1] Dańko J., 2000: Pressurised casting machinery and equipment. AGH University of Science and Technology, Krakow.

[2] Pietrowski S., 2001: Silumines. Lodz University of Technology, Łódź.

For patents:

[3] PN-H-83105:1985. Castings - Breakdowns and Terminology of Defects .

DIGITALIZATION

The 73rd World Foundry Congress
„Creative Foundry"
23 - 27 September 2018, Krakow, Poland

ADAPTATIVE PRODUCT MANUFACTURING TECHNIQUE FOR FOUNDRY INDUSTRY USING IOT TECHNOLOGY: AN INDUSTRY 4.0 INITIATIVE

Yashesh Pandya [1]*, Jozef Suchy [1], Nagaraj S. V. [2]

[1]AGH University of Science & Technology, Faculty of Foundry Engineering, D-8 Reymonta23, Kraków, Poland,
[2] VIT University (Chennai Campus), School of Computing Science & Engineering, Chennai, India,
**Corresponding address: e-mail: ypandya@agh.edu.pl*

Key words: Internet of Things, Multi-agent Systems, Artificial Intelligence Robotics, Goal driven Manufacturing, Production-as-a-Service.

1. Introduction

The main objective of this paper is to bring the principles of the sharing economy to the economic manufacturing space, thereby decreasing dealings prices that are associated with the exchange of underutilized capacity of existing production resources. We tend to refer to this paradigm as IoT adaptive manufacturing (IoT-AM), in which we created a completely automatic production-as-a-service system.

Foundry industrialization efforts have, over the past decades, targeted for the economical production of massive quantities of uniform, standardized product, remarked as production. Consequently, state of the art foundry industry manufacturing styles are based on rather static, inflexible assembly lines that, being designed for lifespans of thirty and more years, specializing in long-term safety and responsible-ness. In fact, such foundries are sometimes simply maintained and only partially refitted over their life, simply because upgrading them would disrupt the assembly method and incur high and arduous to predict cost.

However, driven by globalization and increased access to international information, customers have begun to demand more and more customized merchandise. From cars that customers will customize themselves using the brand's online configuration tool to self-designed jewellery on online marketplaces, mass customization is emerging: customized merchandise has got to be factory-made at the value of the mass-produced product, however in heap sizes of as little as one item. This development challenges the traditional paradigm of today's foundries to efficiently manufacture the same item over and over – instead, it involves the speedy and intensely dynamic reconfiguration of production processes.

As the connectivity of devices on the foundry industry floors will increase, the decomposition of production systems into a less tightly coupled internet of services becomes feasible. this means that technologies from research in ubiquitous Computing and within the internet of Things that enables users to program their smart surroundings [1, 2] currently become applicable to industrial automation scenarios: an equivalent technologies that allow users to make service mashups among heterogeneous smart devices which is able to sense and modify their personal environments, as an example to infer and apply best heating situations among the smart home domain [3], will presently be utilized by operators of foundry industry environments.

In this paper, we'll focus on the on the-fly creation of service mashups in foundry industry environments: here, the easy (re)configuration of production resources is gaining importance to support operators of commercial plants in managing the increasing dynamicity of manufacturing processes [4]. We propose a goal-driven approach that means that we raise potential customers to explain which properties their desired product need to have. Given this linguistics statement of a user's goal, our system determines whether or not the goal may be reached given the set of available production capabilities (e.g., handling robots, production machines, human resources with specific abilities) and derives a concrete production plan to manufacture the item.

2. Transition from Cyber-Physical Systems to Industrial Internet Of Things

The introduction of water and steam powered mechanical production facilities is taken into account as the 1st industrial technological revolution. The subsequent revolution was the deployment of electrically powered manufacturing based on the division of labor, followed by the third industrial technological revolution that introduced electronics and IT to production systems to reinforce automation of production. [5] But, some publications count differently [6]. With electronics, which provides these objects with identification, computing, and communication capabilities. Examples such as basic technology, like radio frequency Identification (RFID) for the identification and tracking of manufactured goods, packets, and pallets in supply chain scenarios, to smart devices, like smart phones and wearables (e.g., smart watches) with extensive computing capabilities and internet connections. This network of ubiquitous smart objects is known as the internet of Things (IoT) and allows novel applications and services, especially within the industrial sector.

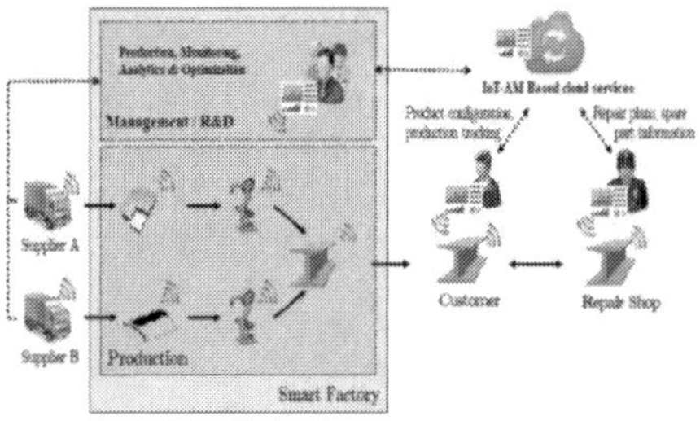

Fig. 1. IoT-AM based Smart Factory Operation

The foundation of foundry industry IoT is Cyber physical systems (CPS) that are computing platforms that monitor and manage physical processes [7]. CPS allows condition monitoring, structural health observing, remote diagnosing, and remote handling of production systems in real time. Further, IoT based CPS are the core basis of smart factories that dynamically organize and optimize production processes with respect to resource-utilization (i.e., costs, accessibility, material, and labor) based on information generated and collected by the underlying CPS, even across company boundaries [8, 9].

In smart foundries, smart product will recognize their own identity, history, specification, documentation, and even manage their own production method (cf. Figure 1). Beyond manufacturing, the smart product is the idea of novel knowledge-based services, referred to as smart services. Specifically, the smart product doesn't solely collect data throughout their production however additionally after they are deployed and utilized by customers. This enables to optimize them with respect to the way they're really used. Smart product are equipped with a digital identity (e.g., stored in barcode or RFID chip) and all data associated with the product is kept in some backend-database. Alternatively, the product is embedded with electronics (e.g., memory and a processor) and stores this data itself. Foundry industry IoT will bring several new challenges with regard to completely different aspects, as well as security, privacy, standardization, legal, and social aspects. Especially increased diversity and enormous numbers of devices in IoT systems need extremely scalable solutions for, e.g., naming and addressing, electronic communication, data management, and service provisioning. Further, most IoT devices have solely limited resources that demands architectures supporting low power, low cost, fully networked integrated devices that are compatible with standard communication techniques.

3. Summarization and Conclusion

The vision of the internet of Things has been around for some years, however, the widespread realization of this vision is just now starting. The field of IoT solutions continues to be a good deal in an exceeding state of flux, as various different industries, research institutes, and completely different entities plan to bring the IoT into reality. This adds another level of complexity to the already pretty tangled mess that the IoT is these days, with its varied technologies, standards, organizations, sectors, and flavors.

The implementation of IoT based adaptive manufacturing (IoT-AM) will change the face of production and will bring new opportunities for growth for the foundry industry, by closing up on the gap between customers and producers either side would be benefitted. Hence, IoT-AM is the rise of the new era of evolution for industries.

References

[1] A. J. Brush, Bongshin Lee, Ratul Mahajan, Sharad Agarwal, Stefan Saroiu, and Colin Dixon 2011. Home Automation in the Wild: Challenges and Opportunities. In Proceedings of the ACM CHI Conference on Human Factors in Computing Systems (CHI). ACM, 2115–2124.

[2] Valérie Issarny, Nikolaos Georgantas, Sara Hachem, Apostolos Zarras, Panos Vassiliadis, Marco Autili, Marco Aurélio Gerosa, and Amira Ben Hamida. 2011. Service-Oriented Middleware for the Future Internet: State of the Art and Research Directions. Journal of Internet Services and Applications 2, 1(2011), 23–45.

[3] Wilhelm Kleiminger, Christian Beckel, and Silvia Santini. 2015. Household occupancy monitoring using electricity meters. In Proceedings of the 2015 ACM International Joint Conference on Pervasive and Ubiquitous Computing (UbiComp). ACM, 975–986.

[4] Jose L. M. Lastra and Ivan M. Delamer. 2006. Semantic Web Services in Factory Automation: Fundamental Insights and Research Roadmap. IEEE Transactions on Industrial Informatics 2, 1 (2006), 1–11.

[5] H. Kagermann, W. Wahlster, and J. Helbig. Securing the future of German manufacturing industry | Recommendations for implementing the strategic initiative Industrie 4.0, 2013.

[6] J. Rifkin. The Third Industrial Revolution: How Lateral Power is Transforming Energy, the Economy, and the World. Palgrave MacMillan, 2011.

[7] J. S. Kumar and D. R. Patel. A survey on internet of things: Security and privacy issues. International Journal of Computer Applications, 90(11), 2014.

[8] Smart Factory | From Vision to Reality in Factory Technologies. International Federation of Automatic Control, 2008.

[9] D. Zuehlke. Smart factory | towards a factory of things. Annual Reviews in Control, 34(1), 2010.

The 73rd World Foundry Congress
„Creative Foundry"
23 - 27 September 2018, Krakow, Poland

Casting EMPS (ERP\MES\PDM\SCADA) Integrated System Architecture and Its Application in Aeronautic Titanium Alloy Foundry Enterprise

Xiaoyuan Ji [1], Hailong Li [1], Jianxin Zhou [1*], Qiuyu Yan [1], Anlong Jiang [2], Qian Zhou [2], Yajun Yin [1], Xu Shen[1]

[1] Huazhong University of Science & Technology, State Key Laboratory of Material Processing and Die & Mould Technology, Wuhan 430074, Hubei Province, People's Republic of China
[2] Baimtec Material Co., Ltd., Haidian district, Beijing 100095, People's Republic of China
**Corresponding address: e-mail: zhoujianxin@hust.edu.cn*

Key words: Foundry, Informatization, EMPS, ERP, SCADA (Supervisory Control And Data Acquisition)

Abstract

ERP, MES, SCADA and other information systems are widely used in foundry enterprises. However, due to the lack of a unified integrated architecture, information barriers , data redundancy and many other problems exist among the systems, which seriously restricting the informatization and industrialization process of foundry enterprises. To this end, the author proposes the casting EMPS integrated system architecture, which defines from four levels: the business value layer, the technology design layer, the manufacturing executive layer and the device data sensor acquisition layer, to conduct seamless integration of ERP, PDM, MES and SCADA systems. In this EMPS framework, EPR is defined to drive the design of PDM, PDM processes guidance for MES, SCADA combined with MES to monitor and warn the PDM production process parameters, information communication and the action mechanism between layers are also defined. Based on the EMPS architecture, InteCAST-EMPS integrated manufacturing platform system was developed and applied to the Titanium alloy research and production base of Beijing Aeronautical Material Research Institute of China Aerospace. After three years of application, the system architecture, on the one hand, satisfies the needs of the first stage of enterprise informatization by eliminating information barriers, reducing data redundancy and sharing global information. On the other hand, the accumulation of industrial big data is realized, which promotes the intelligent development of the second stage of enterprise informatization.

1. Introduction

Foundry enterprises have been actively applying various types of information systems to improve their management level in more than 10 years of informatization and industrialization integration process, such as Enterprise Resource Planning, Product Data Management, Product Lifecycle Management, Manufacturing Execution System, Supervisory Control And Data Acquisition system[1]. Many studies are devoted to exploring new technologies and methods of informatization and industrialization progress in foundry, for example, the production process tracking main-sub card technology and the workshop site production process processing technology[2-4]. However, information barriers exist among these information systems and it is difficult for them to work together. What's more, because of the lack of global planning and unified integrated system architecture, a lot of data duplication, data isolation and data inconsistency problems arise, resulting in system maintenance difficulties, inefficiencies and many other challenges, which seriously restricting the development of foundry enterprises[5]. Integration is an urgent demand and an inevitable trend for the development of foundry informatization[6].

2. Casting EMPS Integrated System Architecture

The casting EMPS Integrated system architecture defines the seamless integration method of ERP, PDM, MES and SCADA in four layers: business value layer, technology design layer, manufacturing execution layer and equipment data acquisition layer. Fig. 1 shows the communication and action mechanism of four layers in casting EMPS. The architecture is based on a highly flexible technology with flexible system integration, the system modules are built and added according to needs. Each part of the four layers in the EMPS integrated system architecture performs a different job in production.

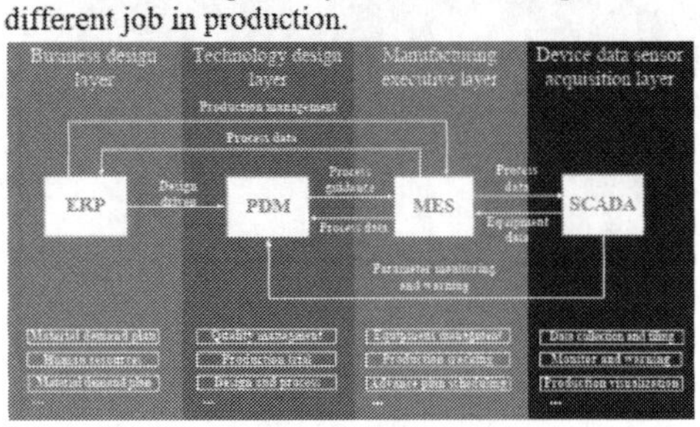

Fig. 1 Communication and action mechanism of various layers in casting EMPS

3. Aviation titanium alloy casting business applications

The InteCAST-EMPS integrated manufacturing system has been used in the Titanium alloy research and production base of Beijing Aeronautical Material Research Institute for three years. Table 1 makes a qualitative analysis of the application of the casting EMPS integrated system from the informational and intelligent aspects.

Table 1 Qualitative analysis of the application of the casting EMPS integrated system

Contrast categories	Compare items	Independent information systems	Integrated EMPS system
informatization	Data redundancy or conflict	A lot	Almost no
informatization	Information sharing scope	Limited to the various systems	Shared among systems
informatization	Enterprise production scheduling efficiency	Low	High
intelligent	Data acquisition, monitoring and early warning	Partly	Completely
intelligent	Industrial big data accumulation	Small amount	A lot
intelligent	Intelligent process optimization, production decision	Can not	Can

During the implementation of InteCAST-EMPS, new technologies such as process tracking main-sub card, production process visualization and other new methods are used. Fig. 2 shows the titanium alloy castings processed by the Baimtec Material Co., Ltd. . Fig. 3 shows the actual production scenario applying the InteCAST-EMPS integrated system. Fig. 4 is a schematic diagram of the main-sub flow card production process. Fig. 5 shows the SCADA system's monitoring interface.

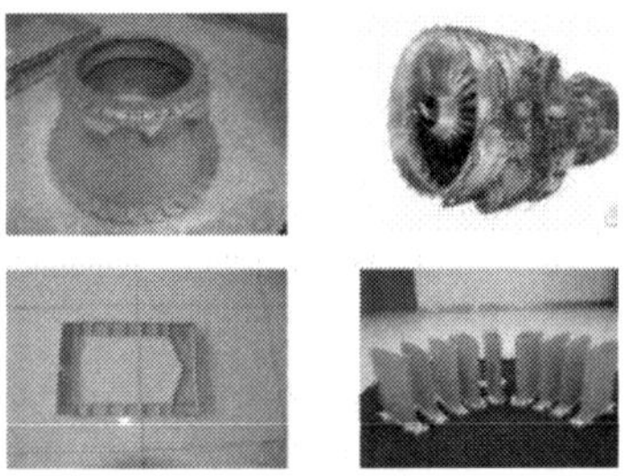

Fig. 2 Titanium alloy casting produced by Baimtec Material Co., Ltd.

Fig. 3 Digital process workshop site

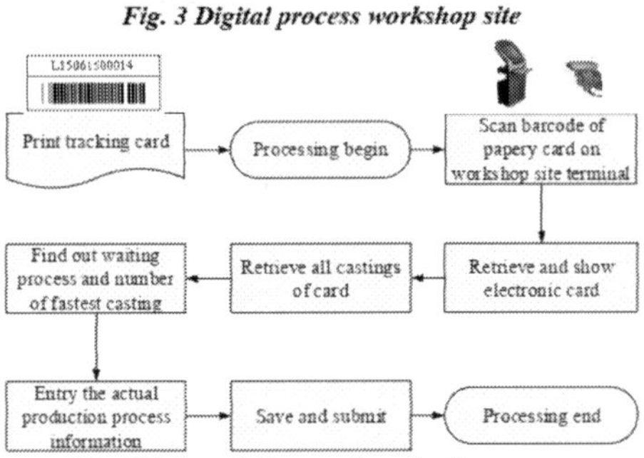

Fig. 4 Main-sub flow card production process

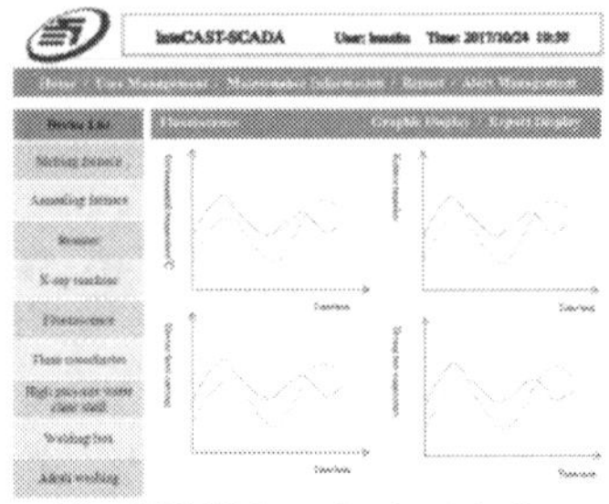

Fig. 5 SCADA monitoring interface

4. Conclusions

In this paper, the author proposes a casting EMPS integrated system architecture, the application results show that the system architecture, on the one hand, eliminates the isolation, duplication and inconformity of information and realizes the global information sharing, which satisfies the demand of the first stage of enterprise informatization. On the other hand, through the integration of SCADA system, the enterprise realizes the collection, monitoring and warning of key data during the operation of foundry equipment, which helps accumulating large industry data and laying a solid foundation for the next intelligent decision and intelligent process optimization.

Acknowledgments

This research is financially supported by the National Science & Technology Key Projects of Numerical Control (2012ZX04012-011), National High-tech R&D Program (863 program) (2013031003), Investment Casting Collaborating Laboratory on Advanced Aeronautical Lightweight Alloy Materials.

References

[1] Xiaoyuan Ji, Hu Ye, Jianxin Zhou, et al. (2016). Digital management technology and its application to investment casting enterprises. China Foundry. 13(5),301-309.

[2] Jianxin Zhou, Xiaoyuan Ji, Dunming Liao, et al. (2012). Construction and case study of "1+N" type digitization platform for foundry enterprises. In 2012 China Foundry Week, 2012/10/22-25 (37-46).

[3] Xiaoyuan Ji, Jianxin Zhou, Yajun Yin, et al. (2014). Research on the digital and intelligent casting technology and its application in casting enterprise. In 2014 China Foundry Week, 2014/10/25-28 (1-14).

[4] Xiaoyuan Ji, Jianxin Zhou, Dunming Liao, et al. 2014. Double chains collaboration model between production management and quality control: three casting factories application cases. In 71st World Foundry Congress, 2014/5/18-23.

[5] Qian Zhou, Weilin Deng, Xiaoyuan Ji, et al. (2015). Research on the main-sub card solution of production processes for titanium precision casting. Special Casting and Nonferrous Alloys. 35(11),1184-1187.

[6] Tingyong Ming, Xiaoyuan Ji, Jianxin Zhou, et al. (2016). A feasibility analysis of the integration of the manufacturing system and the financial supply chain system in the foundry enterprise. Special Casting and Nonferrous Alloys. 36(6),594-597.

The 73rd World Foundry Congress
„Creative Foundry"
23 - 27 September 2018, Krakow, Poland

APPLICATION OF 3D PRINTING TO PRODUCE AN INVESTMENT CASTING MODEL OF HIP-JOINT ENDOPROSTHESIS

Maja Skorupska [1]*, Paweł Leszek Żak [1]

[1]AGH University of Science and Technology, Faculty of Foundry Engineering, Reymonta Str. 23, 30-059 Krakow, Poland,
**Corresponding address: e-mail: skorupska.maja@outlook.com*

Key words: Rapid Prototyping, foundry model, hip endoprosthesis, numerical simulation, investment casting

1. Introduction

The presented work focuses on showing the possibilities of 3D printing casting model of the personalized hip endoprosthesis. During designing of the endoprosthesis model, it is crucial to base on the anatomy of the hip-joint and medical conditions that can be cause of hip-joint replacement [1-2]. The geometry of hip implant was modelled in SolidWorks software on the base of dimensions proposed by Ryan Thomas in [3]. The SolidWorks drawing was then saved as STL file and imported into MAGMA software in order to analyse the casting possibilities and to enable 3D printer to prepare the wax model for the investment casting [4] foundry technique. In the work, it is described with details the process of 3D printing of the hip endoprosthesis elements and briefly analysed the results of numerical simulation of hip implant solidification.

The aim of the work was to develop and verify a procedure of making the personalized hip endoprosthesis with CAD and Rapid Prototyping techniques (RP).

2. Design of hip endoprosthesis geometry

In order to design the hip implant geometry, the SolidWorks software was used. It is one of CAD group programmes, which allows to create 3D geometry projects and perform simple simulations. The designing process consists of adding operations and their dependences.

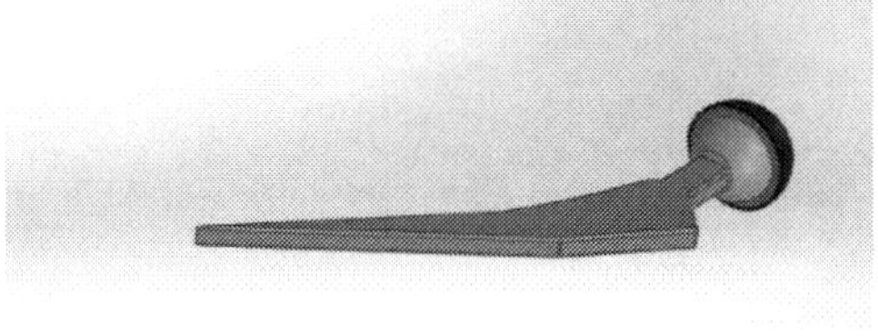

Fig. 1. View of ready designed hip implant geometry

The geometry model was prepared by creating four separate parts, which then were joint to make the assembly. The result of the designing is shown in the Fig. 1. Due to the printer platform size assembly was then sliced into three parts for separate printing.

3. Application of 3D printer to building the wax pattern for investment casting

In the work, the wax model of the designed hip implant was prepared. For this purpose 3D printer which was designed and constructed in the Faculty of Foundry Engineering by student group gathered in "ZGAREK" Student Association was used, Fig. 2.

Fig. 2. 3D printer that was used for the project and wax filament

The base for the printer was Pirx – polish designed simple 3D, FDM technology [5], printer which was equipped with several additional systems as i.e. heating table.

Next stage of the research was to develop the procedure of using a wax filament. The printer has an open working chamber so it was very important to specify the table and nozzle temperature as well as the printing and cooling rates, Table 1.

Table 1. Developed printing parameters for wax filament material

Parameter	Developed value
Table temperature	85 [°C]
Nozzle temperature	145 [°C]
Venting rate	70 – 75 % of max power
Printing speed	maximal possible

To control the printer, free software was used. The Ultimaker CURA was used to generate G-Code, set nozzle and table temperature. Data file prepared in this software was then imported into the Repetier – Host v 2.0.5, program, which was connected to the printer and allows to control the printing even during process run, Fig. 3.

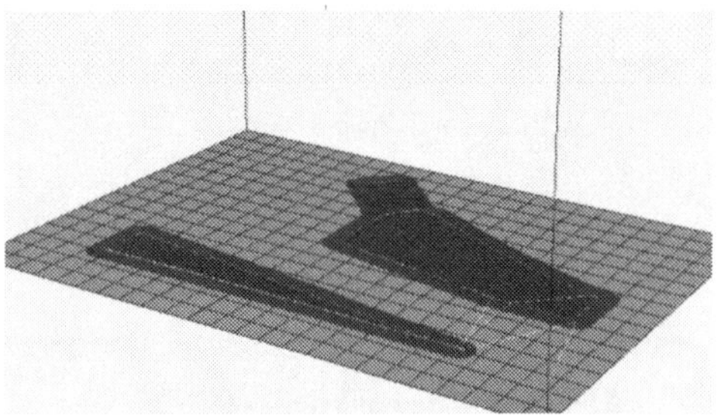

Fig. 3. Repetier – Host window with running printing project

In the figure, Fig. 4, printed part of hip implant is shown.

Fig. 4. Ready printed head of hip implant

During the printing process several problems were noticed connected with part griping to the platform, cracks due to material shrinkage and with the surface quality. Some additional attempts to remove those problems were made, i.e. moving the part on the platform, orientation change, adding gripping medium (office glue, hair spray, melted wax, masking tape) and keeping ambient temperature at constant level. At the end, good quality hip implant model was obtained.

4. Numerical simulation

The MAGMA simulation program was used to analyse hip implant solidification process. During the simulation TiAl6V4 alloy was used. Initial temperature of casting alloy was 1720 [°C]. The mould temperature was 20 [°C]. To simplify the computations, the filling of mould cavity was omitted. Performed simulations took into account the change of mould temperature and adding the feeders.

Simulation results shows that used alloy has high shrinkage and it is likely that some porosity will appear in casting volume, Fig. 5. Change of the temperature did not give significant improvement of this parameter.

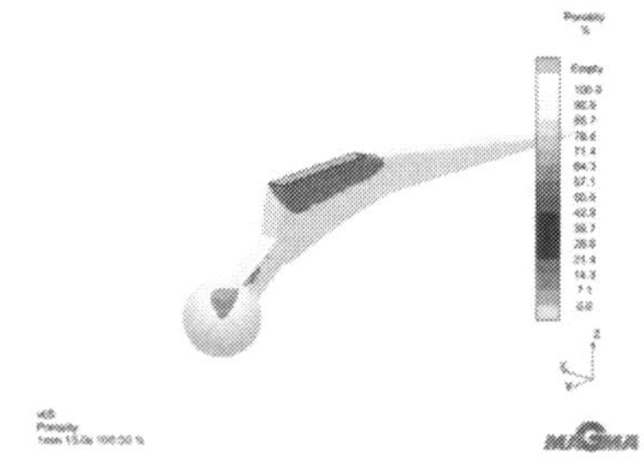

Fig. 5. Porosity distribution – casting without feeders

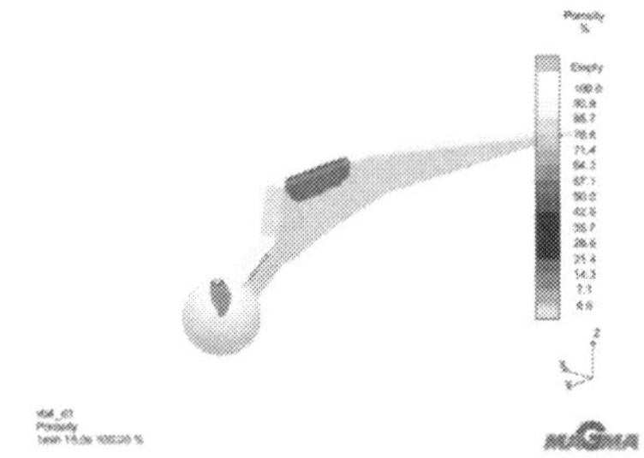

Fig. 6. Porosity distribution after adding feeders

However porosity can be removed from the casting by adding feeders over the porous regions. Some preliminary research was performed on this subject, Fig. 6.

5. Summary

Using the novel computer aided techniques as CAD, RP, simulations it is possible to prepare personalized hip endoprosthesis. Application of wax type filament during the printing allows to build pattern for investment casting. Working with this material is demanding as it is necessary to distinguish and precisely control printing parameters. The casting process can be optimized with numerical simulations.

References

[1] A. Pozowski. (2011). Alloplastyka stawu biodrowego. Wrocław: Górnicki Wydawnictwo Medyczne.

[2] M. Dutton. (2014). Ortopedia Duttona Tom 3. Warszawa: Wydawnictwo Lekarskie PZWL.

[3] R. Thomas. (Sat Sep 26 2015). Hip implant to release osseogenic growth factors. 21.11.2017, https://seelio.com/w/1i4q/hip-implant-to-release-osseogenic-growth-factors

[4] Z. Godlewski. (1986). Technologia Modelarstwo. Warszawa: Wydawnictwo Szkolne i Pedagogiczne.

[5] P. Siemiński, G. Budzik. (2015). Techniki przyrostowe Druk Drukarki 3D. Warszawa: Oficyna Wydawnicza Politechniki Warszawskiej.

The 73rd World Foundry Congress
„Creative Foundry"
23 - 27 September 2018, Krakow, Poland

NUMERICAL MODELLING OF SHSB METAL MATRIX COMPOSITE SOLIDIFICATION

Aleksandra Wojtyla, Michal Szucki

AGH University of Science and Technology, Faculty of Foundry Engineering, Reymonta 23, 30-059 Krakow, Poland
e-mail: aleksandra.wojtyla@interia.eu

Key words: metal matrix composites, SHSB, numerical modelling, in-situ, exothermic reaction

1. Introduction

Composites are widely used in many field of industry because of their unique properties. Nowadays the growing interest in in-situ metal matrix composites (MMC) can be also observed. They are manufactured in single-step process with different technologies. Among others the self-propagating high-temperature synthesis in metal bath (SHSB) should be mentioned here. In this method the ceramic reinforcing phase is created by spontaneous exothermic reaction in powder compact composed of metal and non-metal particles. The compact is surrounded by casting alloy which acts as initiating factor of reaction at the beginning of the process and as a cooling medium during this exothermic reaction. In the result ceramic particles are distributed in the casting [1-3].

From thermal point of view this process is very complex and includes cooling and solidification of metal matrix as well as influence of exothermic reaction on temperature field in the considered system. Therefore besides traditional experimental methods the numerical simulations can be used to predict the process flow and properties of obtained material. Thus, the aim of this work was to develop numerical model of thermal effects accompanying synthesis of titanium carbides during manufacturing of in-situ MMC. Current studies are based on engineer's thesis [4].

2. Mathematical description

Heat transfer in the mold-alloy-powder compact system can be described by the heat equation [4]:

$$c_P \rho \frac{\partial T}{\partial \tau} = div\left(-\lambda \cdot gradT\right) + q_V \qquad (1)$$

where c_p is specific heat (J kg^{-1} K^{-1}), ρ is density (kg m^{-3}), T is temperature (K), τ is the time (s), λ is coefficient of thermal conductivity (W m^{-1}K^{-1}) and q_V is volumetric heat source (W m^{-3}). For reasons of simplification the convection was omitted in the current studies [3,5].

Solidification was in this model included by the method called substitute thermal capacity. This method consist of modification of specific heat independence of temperature. The increase of its value in the liquidus to solidus temperature range will affect the cooling rate [5] (similar to the heat generated during solidification).

Influence of exothermic reaction of synthesis in powder compact on the temperature field was in this model taken into account by modified temperature recovery method. Value of stored temperature θ (K) was determined by equation below:

$$\theta = \frac{E_S}{c_P} \qquad (2)$$

where E_S is total reaction heat (J kg^{-1}) for powder compact. During exothermic reaction the local temperature in powder compact area is corrected by some value taken from this reserve. The reaction will end when $\theta \leq 0$ [4].

Mathematical model requires also definition of system geometry and the initial-boundary conditions. Process was analyzed in two-dimensional space. The model system involves: sand mold, cast alloy and powder compact (Fig. 1). On the internal and external boundaries continuity and Dirichlet boundary condition was applied respectively.

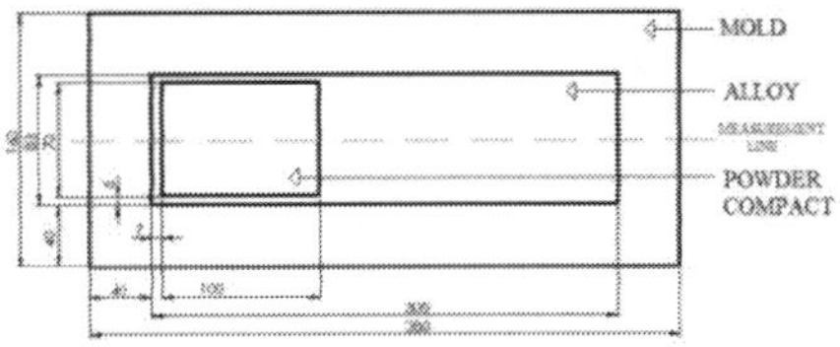

Fig. 1. Geometry of the system

3. Numerical implementation

The differential heat equation was solved by explicit finite difference scheme. This model requires also the time and space discretization. Time step was set manually, wile for the space discretization method proposed in [6] was used.

4. Results

During simulation values of temperature and temperature reserve were recorded for selected time steps. Sample temperature fields are presented in figure 2. Influence of the exothermic reaction can be observed firstly (6 [s]) in the corners of powder compact, because the mesh cells in these areas as the firsts achieve temperature of reaction initiation. Front of the reaction comprises next cells of powder compact and moves to the center.

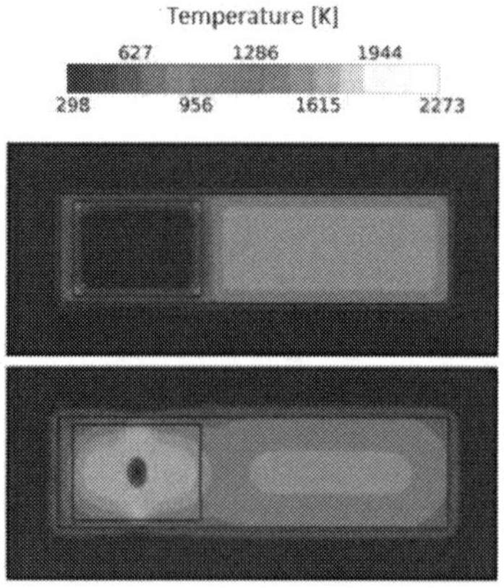

Fig. 2. Temperature field in selected time step

Distribution of temperature on the measurement line (Fig. 1) passing through the center of the system for selected time steps is shown on the figure 3. The blue curve presents temperature at almost initial state (0.6 [s]). Casting alloy has high temperature and remain areas have the ambient temperature. Thermal conditions in the system during reaction of synthesis are shown by the yellow and red curve. It is visible that the reaction moves towards the center of the powder compact. In turn, distribution of temperature during cooling (150 [s]) of system is presented by the black line.

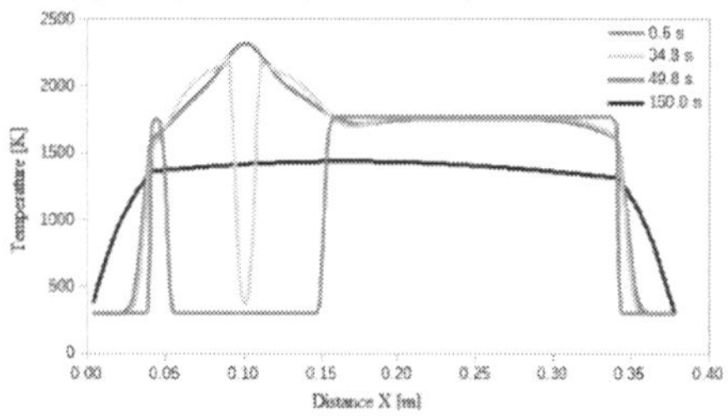

Fig. 3. Distribution of temperature in selected time steps

Computed and real values of the temperature in the compact powder and in the surrounding alloy were compared on the figure 4. Because of some difference between simulation and experiment setup the results were rescaled and presented in the non-dimensional form. The temperature value of 1 means the highest temperature value in the system. When it comes to powder compact the time value of 1 means the moment when system achieves the highest temperature, and in the case of cast alloy time value of 1 means the moment when the system achieves the local lowest temperature before the heat from synthesis reaction is transfer to the surrounding metal matrix. It is clearly visible that simulation results are very similar to the real process progress.

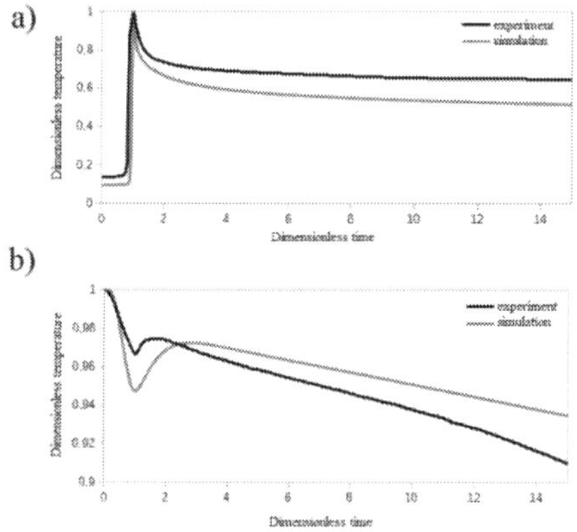

Fig. 4. Comparison of temperature curves inpowder compact (a) and cast alloy (b) [experimental data provided by S. Sobula]

5. Conclusions

Proposed mathematical model and its numerical implementation describes the SHSB process correctly. The results are compatible with the predictions. The synthesis reaction significantly increases the temperature in the system. The similar process progress in simulations and experiments indicates that model includes all important phenomena characteristic for the SHSB technology.

References

[1] E. Fraś, A. Janas, P. Kurtyka, S. Wierzbiński. (2003). Structure and properties of cast Ni3Al/TiB2 composites, Part I. Archieves of Metallurgy, Volume 48 (issue 4), 383-408.

[2] K. Naplocha. (2018). Chapter 8 - Self-propagating high-temperature synthesis (SHS) of intermetallic matrix composites. Eds. Rahul Mitra, Intermetallic Matrix Composites, 203-220.

[3] J. Lelito, P. L. Żak, A. A. Shirzadi, A. L. Greer, W. K. Krajewski, J. S. Suchy, K. Haberl, P. Schumacher. (2012). Effect of SiC reinforcement particles on the grain density in a megnesium-based metal-matrix composite: Modelling and experiment. Acta Materialia, Volume 60 (Issues 6-7), 2950-2958.

[4] A. Wojtyła. (2018). Numerical modeling of thermal effect during manufacturing of in situ composites. Engineer's thesis, Kraków (in Polish).

[5] Mochnacki B., Suchy J. S. (1995). Numerical Methods in computations of foundry processes. Kraków. Polish Foundrymen's Technical Association.

[6] M. Szucki. (2014). Application of the lattice Boltzmann method for modelling of liquid metal motion. PhD thesis, Kraków (in Polish).

The 73rd World Foundry Congress
„Creative Foundry"
23 - 27 September 2018, Krakow, Poland

IMPLEMENTATION OF 3D PRINTING IN DENTISTRY TO IMPROVE THE PRODUCTION OF DENTAL PROSTHETICS RESTORATION

Daniel Dulian [1*], **Janusz Lelito** [1]

[1] *AGH University of Science and Technology, Faculty of Foundry Engineering, al. Mickiewicza 30, 30-059 Krakow Poland*

**Corresponding address: e-mail: duliandan@gmail.com*

Key words: 3D printing, 3D scanning, dental prosthetics, medicine, casting

1. Abstract

The aim of the below paper is to describe the techniques using 3D printing and CAD-CAM technologies, which allow for the improvement or even the replacement of the traditional production processes in dental prosthetics. What follows is the research into the achieved results of the techniques' application, with regards to the most important parameters of the example objects.

2. Introduction

Dental crowns form a permanent prosthetic replacement in a patient's oral cavity. They are used on severely decayed, discoloured or broken teeth. The traditional treatment process, with the use of a porcelain dental crown, involves the rough-grinding of the decayed tooth, followed by drawing a denture impression as the basis for the creation of a plaster casting, which is a precise model of the patient's set of teeth. With the plaster casting, a wax pattern is prepared, which in turn can be used to cast a metal sub-structure for the porcelain crown, by the means of the lost-wax technique. The last step involves a dental technician positioning the porcelain tooth replacement layer and affixing it permanently to the patient's own by the dentist, previously rough-ground tooth. The above process is extremely time- and material-consuming [1, 2].

Thanks to the 3D printing technology, supported in this case by the use of a 3D scanner, the process at hand can be both shortened and made more resource-efficient. A 3D lab scanner reads the dental plaster casting or the denture impression of the patient, and the accompanying CAD software creates a precise virtual model of the patient's oral cavity. This digital object can be processed in specialised stomatological CAD software, to easily design the crown's sub-structure. Such a process lasts as short as 15-30 minutes, which makes a great difference in time required, considering the fact that the traditional method would require about a full day to reach this stage. When the virtual model of the sub-structure is ready in STL format, it is time to choose the optimal type of 3D-printing support.

3. Experimental method

The first of the already tested technologies involves using the 3D printer to create the wax pattern from a suitable material, which can be burnt without trace in a high temperature. This model is then used to finalise the moulding process. The printing process used the DDS 3DPrinter, provided by the Poland-based company Natrodent. This device works in the 'DLP' Digital Light Processing technology - a 3D printing method based on hardening a photopolymer layer by layer with the light projector beam, as opposed to the, otherwise similar, SLA (stereolithographic) technology, in which resins are hardened with a laser beam. This process is limited to improvements in the traditional method of creating the metal prosthetic sub-structure, as the final element is still produced by casting. Even though this difference seems insignificant, the results are considerable. Not to go into detail of higher levels of precision and adjustment to the ground tooth, the very time required to produce the medical component in question is greatly shortened [3, 4].

Another researched technology is known as Direct Metal Laser Sintering or DMLS. This method uses a powdered metal alloy dosed from a dispenser, formed into layer over a detail-building platform by a special blade. What follows is the process of selective remelting of the layers of the printed object with a laser beam. This must be conducted in a special chamber, where the required protective atmosphere can be created (usually using nitrogen or argon). By using the STL data from the CAD software, the printer can produce ready-to-use, high-quality, and perfectly fitting components of the dental replacements. These are characterised not only by their superb precision and high leak tightness, but also by the very little amount of final processing they require, before being combined with the porcelain crown. We can say without fear of contradiction, that this method is not a mere improvement, but a full replacement for the traditional method, which seems quite obsolete when compared to DMLS printing [3, 4].

Its most prominent advantage is the even greater time-efficiency, well surpassing other methods, DLP printing included. The other is the elimination of the costs of the materials required in the traditional casting process. As a consequence, we receive a final product of the highest quality, while using less resources and time. This is an innovative and aesthetic process of much greater level of environment-friendliness, as fits the global trends in production. The material used during the test research was a powdered cobalt and chrome superalloy.

After creating a pair of metal crown sub-structures from the same STL file with the two different methods above, the objects were scanned in the 3D lab scanner, allowing for the comparison of their significant parameters, and their renderings, to the original virtual master.

4. Summary

The research results lead us to believe that 3D printing is the future of the medical branch, with special regards to stomatology. This technological innovation will allow specialists to work in much more sterile conditions, save their precious time and materials, while offering an unprecedented quality. Thanks to the 3D printing technologies it had been made possible to reach higher levels of precision than in the case of the traditional lost-wax method, which in itself has long been greatly valued in many branches of science and art, where precise rendering was of key importance.

Acknowledgments

This project would have been impossible without the support of two Polish companies - Natrodent from Łódź and pro.orto.dent sp. z o.o. from Cracow which provided the specialized equipment for research

References

[1] Ian Gibson. (2006) *Advanced Manufacturing Technology for Medical Applications*. John Wiley & Sons, Ltd.

[2] Grzegorz Pawlicki. (1997). *Podstawy inżynierii medycznej*. Oficyna Wydawnicza Politechniki Warszawskiej.

[3] Anna Kaziunas France. (2014). *Świat druku 3D. Przewodnik*. wydawnictwo: Helion

[4] Przeysław Siemiński, Grzegorz Budzik. (2015). *Techniki przyrostowe*. Oficyna Wydawnicza Politechniki Warszawskiej

The 73rd World Foundry Congress
„Creative Foundry"
23 - 27 September 2018, Krakow, Poland

NUMERICAL SIMULATION OF THE IMPACT OF CASTING WALL THICKNESS AND INITIAL MOULD TEMPERATURE ON THE DUCTILE CAST IRON MICROSTRUCTURE

Grzegorz Witek [1]*, Paweł Leszek Żak [1] , Marcin Górny[1]

[1]*AGH University of Science and Technology, Faculty of Foundry Engineering, Reymonta Str. 23, 30-059 Krakow, Poland,*
**Corresponding address: e-mail: grzegorzw.mail@gmail.com*

Key words: numerical simulation, ductile cast iron, wall thickness, microstructure

1. Introduction

Thin wall ductile iron (TWDI) is modern material with very good mechanical properties, and dumping capacity. From an economic point of view, it has lower production costs than aluminium. TWDI can be lighter than their substitute made of aluminium alloys and to successfully replace them.

The purpose of this work is an analysis of the impact of the wall thickness and initial mould temperature on the microstructure in ductile cast iron, using numerical simulation [1][3]. For this purpose, a special geometry with different wall thickness was modelled in SolidWorks software. Next the model was saved as STL file and imported to MAGMA 5.3 software to simulate a casting process. To confirm the simulation, results were compared with experimental casting. The aim of the work was to check the possibility of using numerical simulation to design thin-wall ductail iron castings [4].

2. Geometry design

To design a testing casting geometry, the SolidWorks software was used. The program allows to create 3D geometry, assembly and basic simulation, Fig. 1.

Fig.1. Orientation of plates, relative to each other

Casting geometry was prepared by modelling separate part. For in-depth reporting a problem, geometry has three thickness of wall, 2,3 and 5 mm. Final step was prepare assembly with all prepared geometry. Orientation of plates was show on Fig. 1.

3. Mesh generation and simulation setup

For analyse casting process and microstructure MAGMA simulation program was used. First step was dividing model geometry with discrete mesh. Due to thickness of the wall in model, it was necessary to use Minimal Wall Thickness option to generate mesh. After this procedure, the number of cells was 44 million [2].

The alloy used in simulation was ductile cast iron its chemical composition is show in Table 1.

Table 1. Chemical composition of using alloy

C	Si	Mn	P	Mg	S
3.7	2.7	0.2	0.016	0.036	0.02

For calculation pouring time was use option Automatic Filling Control with Pouring Basin Fill Level set to 70%

For check the impact of mould temperature for microstructure, two simulations were conducted with the same chemical composition and pouring time, but with different inertial mould temperature.

The simulations had following parameters, Table 2.

Table 2. Simulation parameters using in simulation

Parameter	Developed value
Cast alloy initial temperature	1425 [°C]
Pouring time	5 [s]
Initial mould temperature in first simulation	160 [°C]
Initial mould temperature in second simulation	20 [°C]

4. Experimental Casting

To validate the simulation result, experimental casting was conducted in the Faculty of Foundry Engineering. Casting parameters were the same as set during simulation. The form was made of green sand and heating before pouring g in 200°C. Fig. 2.

Fig. 2. Mould use to experimental casting

5. Microstructure analysis

Simulation results shows differences in microstructure between each wall thickness and mould temperature.

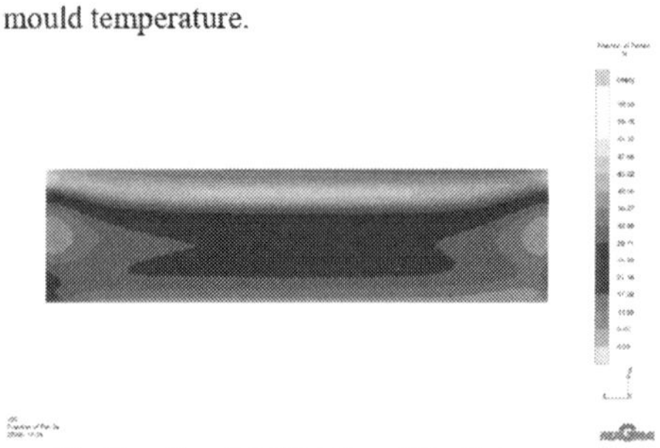

Fig. 3. Simulated fraction of ferrite in normal temperature mould- 2mm plate

Comparing plates with each other we can see that with increasing of wall thickness, the percentage share of ferrite is increasing and nodule count is decreasing, Fig. 3, 4.

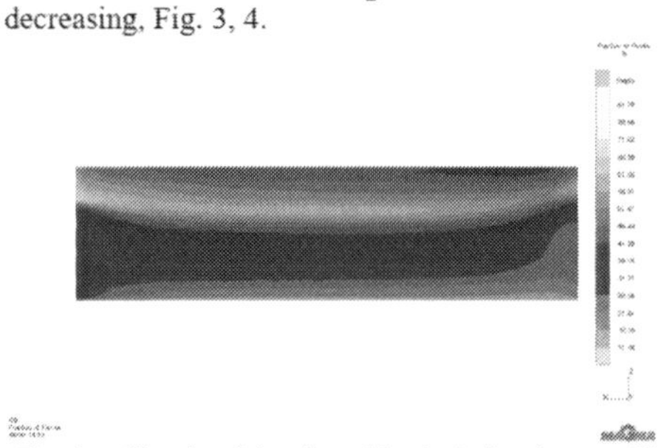

Fig. 4. Simulated fraction of ferrite in heating mould - 2mm plate

On simulation we can also saw an impact of initial mould temperature to microstructure. Higher temperature translates into bigger share of ferrite in microstructure. Also nodule count are lover in heating mould.

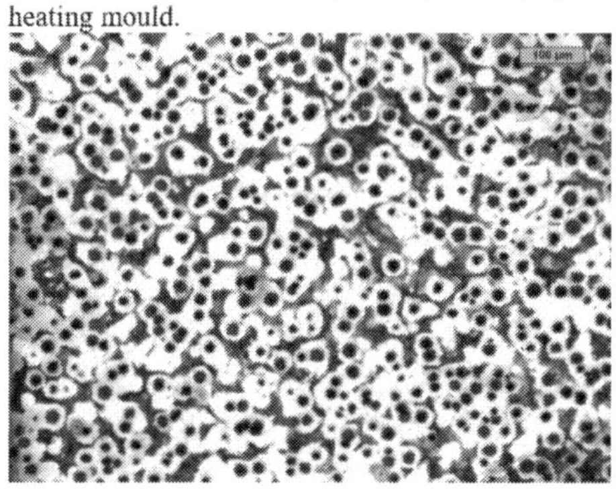

Fig. 5. Microstructure of 2mm plate - heating mould. Magnification 100x

From experimental casting were made samples. Then using Nital, samples was etching and observed with microscope, Fig. 5.

Compering simulation results to experimental casting, we can see comparable microstructure and nodule count. This may suggest the correctness of the results from the simulation.

6. Summary

Many parameters has impact to microstructure in casting. It increases the difficulty of constructing thin wall casting and production process of them.
The use of MAGMA simulation software allows one to predict the microstructure type and nodule count created in the thinwall casting.

References

[1] Podrzucki Cz.(1991). Żeliwo:Struktura właściwości zastosowanie .Kraków. Polish Foundrymen's Technical Association

[2] Mochnacki B., Suchy J.S.(1995): Numerical methods in computations of foundry processes. Kraków . Polish Foundrymen's Technical Association.

[3] Guzik E.(2001): Procesy uszlachetniania Żeliwa, wybrane zagadnienia. Katowice. Archives of Foundry.

[4] Witek G.(2018): Numerical simulation of the impact of casting wall thickness on the spheroidal cast iron microstructure. Kraków. Engineer's thesis

The 73rd World Foundry Congress
„Creative Foundry"
23 - 27 September 2018, Krakow, Poland

VISION OF THE DIGITALIZED FOUNDRY

WITH CYPER-PHYSICAL SYSTEMS TO FULLY AUTOMATIC CUTTING PROCESS FOR CASTING

Füglister Matthias, Max Edelmann, Dr. Roland Anderegg

University of Applied Science and Arts Northwestern, School of Engineering, Institute of Automation, Klosterzelgstrasse 2, 5210 Windisch, Switzerland

**Corresponding address: matthias.fueglister@fhnw.ch*

Key words: Digital Twin, Robotic, Automation, Optimal Clamping, Foundry

1. Introduction

High quality standards and high wages demand for rationalization and automation in the small series foundry production. The first automation approaches in the field of molding and casting processes were already pushed forward in the 1980s and are now part of the basic equipment of the modern foundry industry. However, fettling of casting was excluded from innovations, although this area is responsible for a major part of the costs. In particular, the process of cutting feeders and sprue material has great potential for improvement.

In this area major issues prevent increasing level of automation. One of the biggest problems is robot path planning. It involves a great deal of effort for automated fettling and is only profitable for large series. In addition, well-founded knowledge in the field of robots is required, which is not available in foundries with small series production.

This issue leads to the aim of the project, which is to simplify the entire process of automated fettling in such a way that a minimum knowledge in the field of robotics is required. In addition, robotics can be introduced into the small series foundry industry.

To achieve this ambitious goal, the primary focus is set on optimized clamping of castings (free-form objects) and an algorithm to generate a tool-path.

2. The path of digial twin

To achieve the optimal clamping and robot path planning the concept of digital twin is introduced. The basic geometric model of the cyber-physical system (see figure 1) is based on data from the industrial production 3D sand mold printer. The 3D data contains the geometry of the part as well as the additional sprue parts and is the foundation for the digital optimization of the clamping and the path planning for separating the gating system.

A) Optimal Clamping algorithm

To solve the first problem (optimal clamping of castings) three or more universal clamping systems (16 tappets per system) are used for clamping cast parts. This hardware clamping system allows sturdy, positive-fitting and friction-locked connections. Since the geometry of castings may arbitrary, the exact position for proper clamping is unknown. The optimization of clamping is implemented as a geometric surface analysis and uses curvature as a quality criterion - the more curved the surface, the better the form fits. According to this, the optimum clamping of a casting is achieved at the maximum curvature H of the surface, which can be solved by the following optimization task with the quality criterion J (see equations 1,2).

$$J(\varphi_z, z) = \sum_{i=1}^{n} \|H_i\|_2 \qquad (1)$$

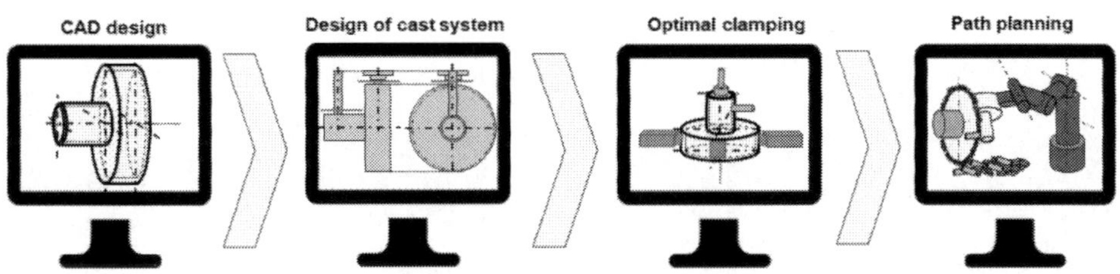

Fig. 1. Path of the digital twin: From CAD Design over generate a tool-path and optimal clamping to the virtual final product

$$\varphi_{z\,max}, z_{max} = \max J(\varphi_z, z) \quad (2)$$

B) Automated adaptive path planning and simulation

To solve the second problem a combination of the 3D data of the casting and the sprue systems creates slice planes which can be used for cutting. Based on this, the path planning is automatically created which takes into account the positions of the universal clamping systems and ensures that no cuts damage the casting or peripheral equipment (clamping table, robots, etc.). This is achieved by end-to-end digitization of the robot cell and enables complete simulation of the fettling process.

3. Path of physical casting

In the previous section, we considered how optimal clamping and robot path planning for the separation of sprue systems is calculated using the digital twin. In the following, the path of the physical casting (see figure 2) is described.

After the casting has been designed, the sand casting mold is created using a 3D sand printer for industrial production. This enables the production of molds in batch size 1 and more complex geometries. After the casting, the employee clamps the casting using the calculated optimum clamping position based on the digital twin calculation and initiates the fettling process. During the process, small position errors are recorded and used for automated compensation of path planning to ensure accuracy. The sprue system is removed from the casting and is ready for further manufacturing.

4. Results

This work shows possibilities to calculate an optimal clamping of arbitrary castings and the requirements of an automatically generated tool-path for an robotic sprue system remover only using the digital twin and without the know-how of a robotics specialist. Furthermore the effort to create a valid path will be significantly reduced. In addition, it enables the user to carry out lot size 1 path planning for automated fettling through a familiar CAD programming interface.

Acknowledgments

I would like to express my deep gratitude to Professor Dr. Roland Anderegg and Max Edelmann, my research supervisors, for their patient guidance and enthusiastic encouragement.

I would also like to extend my thanks to the technicians of the laboratory of the Institute of Automation [1] and Christenguss AG [2] for their help in offering me the resources in running the program.

This Masterthesis was carried out within a research and development project of Christenguss AG in collaboration with University of Applied Science and Arts Northwestern and granted by Swiss Commission of Technology and Innovation CTI [3].

References

[1] Institute of Automation, FHNW (29.05.2018). https://www.fhnw.ch/ia

[2] Christenguss AG (29.05.2018). Company Homepage http://www.christenguss.ch/

[3] CTI (29.05.2018). https://www.sareco.org/institution/cti-commission-for-technology-and-innovation/

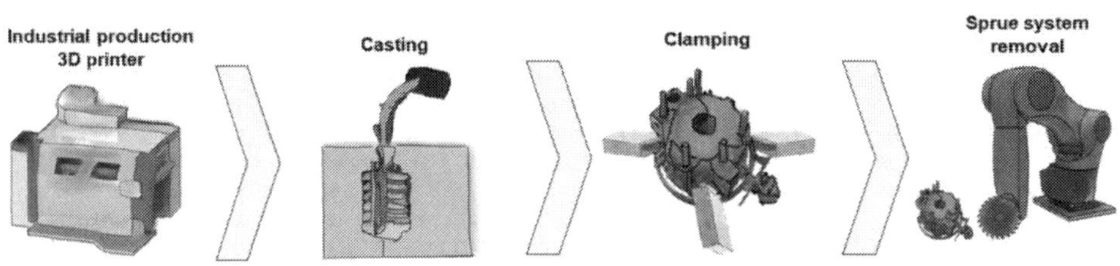

Fig. 2. Path of the casting: From 3D sand printer for industrial production to the final-cut product

ECOLOGY

The 73rd World Foundry Congress
„Creative Foundry"
23 - 27 September 2018, Krakow, Poland

THERMAL RECLAMATION OF FOUNDRY SANDS USING REPURPOSED SAND DRYER EQUIPMENT

Tommi Sappinen [1]*, Juhani Orkas[1], Timo Kronqvist[2]

[1]*Aalto University, School of Engineering, Otakaari 4, Espoo, Finland*
[2]*Timexcon Oy, Seppäsentie 8, Kempele, Finland*
**Corresponding address: e-mail: tommi.sappinen@aalto.fi*

Key words: molding sands, sand reclamation, industrial symbiosis, circular economy, sustainability, environment

1. Background

In Finland, foundries are experiencing rising pressure to properly recycle surplus foundry sands, as environmental legislations get tighter and sands become more difficult to landfill. While most Finnish foundries have different methods of mechanical reclamation, thermal reclamation is still much underutilized, as only one foundry currently reclaims their used molding sand this way. This changed in the spring of 2017, when a foundry raw material supplier acquired an out-of-operation industrial plant with the idea of converting its sand drying facility into a thermal reclamation plant. This paper shortly describes the development of the equipment, the theory of three foundry sand binder systems, test runs of reclaiming sand in the new plant and results of analyzing the reclaimed properties of the sand samples.

2. Sands and Equipment

Six Finnish foundries took part in the project by providing some of their used sands for the reclamation tests. One foundry has a furan no-bake system (Foundry 1), one has a green sand system (Foundry 2) and the rest have an ester-cured alkaline phenolic system (Foundries 3-6). The sands were treated in as received condition. For most foundries, this meant mechanically reclaimed sand from circulation. However, foundries 3 and 5 sent surplus sand, which was in an intermediate storage. This did not affect the reclamation process significantly, as the raw material was only dried beforehand.

Thermal reclamation of green sands is not fully beneficial because the heat deactivates active bentonite. Burned bentonite also does not break from the sand grain so a mechanical treatment afterwards should be required, which this plant did not have. [1]

Furan bonded sands are one of the easiest foundry sand types to reclaim as it consist of only organic substances that combust efficiently in even lower temperatures. The reclaimed sands are usually controlled by their acid demand value. [2]

The alkaline phenolic system has an interesting feature that holds back efficient thermal reclamation. Combustion of binders leaves remains of alkaline salts in the sand that can be a problem both for the reclamation process and for the rebonding characteristics. This problem is faced by mixing a special liquid into the sand before charging it to the furnace, which reacts with the salts in high temperatures. [3]

The plant equipment consists of a charge silo with a conveyor system. The raw material is fed through these into a rotating drum type reclamation unit with a gas burner. After burning, the sand is cooled. The material is then fed to a storage silo system with dust separation units.

3. Test runs

Test runs with the machinery proved that automated temperature control enabled constant reclamation temperatures. The collection of two samples could be done in a single test run, as seen in figure 1. Samples for 550 and 650 degrees were made during these constant temperature periods of around half an hour. Frequent testing also demonstrated the capacity and capability of the rotating dryer as well as the parts of the process that required more for example insulation and heat shielding.

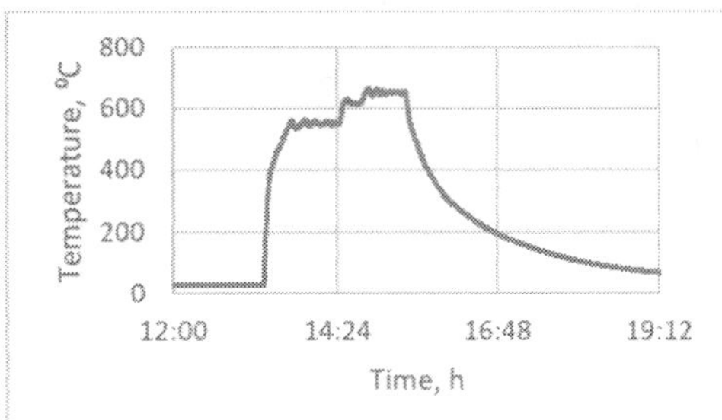

Fig. 1. Temperature data from collection of two samples in 550 and 650 degrees Celsius, respectively

4. Analysis of samples and results

All the samples were analyzed for loss on ignition (LOI) and grain size distribution, as well as the alkaline phenolic sands were tested for their pH (not present in this abstract). In addition, micrographs were taken from the as received sands compared to the reclaimed ones. An example of the comparative micrographs are presented in figures 2 and 3.

Fig. 2. Foundry 4 sand, as received

Fig. 3. Foundry 4 reclaimed sand

The loss on ignition results for as received sands compared to three different reclamation temperatures when applicable is in figure 4. The average grain sizes of as received, reclaimed and new sands are also presented in table 1. Dust percentages and detailed grain size distribution curves are in a more thorough article.

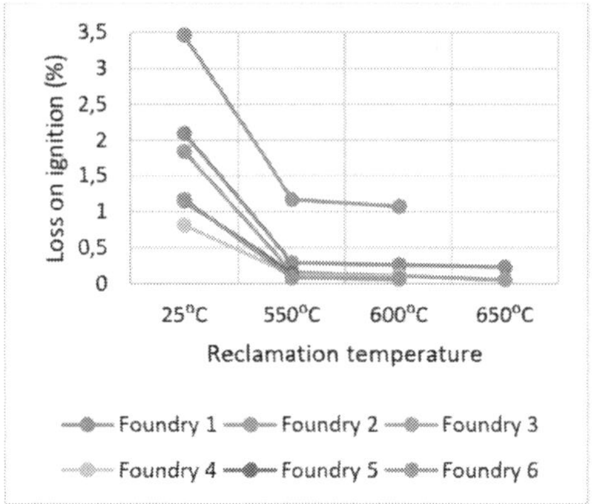

Fig. 4. LOI for all samples.

Table 1. Average grain size of samples in mm according to standard DIN 4188

	Reclaimed	As received	New
Foundry 1	0,353	0,373	0,31
Foundry 2	0,319	0,297	-
Foundry 3	0,363	0,36	0,32
Foundry 4	0,247	0,278	0,27
Foundry 5	0,326	0,369	-
Foundry 6	0,355	0,36	0,31

5. Discussion

The micrographs and general visual inspection showed that the alkaline phenolic sands had an appearance very similar to virgin quartz sand. Loss on ignitions were lower than 0,3% for all samples except the green sand, which was expected. Grain size distribution did not change dramatically in any case, but most samples showed that the impure sand grains settle on the higher sieves, while the reclaimed sands have those grains cleaned so they can settle on their actual lower sieves.

The plant equipment fit the thermal reclamation needs well. Initial testing and co-operation with foundries and suppliers has shown interest to continue on developing the facility. While the laboratory tested samples showed good results, the most important result of this work was the successful initiation of a central reclamation facility in the Finnish foundry business. This does not only advance sustainability and circular economy, but also the idea of industrial symbiosis, where companies exchange material streams for efficient waste management and use of natural resources.

References

[1] Granlund, M. & Silsby, D.V. (2008). Thermal-pneumatic reclamation of molding sands for reuse in core work. Transactions of the American Foundry Society. Volume 116, pages 309-328.

[2] Carey, P.R. & Lott, M. (1995). Sand binder systems part V – Furan no-bake. Foundry management & technology. Volume 103 (issue 7), pages 26-30

[3] Busby, A.D. (2015). Patent nr. 0246386 A1. US. Reclamation of ester-cured phenolic resin bonded foundry sands.

MANAGEMENT

The 73rd World Foundry Congress
„Creative Foundry"
23 - 27 September 2018, Krakow, Poland

A SIX SIGMA FRAMEWORK FOR ELIMINATING DEFECTS IN SAND CASTING FOUNDRIES

Cindy Sithole[1]*, Kasongo Nyembwe, Peter Olubambi

[1]University of Johannesburg, Faculty of Engineering and the Built Environment, 37 Nind Street, Doornfontein, Johannesburg, 2094, South Africa,

**Corresponding address: e-mail: cindy.sithole@ymail.com*

Key words: Sand casting, Six Sigma (DMAIC), Sand moulding, Defects, Quality Improvement

1.Introduction

A number of foundries in South Africa (SA) are jobbing foundries, meaning that they produce a large number of different castings based on received orders [1]. Due to the nature of business, the success and sustainability of local foundries depend on their ability to provide castings of high quality at the minimum cost. This means that SA foundries need to adopt quality systems and strategies that are cost-effective and allow the production of casting of high integrity [2]. Production of good quality castings will enable South African foundries to gain operational competitiveness and access to worldwide market. The traditional quality control in sand moulding foundries based on the sand testing employed in most SA foundries have limitations in terms of efficacy to avoid casting defects.

This paper presents an alternative quality control philosophy based on Six Sigma framework that is proposed to be used in sand moulding foundries for reduction of casting defects related to moulding sand. It is designed to explore the fundamentals of Six Sigma tools for the sand moulding process in SA foundries and to establish a continuous improvement system within the sand moulding plant by eliminating variation using Six Sigma principle shown in figure1.

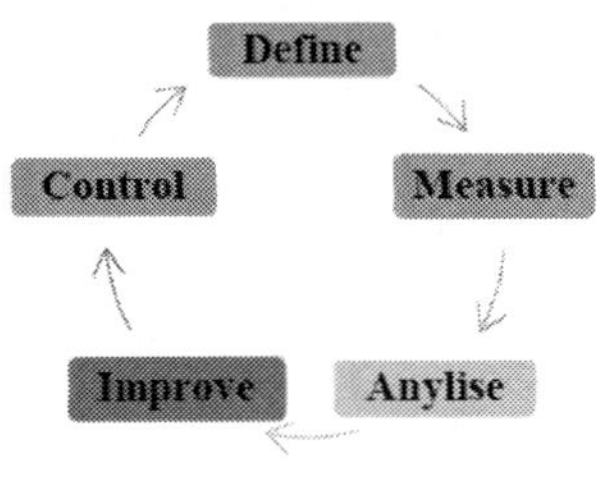

Fig. 1. Six Sigma DMAIC (Define, measure, analyze, improve and control) adopted from [5]

2. Methodology

Six Sigma is a data driven method to process improvement that aims to eliminate variation in business processes using statistical control methods [3]. Six Sigma focuses on business performance improvement through quality initiative using the basic tools of quality. The objectives of Six Sigma are to eliminate features that affect productivity, profit and meeting customers' demands [3]. The principle applied in Six Sigma is refered to as DMAIC explained in Table 1 as a framework for sand moulding foundries. The activities outlined in the Table 1 are compulsory for the foundry to execute. This tool could asssist foundries in solving the problem.

In today's global competition, Six Sigma has become one of the most important tools for attaining competitive advantage [4]. Competitive advantage for sand casting foundries in South Africa means improved plant maintenance processes, improved product quality and proper utilization of resources especially scarce resources.

Table 1. The Six Sigma (DMAIC) framework and applicable tools [5]

SIX SIGMA PHASE	ACTIVITIES	TOOLS
Define	Define the Customer (internal and external). Define what the customer require. Define the critical to quality elements (CTQ). These are the properties of the moulding sand. Define the problem in the foundry. Define core foundry process. Define the goal of the project and the deliverables.	Voice of customer. Voice of the business. Process map SIPOC Project plan
Measure	Measure the current products in terms of the CTQ. Measure the current performance of the process. Calculate the sigma level of the process.	CTQ Tree Detailed process map. Data collection sheet. Sigma level calculation sheet Process capability

	Collect data for improvement.	calculation Measurement system analysis
Analyze	Analyze the existing variation in data. Analyze the relationship between the output and input. Analyze the root causes of the problem.	Histogram Scatter plot Cause and effect diagram Failure mode effect analysis ANOVA Pareto analysis
Improve	Identify possible solutions. Assess the failure mode. Prioritize the solutions. Verify the suitability of the possible solutions and eliminate the variation causes.	Brainstorming Failure mode effect analysis Multi voting Suitability Matrix
Control	Developing continuous process standard and procedures. Verification of the outcomes. Statistical process control and implementation.	Control Chart Control plan Standard operating procedure TPM

Every step in the sand moulding process is very critical for production of good quality castings. Defects produced from the sand moulding process are only visible in the final product (casting) which results in increased number of scrap, reduced sales and increases cost in the foundry . In most South African foundries, casting defects are classified however,their root causes in sand casting foundries are not found to prevent future occurrences.

It is also observed that documented possible defect causes and collected data in the sand moulding plant are not often reviewed to enable quality improvement within the foundry. Six Sigma should force the foundry to collect and review data for improvement. Inefficiencies in sand moulding process, material preparation and both process as well as material parameters influences defects formation in SA foundries. Defects such as flashes, metal penetration, sand inclusions and burn-on are some of the defects that arise from the above mentioned foundry process and are observed in the casting surface during fettling and or after heat treatment.

The appearance of these defects in a casting affect the profitability as well as the competitiveness of the foundry [1]. Defects in a foundry increase waste in terms of time, resources and money, improved yield means productivity and profitability the foundry.

For increased productivity and enhanced competitiveness, foundries should focus on scrap and rework reduction as a practical technique by using the Six Sigma framework [4]. Six Sigma deployments can efficiently reduce defects and rework in production environment and in a foundry specifically.

3. Expected Results

This framework is to be implemented as a case study in a sand moulding foundry. After Six Sigma deployment, the expected results are increased customer satisfaction, improved profitability, quantified cost savings and reduced lead times as well as inventory [5]. With Six Sigma deployment, castings will be produced according to the specified specification of the customer and the critical to quality elements (properties of the moulding sand) will be met at all times. Due to this, the customer reliability in terms of quality will increase. Good relationship with customers continuously influences business expansion (access to new markets) and increases profitability for the foundry [5].

The production cost of the foundry will be reduced due to the improved process and profitability will increase. The deployment of this technique by a foundry will also reduce cost in the foundry by improving the yield. The reduction of scrap will allow optimum use of resources within the foundry which in return enhances the production rate and increases the foundry's productivity [6].

Six Sigma also increases employee morale. The biggest benefit of Six Sigma technique is the removal of partiality in decision-making by forming a system that allows everyone in the organization to collect, evaluate and present data in a constant and reliable manner to reduce variations [7].

4.Conclusion

Six Sigma is proposed as a quality improvement framework for sand foundries. A case study would be used to demonstrate the benefits of this tool in a local foundry.

References

[1] Davies, J. (2015). The impact of IPAP on the foundry industry, Gauteng: South African institute of Foundry Men.

[2] Singh, B.J & Khanduja, D. (2015). Wrap the scrap with DMAIC. Strategic deployment of Six Sigma in Indian foundry SMEs. Anchor Academic publishing. Hamburg.

[3] Pyzdek, T. (2003). The Six Sigma handbook. United States of America. The McGraw-Hill. 55

[4] Anthony, J. and Buñuel, R. (2002). Design for Six Sigma: a breakthrough improvement business strategy for achieving competitive advantage. Manufacturing Engineering. 81 (1), 24-6.

[5] Gopalakrishnan, N. (2013). Simplified Six Sigma: Methodology, Tools and Implementation. PHI learning Private limited, Delhi.

[6] Goh, T.N &Xie, M. (2004). Improving on the Six Sigma paradigm. The TQM Magazine. 16 (4), 235-240

[7] Maleyeff, J. and Kaminsky, F.C. (2002) 'Six Sigma and introductory statistics education', Education and Training, Vol. 44 (2), 82-89

World Foundry Organization Ltd
Winton House, Lyonshall
Kington, Herefordshire
HR5 3JP, United Kingdom

ISBN 978-1-5108-8042-9
9 781510 880429

CURRAN ASSOCIATES INC.
proceedings.com